Green Energy and Technology

Climate change, environmental impact and the limited natural resources urge scientific research and novel technical solutions. The monograph series Green Energy and Technology serves as a publishing platform for scientific and technological approaches to "green"—i.e. environmentally friendly and sustainable—technologies. While a focus lies on energy and power supply, it also covers "green" solutions in industrial engineering and engineering design. Green Energy and Technology addresses researchers, advanced students, technical consultants as well as decision makers in industries and politics. Hence, the level of presentation spans from instructional to highly technical.

Indexed in Scopus.

Indexed in Ei Compendex.

Ramchandra Pode • Boucar Diouf

Decarbonization of Road Transportation

The Electric Vehicle (EV) Transition

Springer

Ramchandra Pode
Department of Physics
Kyung Hee University
Seoul, Korea (Republic of)

Boucar Diouf
Department of Information Display
Kyung Hee University
Seoul, Korea (Republic of)

ISSN 1865-3529 ISSN 1865-3537 (electronic)
Green Energy and Technology
ISBN 978-3-032-11387-0 ISBN 978-3-032-11388-7 (eBook)
https://doi.org/10.1007/978-3-032-11388-7

This Springer imprint is published by the registered company Springer Nature Switzerland AG
The registered company address is: Gewerbestrasse 11, 6330 Cham, Switzerland

Preface

The global road transport sector stands at the threshold of a historic transformation. For more than a century, the internal combustion engine has dominated our mobility systems, shaping economies, lifestyles, and even geopolitics. Yet today, this paradigm is being challenged by an urgent imperative: **decarbonization**. The transport sector accounts for a significant share of global greenhouse gas emissions, and without decisive action, the goals of the Paris Agreement will remain out of reach. This book was born out of a deep concern for the environmental impact of our mobility choices and a conviction that technological innovation, policy reform, and behavioral change can together drive a sustainable transformation.

The electric vehicle transition, however, is not a linear path. It unfolds amid complex dynamics: progress in energy storage technology, volatility in critical mineral supply chains, evolving policy frameworks, and shifting consumer expectations. Questions abound. How fast can EV adoption scale? What role will charging infrastructure, renewable energy integration, and digital technologies play? How do we ensure that this transition is inclusive, equitable, and resilient against future shocks?

This book is not only about vehicles: it is about systems, policies, and human choices. It seeks to provide a comprehensive understanding of the transition from the enabling technologies and the role of regulation in the dynamics of consumer adoption and the infrastructure backbone that will support the future of mobility.

This book explores the complex and dynamic transition from conventional road transport to a low-carbon, electrified future. The **electric vehicle (EV) revolution** is not just a technological shift; it represents a convergence of multiple forces: technological innovation, market evolution, regulatory ambition, and societal change working in concert to redefine mobility.

Several key drivers shape this transformation:

- **Technological innovation**: Advances in battery chemistry, power electronics, and charging infrastructure have dramatically improved the performance and affordability of electric vehicles, challenging long-standing barriers such as range anxiety and high costs.

- **Policy leadership**: Governments worldwide are enacting policies that set ambitious emission targets, incentivize clean vehicle adoption, and penalize polluting alternatives. Without these frameworks, ranging from zero-emission vehicle mandates to subsidies and carbon pricing, the pace of change would be far slower.
- **The second-hand EV market**: As the first generation of electric vehicles enters the resale market, accessibility will expand beyond early adopters to mainstream consumers. This segment will play a critical role in democratizing the transition, particularly in emerging economies.
- **Grid readiness and renewable integration**: The widespread adoption of EVs brings both opportunities and challenges for electricity systems. Vehicle charging will significantly increase demand on the grid, requiring smart infrastructure, robust storage solutions, and flexible demand management. At the same time, EVs offer an unprecedented opportunity to integrate **renewable energy**, turning cars into active participants in the energy ecosystem through vehicle-to-grid technologies.
- **A complete shift to electric vehicles**: The global shift to electric vehicles (EVs) represents a pivotal strategy in combating climate change and modernizing transportation. Governments and automakers are accelerating the shift through policy incentives, technological innovation, providing financial incentives to autoworkers for skill upgradation, and so on. As battery technology improves and renewable energy integration grows, EVs are poised to dominate new car sales by 2030.

The decarbonization of road transport is not a single leap but a journey, one that involves innovation, collaboration, and foresight. This book is intended for policymakers, industry leaders, researchers, students, and anyone interested in understanding the intricate web of changes that will define the future of mobility. Our aim is to provide a comprehensive, evidence-based analysis of the strategies, technologies, and market forces driving this evolution while addressing the obstacles that remain.

The transition to electric mobility is inevitable, but its speed, equity, and sustainability depend on the choices we make today. It is our hope that the insights offered in this volume will inform those choices and contribute to shaping a cleaner, more resilient transport future.

As we turn the page on the fossil fuel era, the road ahead is both challenging and full of promise. The decarbonization of transport is not just a technical necessity; it is a societal transformation, redefining what progress and sustainability mean for generations to come.

This book has developed from a couple of years of intensive research on renewable energy and zero-carbon emission mobility systems. The advancements in literature and knowledge over the past few years have prompted the need for a comprehensive presentation of this topic. A large part of the book is based on the practical experience and research related to electric vehicles and clean energy sources powering them.

We express our appreciation for the open and intense discussion with colleagues from the Physics and Information Display Departments at Kyung Hee University,

Seoul. Our special acknowledgement is given to **Professor Jang Hyuk Kwon** from the Department of Information Display at Kyung Hee University, Seoul, for his continuous support and encouragement throughout the duration of this project. His insightful suggestions and recommendations have contributed significantly to the preparation of research articles, which are now part of this book.

We believe that the full transition to EVs is seen as an opportunity not only for environmental benefits but also for a redefinition of mobility, making it smarter, cleaner, and more interconnected. It is hoped that this book will serve as a valuable resource and reference for academicians, businesses, policymakers, climate researchers, and professionals involved in zero-emission mobility engineering.

Seoul, Korea (Republic of) Ramchandra Pode
Seoul, Korea (Republic of) Diouf Boucar

Competing Interests The authors have no competing interests to declare that are relevant to the content of this manuscript.

Contents

About the Authors

Ramchandra Pode is a professor in the Department of Physics at Kyung Hee University in Seoul, South Korea. He has received the European Union-DST (India) Fellowship (ENEA, Frascati, Rome, Italy, in 1995) and the Brain Pool, KOFST Fellowship (South Korea, in 2003–2004 and 2005–2006). His research focuses on organic light-emitting diode (OLED) devices and renewable energy technologies. He has authored numerous publications in international journals and has guided many research and postgraduate students. Prior to his tenure at Kyung Hee University, he was affiliated with Nagpur University in India. Ramchandra Pode is a distinguished physicist known for his work in **organic light-emitting diodes (OLEDs)** and **renewable energy technologies** and has made notable contributions in these fields. He has also contributed to several influential books in the fields of **organic electronics** and **renewable energy**, often as a co-author or chapter contributor. He co-authored a book titled ***Solar Lighting***, which is part of the *Green Energy and Technology* series, in 2011. Key aspects of his work include the following:

- **OLED research**: He has extensively worked on improving the efficiency and structure of OLED devices, including phosphorescent OLEDs and transparent conducting electrodes.
- **Renewable energy**: His research includes exploring the potential of lithium-ion batteries in renewable energy systems and the applications of rice husk ash waste from biomass power plants.

- **Sustainable lighting**: He has been involved in developing energy-efficient lighting technologies, such as solar-powered LED systems, aimed at enhancing accessibility in developing regions.

His research has produced several practical applications that tackle real-world challenges:

1. **Energy-efficient lighting**: His work on organic light-emitting diode (OLED) devices has led to the development of energy-efficient lighting solutions, which are widely used in displays and general lighting.
2. **Renewable energy systems**: His investigations into lithium-ion batteries and rice husk ash waste have applications in renewable energy storage and sustainable waste management, promoting cleaner energy solutions.

These contributions not only advance technology but also aim to improve accessibility and sustainability across various sectors.

Boucar Diouf is a scholar and researcher specializing in sustainable technologies and energy systems. He is currently a professor at Kyung Hee University in Seoul, South Korea, where he teaches physics and conducts research in advanced energy solutions and electric mobility. Dr. Diouf earned his Ph.D. in Physics from INSA Toulouse in France, where he developed a strong foundation in applied physics and later in energy systems. Over the past 15 years, his work has focused extensively on renewable energy integration, electric vehicle technologies, and innovative strategies for energy efficiency.

Professor Diouf has made noteworthy contributions to academia, particularly in the field of renewable energy technologies. One of his standout projects involves repurposing discarded smartphone batteries into solar-powered LED lighting systems for rural communities, combining environmental sustainability with social impact. He developed a model for recycling discarded lithium-ion smartphone batteries into solar-powered LED lighting systems. Some of his remarkable achievements are the following:

- **Electronic waste reduction**: Instead of letting old batteries pile up in landfills, they are tested, refurbished, and reused.

- **Energy access**: These recycled batteries are paired with small solar panels to create affordable lighting systems for rural households without electricity.

He has authored numerous academic papers, engaged in international research collaborations/projects, and served as an advisor for industry initiatives and policies aimed at accelerating the transition to clean energy. Professor Diouf is committed to bridging the gap between scientific research and real-world applications, ensuring that innovative ideas translate into tangible solutions for society. Through groundbreaking academic contributions and active involvement in global initiatives, he promotes sustainable mobility solutions that address the pressing challenges of climate change and urbanization. His research is a great example of how technological ingenuity can be harnessed to address both environmental and social challenges. Professor Diouf's work is a game-changer for the following:

- **Sustainable development**: It aligns with global goals for clean energy and responsible consumption.
- **Social equity**: It empowers communities that rely on harmful, expensive lighting like kerosene lamps.
- **Circular economy**: It turns waste into a resource, proving that innovation does not always require new materials, just new thinking.

Chapter 1
Electric Vehicles in the History of Road Transportation

Contents

1.1 Introduction

Humans are essentially born to move like animals. In fact, migration or movement from place to place is an intrinsic characteristic of human beings and not the sedentary lifestyle. They are endowed with two skills to attain perfection in life: intelligence and physical activity. With these two skills, they accomplished perfection in life. The importance of physical activity is far greater. As a result, the need for transportation since the beginning of human civilization has been vital to survival, trade, and exploration. In order to exchange goods, ideas, traditions, and cultures, people used to interact frequently by connecting distant communities. The Sumerian Mesopotamia region, the cradle of the ancient civilization, emerged as a trading hub, and people around the world used to travel to that region, mostly to trade goods [1]. The Mesopotamian region was the converging point and used to be called the

R. Pode, B. Diouf, *Decarbonization of Road Transportation*, Green Energy and Technology, https://doi.org/10.1007/978-3-032-11388-7_1

Fertile Crescent. From the Indus Valley to Persian Iran and from China to India, people used to interact to trade species and exchange ideas and vice versa. All regions and civilizations benefited and flourished due to the human interaction. The ancient Silk Road, a network of trade routes linking Asia, the Middle East, and Europe from second BC to the mid-fifteenth century, was not merely used by traders, intellectuals, and armies to travel from east to west and from west to east, but ideas, religions, and cultures were also exchanged [2]. Transportation has been a vital part of human life for thousands of years in the past and continues to be so. Undoubtedly, transportation tools played a significant role and influenced the evolution of society. Long before the invention of modern vehicles, our ancestors relied on simple and ingenious techniques to travel from one place to another. The earliest forms of transportation were walking across vast terrain, riding on animals, or making primitive rafts to cross bodies of water.

1.1.1 Animal-Driven Transportation

Simple walking, using the strength of human muscles, was the earliest mode of transportation. However, walking was not the most efficient way of transportation, as it covered only limited distances. Later, animals like horses, oxen, donkeys, etc. were used as primary forms of transportation for carrying goods over tracks to cover long distances, as shown in Fig. 1.1 [3, 4]. The domestication of animals allowed humans to travel greater distances more effectively. These animals were highly valued for their ability and strength to move people, building materials, and merchandise. Even today, these animals are used in Africa, Asia, and Latin America to transport water, cut-and-carry fodder for food production, and move harvested crops from the fields into storage sites [5]. Animals are also used for food security at the household level.

Horses were domesticated around 4000 BC and commonly used for riding (see Fig. 1.1), whereas camels became the main mode of transportation in arid regions. Buffaloes were extremely valued due to their strength and endurance to carry goods across rugged terrain. Although they were not as fast as horses, donkeys and mules were particularly important in agricultural societies, where they were used to transport crops and other products over long distances. In more settled agricultural societies, oxen were frequently used to pull carts and plows, which in turn allowed for the development of early infrastructure.

In summary, the earliest means of transporting items were on human heads and backs; nevertheless, the usage of animals evolved during the Stone Age (before 4000 BC). It is thought that a travois, a frame used to drag loads, was the first vehicle. Around 5000 BC, sleds were created because they were simpler to move on level terrain. A larger roadway and more clearance were needed for pack animals,

Fig. 1.1 Early primary modes of transportation for carrying goods over unpaved tracks caused by human or animal movements

goods ridden on horses, and bullocks pulling sleds or travois than for walkers. Consequently, better tracks were created [4]. Also, roads were developed along the tops of ridges in England, including the Ridgeway, to avoid crossing rivers and bogging by about 5000 BC [4]. The evolution of road transportation is depicted in Fig. 1.2. From horse-drawn carriages to electric vehicles, the history of transportation has seen an incredible amount of innovation [6, 7].

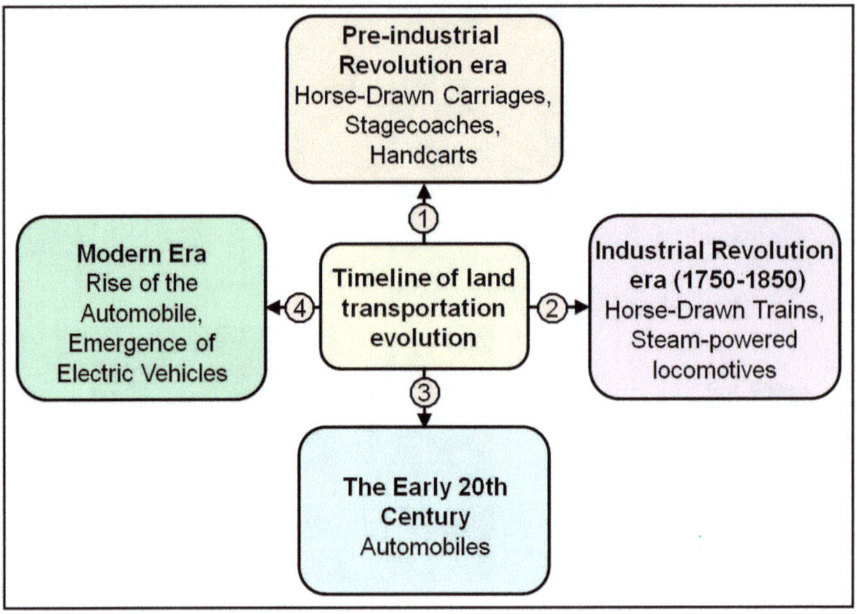

Fig. 1.2 Evolution of road transportation

1.2 Advancement in Road System

In the Stone Age, people lived in open land without the need for constructed tracks/routes. The first improved trails would have been at passages, mountain passes, and through swamps [4]. The first improvements would have involved removing huge stones and trees off the path. The tracks were frequently enlarged or leveled to accommodate both human and animal traffic as commerce and trade increased. Some of these dirt pathways evolved into rather vast networks that enabled trade, governance, and communication across large regions.

In ancient times, people made mud tracks to transport products and used these routes for several years. Often, tracks were naturally formed in areas with heavy traffic. Also, animals were utilized for transportation all around the world till the eighteenth century. However, in the nineteenth century, as trade increased, the movement of people and products became increasingly costly as animals required significant amounts of food and water. Therefore, various types of tools like wagons were devised to meet the growing transportation demands. Correspondingly, the needs of comparable roads for such transportation were anticipated too. As a result, better roads were constructed to enable efficient movement between far-off regions within a vast state or empire, as well as to serve military objectives.

1.2.1 Mesopotamian Roads

Roads structures in ancient times are illustrated in Fig. 1.3. Mesopotamian road systems around 400 BC were primitive and constructed using materials like stones and other local resources and construction techniques available at that time, as shown in Fig. 1.3a [8, 9].

1.2.2 Harappan Roads

Street paving has been found from the first human settlements around 4000 BC in cities of the Indus Valley Civilization on the Indian subcontinent, such as Harappa and Mohenjo-daro [10]. Roads in the Indus civilization region were primarily paved with baked bricks (see Fig. 1.3b).

1.2.3 Ancient Roman Roads

Ancient Roman roads were built from about 300 BC through the expansion and consolidation of the Roman Republic and the Roman Empire. These roads were used for the efficient movement of armies, people, carriages, and traded goods. Almost all major roads were often paved using stones (see Fig. 1.3c). Stone, gravel, and sand layers to create a strong, lasting surface were employed. The road construction process typically began with digging a trench to lay foundations [11, 12].

Fig. 1.3 Roads in (**a**) Mesopotamian, (**b**) Harappan, and (**c**) Roman eras

1.2.4 Early Tar-Paved Roads

The first recorded use of asphalt as a road-building material was in Babylon, Mesopotamia (625 BC). In 1595, Europeans discovered natural deposits of asphalt on the island of Trinidad, near Venezuela. With the beginning of the eighteenth century, road technologies began to improve as engineering was more greatly utilized. A tarmacadam road basically is a macadam road having a tar-bound surface. The first tarmacadam pavement was placed outside of Nottingham (Lincoln Road) in 1848 [13]. Paved roads constructed using sheet asphalt on a concrete foundation became common during the mid-1800s. One of the first "tar" roads was built in Paris in 1858. Tarmac was employed to pave roads for the first time in 1901. Tar was mixed with aggregate material to create a more mixed product than pure tar from bitumen [14]. Today, 96% of all paved roads and streets in the U.S., almost two million miles, are surfaced with asphalt. All the paving asphalts used today are acquired by processing crude oils. Roads paved with asphalt are very popular throughout the world because of their smooth surface, increased tire contact, and hence improved skid resistance, although they have higher maintenance costs.

Regardless of the road structure and construction techniques used, the original intention for building roads and transport systems has remained fundamentally the same throughout time and has been important for the advancement of many cultures.

1.3 Wheeled Transportation

The evolution of wheel inventions and transportation tools is depicted in Table 1.1. About 5000 years ago, simple 2-wheeled carts were constructed and used. Soon after, chariots were developed. The most significant advancement in wheeled transportation was the construction of the chariot. Chariots, developed around 2000 BC, were a faster, more transportable option compared to earlier vehicles. Usually, chariots were constructed using two wheels, pulled by horses, which allowed quick movement of goods. Subsequently, four-wheeled wagons pulled by oxen appeared. With time, the improvement in the road system followed the changes in transport tools. Chariots, horse-drawn coaches, and Macmillan's Bicycle (in 1839) followed. Around 1000 BC, wheels with iron rims around them were first used on chariots employed in warfare. After that, there were no major changes in the advancement of wheels till spoked wheels in 1802. Today's wheels are very efficient and light, look different from the simple wooden wheels of the past, but the basic concept and principle remain the same.

G.F. Bauer produced and patented the first metal-spoked wheel (also known as a wire-spoked wheel) in the United States in 1802 [15]. Later, the iron-spoked wheel *for vehicles* was patented by Theodore Jones in 1826 (London, England). The early Ford Model T used wooden spoked artillery-style wheels (around 1908–1925) but changed to steel welded-spoke wheels in 1926. The wheel seems to be one of the greatest mechanical inventions of all time. The very first automobile wheels were

Table 1.1 Timeline of wheel inventions and transportation tools

Year	Breakthrough	Description
3500 BC	Invention of the first wheel	Ancient Sumerians of Mesopotamia invented solid mud and wooden wheels
~3402–2800 BC	Invention of two-wheeled cart	A two-wheeled cart was illustrated at the end of the fourth millennium BC, at Lohne-Engelshecke, Germany
3000–2000 BC	Invention of 2 wheeled carts	Domesticated horses were used for transportation. Two wheeled chariots were pulled by horses/oxen or buffaloes
First century BC	Invention of 4 wheeled wagons	4 wheeled wagons pulled by oxen appeared. Four-wheeled wagons with a rectangular vehicle body fitted with solid disc wheels. Animals such as horses, mules, or oxen usually pull wagons
Between sixth to fourth century BC	Invention of one wheeled vehicle	A hand drawn vehicle with one wheel was first invented in Greece
2000 BC	First known spoke wheels; wooden spoked wagon wheel	Easier to carry, more flexible, and had a controllable wobble. The lighter spoked wheel enabled much higher speed. The spoked wheel was invented around 2000 BC in Anatolia. A wheel with spokes first appeared on Egyptian chariots around 2000 BC
1802	Metal spoked wheels	The first metal-spoked wheel was manufactured and patented by G.F. Bauer in 1802
1826	Invention of iron-spoked wheel for vehicles	The invention of the iron-spoked wheel for vehicles is credited to Theodore Jones in 1826 in London, England
1839	Invention of Macmillan's Bicycle	Macmillan was the first to invent the pedal-driven bicycle. A pedal driven bicycle of wood included iron-rimmed wooden wheels, a steerable wheel in the front and a larger wheel in the rear which was connected to pedals via connecting rods
1885	First automotive wheels patented-wire-spoke wheels with rubber tyre	In 1886, Carl Benz submitted the patent for his motor car at the Imperial Patent Office in Berlin. The patent "DRP 37435" was on 29 January 1886 for the first vehicle to be propelled by a motor

remarkably similar to the bicycle wheels of the time. The three-wheeled Patent Motorwagen, owned by Karl Benz in 1885, featured wire-spoked wheels and a simple hard rubber tire. Together with a crude suspension, these very early wheels had very little cushion and produced rather a bumpy ride.

1.4 Mobility and Transportation Transformed Society

One of the most basic features of human activities is the movement from place to place. The international migration since the late 1940s, investigated by Stephen Castles, observed that the cheaper transport and the growth of new media drove the growth of international migration at all skill levels [16]. Movability and freedom of movement facilitate social, cultural, political interactions, and economic activities.

Changes in mobility due to technological developments improved societies and the quality of life of populations.

The evolution of transportation goes much beyond mere travel from one point to another. Transportation summarizes an amazing account of human development, ingenuity, and the continuous pursuit of efficacy [17]. It has had a significant impact on economies, civilizations, and daily life. Transportation has influenced society, the economy, and our worldview throughout history, with a major impact in shaping our past. The progression of transportation displays human invention and persistence.

Surface transportation played a critical part in the growth of ancient civilizations. The role of transportation and infrastructure in ancient civilizations supported the vast imperial ambitions and facilitated cultural exchange and economic prosperity throughout the centuries. The remarkable wheel innovation not only revolutionized transportation in ancient Mesopotamia but also transformed agriculture and trade, laying the groundwork for the rise of complex societies. Although the Indus Valley was mostly an agricultural society, trading of goods for them was very important. Probably, the Indus Valley was the first place in the world to use wheels for transportation, such as carts. Technology enabled advances in trade. Transportation in ancient China was vital to maintaining the unity and functionality of this remarkable civilization. In Chinese dynastic society, transportation facilitated communication, trade, and cultural exchange throughout the vast empire.

In the pre-industrial age, transport modes were restricted to an agricultural economy based on local production and consumption with limited long-distance trade. The need for more rapid and cheaper transport modes was felt due to the advancements in agriculture, industrialization, and the consequent increase in medium- and long-distance traffic. During the Industrial Revolution, transportation saw substantial advances that reshaped society and the economy. Urbanization was accelerated by these advancements in transportation. More and more workers from rural areas migrated as industrial units in cities grew, resulting in the expansion of cities. Consequently, improved transportation facilitated fast delivery of raw materials and finished products, resulting in economic growth. Increased mobility also contributed to social changes, facilitating the spread of ideas and cultures [18].

In brief, the transport revolution allowed significant societal changes, such as the dimension and structure of human settlements, tradable goods and routes, the production and distribution of goods, personal and societal lifestyles, and the physical environment [19]. The structure of society can be affected significantly and often unpredictably in a relatively short period of time by the transport revolutions, which are produced by a combination of technological innovations and related organizational ones.

1.5 Transition to Gasoline Automotive Transportation

One of the most remarkable prime movers in the world, the horse very much ruled urban and rural life and culture till the nineteenth century in Asia, Europe, and North America. It took several decades to completely switch from horse-drawn to gasoline-powered transportation. Dislodging the horse from the public transportation and wagon delivery systems took almost half a century. The horse and carriage era ended as it was disrupted by a new superior technology, the internal combustion engine. They became obsolete for the purposes of transportation. The transition was driven by several factors, as displayed in Fig. 1.4.

In 1850, the steam engine was ranked as one of the greatest inventions in the world. With the advent of the electric motor and internal-combustion engine by 1950, an older invention, the steam engine, became obsolete and disappeared from transportation. This led to the spread of automobiles, trucks, and buses, including grocery carts, bicycles, roll-aboard luggage, etc. [20]. Moreover, the popularity of vehicles in 1916 was spurred by a notable decline in car prices and an increase in reliability.

The first automobile powered by gasoline was designed and built by a German engineer, Karl Benz, in 1886. Around the same period in 1886, Gottlieb Daimler and his colleague Wilhelm Maybach developed a carburetor that allowed gasoline to be used as fuel and patented one of the first successful high-speed internal combustion engines. Henry Ford later invented the Model T, sometimes known as the Tin Lizzie (1908). In the 1890s, the first automobiles were brought to Australia.

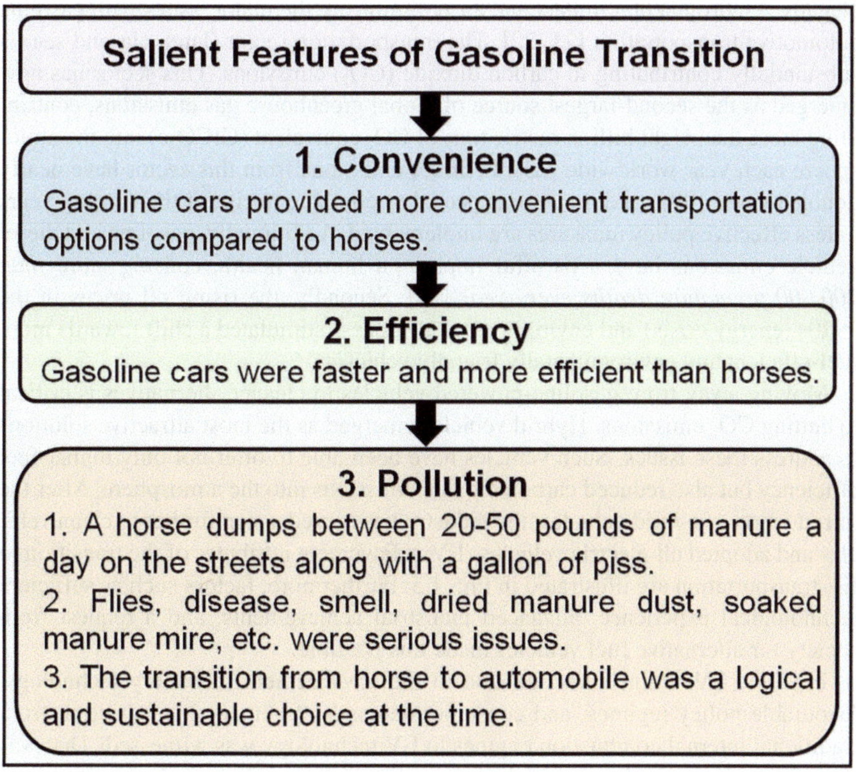

Fig. 1.4 Favorable attributes for transitioning to gasoline automotive

Gasoline and diesel engines became the dominant power in the automobile sector throughout the twentieth century. The prosperity of the automobile industry stimulated economic growth and technological innovation. These automotive innovations (engine fuel injection, turbocharging, catalytic converters, etc.) have improved engine performance, fuel efficiency, and emissions control while increasing the influence of automotive production on the economic and political arenas of numerous nations. The infrastructure for gasoline is well-established, with gas stations readily available in even the most remote areas. As a result, the reliance on and acceptance of gasoline vehicles will remain for years to come. However, gasoline-based automobiles generally have detrimental impacts on the environment, which is why hybrid car technologies were thought of and developed. By combining internal combustion engines with electric motors, an environmentally friendly transitional solution evolved.

1.6 Twenty-First Century Transportation: Transition to Electric Vehicles

Negative environmental impact and air pollution are the major issues with gasoline automotive transportation [21, 22]. The transportation sector (land, air, and sea) is substantially contributing to carbon dioxide (CO_2) emissions. This sector has now emerged as the second-largest source of global greenhouse gas emissions, contributing more than eight billion metric tons of CO_2 equivalent ($GtCO_2e$) into the atmosphere each year worldwide [23]. In 2023, emissions from this sector have nearly doubled since 1990, and they are expected to continue to rise in the coming years unless effective policy measures are implemented. Additionally, gasoline and diesel vehicle emissions have a harmful impact on human health, *causing more than 200,000 premature deaths every year* [24]. Secondly, the rising oil prices in the 1970s (energy crisis) and environmental awareness stimulated a shift towards more fuel-efficient and environmentally friendly vehicles.

Moving away from gasoline-powered vehicles to cleaner alternatives is critical to cutting CO_2 emissions. Hybrid vehicles emerged as the most attractive solutions to address these issues. Such vehicles have been able to offer not only higher fuel efficiency but also reduced carbon dioxide emissions into the atmosphere. After the introduction of hybrid vehicles, the industry has moved a step further to clean vehicles and adopted all-electric vehicles (EVs). Favorable attributes of the transition to EV transportation are illustrated in Fig. 1.5. Furthermore, factors such as sufficient technological experience, advanced industrial achievements, and a request from society for alternative fuel vehicles made this possible.

In 2010, EVs gained traction due to the advancement in battery technology, favourable policy regimes, and consumer demands. A remarkable transition from traditional internal combustion engines to EV technology was witnessed. That was a turning point in the history of the automotive industry, as many leading automotive brands and start-up companies introduced more electric vehicles to the market. This transition has revolutionized the transportation sector and has significant

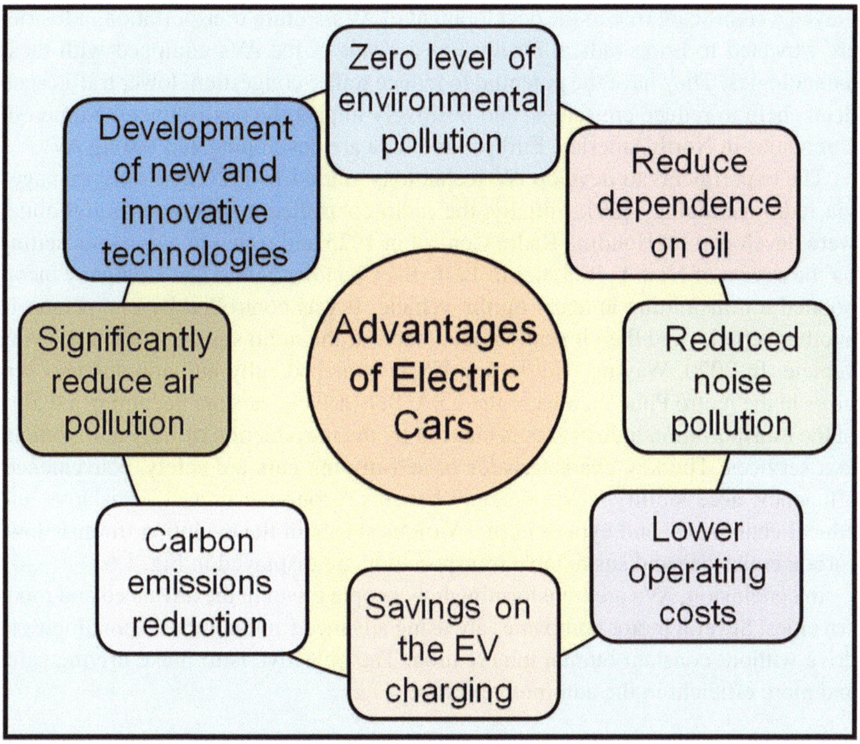

Fig. 1.5 Some undeniable advantages of electric cars

implications for business, information technology, outsourcing, and shaping new industries worldwide. The goal of having zero net emissions from the transportation sector by 2050 has also been established by a number of countries and policymakers [25]. From this perspective, the key component of the decarbonization of the transport sector and hence the economy is the use of clean vehicles, i.e., EVs [26, 27]. The development of advanced Li-ion batteries has emerged as the main instrument to achieve the successful penetration of EVs in our societies to date [28]. More and more innovation to produce eco-friendly long-range batteries for a global and sustainable EV transition is necessary to completely decarbonize the transport sector.

1.6.1 Autonomous Vehicle

With the arrival of self-driving technology, a whole new range of opportunities in the field of autonomous vehicles (AVs) may be possible in coming years. This is a new type of technology in the automotive industry. Herein, AV will be able to select the best possible route without human input or control and reach the predetermined destination. Technological advancements in sensors, cameras, radars, LiDAR systems, artificial intelligence (AI), deep learning algorithms, or a combination thereof played a significant role in the development of AVs. Future transportation industries are expected to bring radical changes as a result of the AVs equipped with these technologies. They have the potential to reduce traffic congestion, lower traffic accidents, help to reduce emissions, and positively impact the environment. Numerous companies in North America, Europe, and Asia are developing and testing AVs.

The experiments to develop AV technology started in 1920 and were managed via radio technology [29]. Initially, the radio-controlled autonomous automobiles were developed by Houdina Radio Control in 1925 and tested in a dynamic setting on the streets of New York. Later in 1926, the Chandler Motor Car Company incorporated a transmitting antenna on the vehicle. It was controlled by one person in another vehicle, and then it sent out and received the radio signals of the following vehicle. In 2020, Waymo (California, USA) launched fully autonomous taxi services in the metro Phoenix area of the USA. Self-driving cars are becoming a reality in the transportation industry, as evidenced by the introduction of fully autonomous taxi services. The key characteristics of self-driving cars are safety, convenience, efficiency, accessibility, environmental benefits, economic impact, legislative and ethical challenges, and cybersecurity. Various stages in the evolution towards low-carbon emissions and sustainable transportation are displayed in Fig. 1.6.

In conclusion, AVs are transforming how people travel in the advanced and modern cities. Several motor companies are using advanced autonomous technologies to drive without constant human intervention. The objective is to make driving safer and more efficient in the automotive sector.

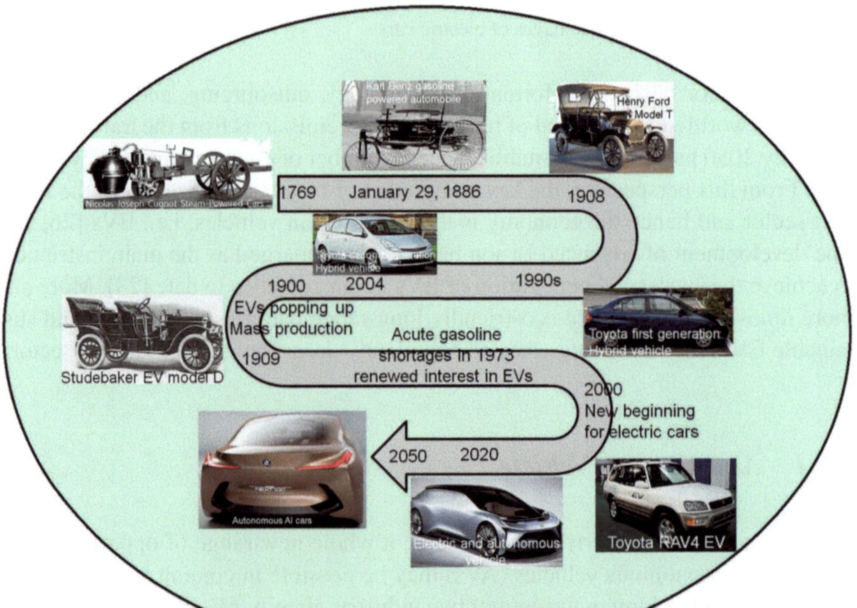

Fig. 1.6 Evolution in the automobile sector on the road to achieving clean and sustainable transportation

1.7 Inevitable Impacts of EV Transition

Transitioning to EVs is important to decarbonize the transport sector and to address climate change and environmental concerns. Future modes of transportation, such as electric and driverless cars, are predicted to drastically transform society by changing how we get around. This may have an influence on job markets, urban planning, and even on the ownership of a car. It will also make the transportation system more accessible, effective, and possibly less congested. The transition to autonomous EV transportation is probably upending a number of industries, including the labor and service market, the gasoline automotive sector, the oil (energy business), etc. In the autonomous EV age, the existing mode of transportation will simply become outdated. In essence, this could change how we live and interact with others, work, and the environment. The clean disruption of energy and transportation is inevitable by 2030 or maybe even before [30, 31]. Further negative consequences of transitioning to 100% EV transportation could include a decrease in or total cessation of demand for gasoline and diesel, strain on the electric power grid and distribution, and unemployment among autoworkers.

In brief, transportation means have played a pivotal role in shaping society as we know it today. From ancient times to the present day, advancements in transportation systems have transformed the way we live, work, and interact with one another [32]. Transportation had profound social, economic, and cultural impacts on society. Overall, the invention of the wheel and human mobility facilitated economic growth, cultural exchanges, and community relationships, leaving a lasting legacy that continues to influence modern society. It signalled the beginning of a new chapter in human civilization, characterized by enhanced economic prosperity, cultural exchanges, and mobility. Innovations in transportation have transformed our society in the past and will surely do so in the future.

References

1. Kriwaczek P (2012) Babylon: Mesopotamia and the birth of civilization, illustrated. St. Martin's Griffin Publisher, New York
2. Liu X (ed) (2024) The world of the ancient silk road (Routledge Worlds), 1st edn. Routledge, London
3. Simon E. Oldest means of transportation in the world. Website: Oldest means of transportation in the world—school drillers. Accessed 8 Jan 2025
4. Lay MG (1992) Ways of the world—a history of the world's roads and of the vehicles that used them, 1st edn. Rutgers University Press
5. Pritchard JC (2010) Animal traction and transport in the 21st century: getting the priorities right. Vet J 186(3):271–274. https://doi.org/10.1016/j.tvjl.2010.08.004
6. Sharma M (2024) The journey of transportation: from horse carts to EVs. https://1charging.com/the-journey-of-transportation-from-horse-carts-to-evs/. Accessed 9 Jan 2025
7. Illuminating Facts (n.d.) Timeline of transportation evolution: from horse-drawn carriages to electric vehicles. https://illuminatingfacts.com/timeline-of-transportation-evolution-from-horse-drawn-carriages-to-electric-vehicles/#:~:text=%EE%80%80Horse-drawn%20carriages%EE%80%81%20were%20an%20essential. Accessed 9 Jan 2025

8. Casson L (1994) Travel in the ancient world. Johns Hopkins University Press
9. Mark JJ (2022) Trade in ancient Mesopotamia. World History Encyclopedia. https://www.worldhistory.org/article/2114/trade-in-ancient-mesopotamia/
10. Possehl G (2002) The Indus civilization a contemporary perspective. Alta Mira Press
11. Becker L (2024) The Roman Road system (Ancient Engineering Marvels). https://ancientengineeringmarvels.com/the-roman-road-system/. Accessed 12 Jan 2025
12. Forbes RJ (1993) Studies in ancient technology, vol 2, 3rd edn. E.J. Brill, Leiden
13. Collins HJ, Hart CA (1936) Principles of road engineering. Edward Arnold, London
14. Ralston A (2005) Between a rock and a hard place. Pocket, London
15. Delgado J (2018) History of the wheels. OEM Wheel Shop, Anaheim, CA. https://www.oemwheelshop.com/blogs/oem-wheels-and-rims/history-of-the-wheel
16. Castles S (2018) Social transformation and human mobility: reflections on the past, present and future of migration. J Intercult Stud 39(2):238–251. https://doi.org/10.1080/07256868.2018.1444351
17. Leaders in Law (2024) How transportation has evolved over time and shaped our society. Legal News. https://www.leaders-in-law.com/legal-news/how-transportation-has-evolved-over-time-and-shaped-our-society/. Accessed 19 Jan 2025
18. Cars Trucks Roads (2024) Transportation during the industrial revolution: key innovations that changed society. Spencer Gasselin, WY. https://carstrucksroads.com/transportation-during-industrial-revolution/. Accessed 19 Jan 2025
19. Cascetta E, Henke I (2023) The seventh transport revolution and the new challenges for sustainable mobility. J Urban Mobility 4:100059. https://doi.org/10.1016/j.urbmob.2023.100059
20. Richard B (2016) The wheel: inventions and reinventions. Columbia University Press, New York. https://doi.org/10.7312/bull17338
21. Azhar U, Yaseen S, Arif M, Babar M, Sagir M (2024) Emission of greenhouse gases from transportation. In: Rahimpour MR, Makarem MA, Meshksar M (eds) Advances and technology development in greenhouse gases: emission, capture and conversion. Elsevier, pp 147–163. https://doi.org/10.1016/B978-0-443-19231-9.00016-8
22. Fei Y, Qin P, Chu Y, Zheng H, Tan-Soo J-S, Zhang X-B (2025) Does high gasoline price spur electric vehicle adoption? Evidence from Chinese cities. Energy Econ 142:108188. https://doi.org/10.1016/j.eneco.2025.108188
23. Statista (2025) Transportation emissions worldwide—statistics & facts. Statista Research Department. https://www.statista.com/topics/7476/transportation-emissions-worldwide/#topicOverview. Accessed 23 Jan 2025
24. Shultz D (2020) The climate and health impacts of gasoline and diesel emissions. Eos. https://doi.org/10.1029/2020EO142849
25. Alsauskas O, Dasgupta A, Connelly E, Gouy A, Le Marois J-B, McDonagh S, Pales AF, Petropoulos A, Teter J (2023) Global EV outlook 2023. IEA, Paris. https://www.iea.org/reports/global-ev-outlook-2023. Accessed 27 Jan 2025
26. Glitman K, Farnsworth D, Hildermeier J (2019) The role of electric vehicles in a decarbonized economy: supporting a reliable, affordable and efficient electric system. Electr J 32(7):106623. https://doi.org/10.1016/j.tej.2019.106623
27. Kumar RR, Alok K (2020) Adoption of electric vehicle: a literature review and prospects for sustainability. J Clean Prod 253:119911. https://doi.org/10.1016/j.jclepro.2019.119911
28. Boucar Diouf B (2024) The electric vehicle transition. Environ Sci Adv 3(2):332–345. https://doi.org/10.1039/d3va00322a
29. Editorial, RoboticsBiz (2021) History of autonomous vehicles—timeline. https://roboticsbiz.com/history-of-autonomous-vehicles-timeline/. Accessed 16 Jan 2025
30. Seba T (2014) Clean disruption of energy and transportation: how silicon valley is making oil, nuclear, natural gas, and coal obsolete by 2030. Beta Edition
31. Jeong JH, Kim C, Jo HJ (2024) Three major challenges in the shift to electric vehicles: industrial organization, industrial policy, and a just transition. Sociol Compass 18(5):e13218. https://doi.org/10.1111/soc4.13218
32. Wray J (2024) How has transportation impacted society? NextGen Horizons Inc. https://www.ncesc.com/how-has-transportation-impacted-society/. Accessed 19 Jan 2025

Chapter 2
Decarbonizing Road Transportation

Contents

2.1 Introduction: Global Warming and Greenhouse Gas Emissions

Carbon dioxide (CO_2) and other greenhouse gases (GHG) are the leading factors driving the current increase in global temperatures and climate change. The CO_2, is contributing more to the warming effect than other gases that are present at smaller levels in the atmosphere. According to the International Energy Agency report, global energy-related CO_2 emissions reached a new record high of 37.4 billion tonnes (or gigatonnes, Gt) in 2023 [1]. Figure 2.1 shows the global energy-related CO_2 emissions from various sectors in 2021. Here, fugitive emissions signify the direct discharge of GHG to the air from different kinds of equipment and processes. In 2022, more than 21.7% of global CO_2 emissions originate from the transportation sector [2]. The transportation sector is contributing significantly to the total global CO_2 emissions (21.7%), with road transport accounting for 74% in 2018 [3]. In 2022, emissions from transportation surged by 254 tonnes, compared to 7.73 Gt in

R. Pode, B. Diouf, *Decarbonization of Road Transportation*, Green Energy and Technology, https://doi.org/10.1007/978-3-032-11388-7_2

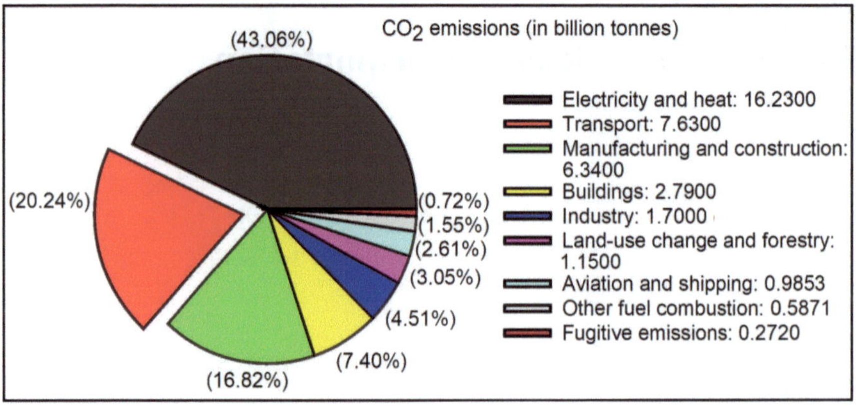

Fig. 2.1 Sector-wise global CO_2 emissions in 2021 (in billion tonnes, Gt)

2021. The World Energy Outlook reported that globally road vehicles emitted 5.9 Gt CO_2 in 2021, with heavy-duty trucks contributing 1.8 Gt, which is 4% of global CO_2 emissions [4]. Nearly 25% of the world's energy-related CO_2 emissions were generated by the transport sector each year. In 2019, GHG emissions from the transport sector reached 8.7 $GtCO_2$-eq (up by 3.7 $GtCO_2$-eq compared to 1990 levels) and accounted for 23% of global energy-related CO_2 emissions [5]. Nearly 70% of direct transport emissions originated from road vehicles. The Intergovernmental Panel on Climate Change (IPCC) has identified the transport sector as the fourth-largest contributor to global carbon emissions, accounting for 15% of total emissions directly [5].

Recently, Alam et al. reported that Bangladesh had registered a total of 5,982,765 vehicles by January 2024, according to the Bangladesh Road Transport Authority (2023) [6]. Interestingly, over 80% of these vehicles are responsible for emitting excessive black smoke, primarily from petrol and diesel engines. The transport sector contributed 9.92% to Bangladesh's GHG emissions, which are anticipated to increase to 36.28 million tonnes by 2030, with road transport being the largest contributor. As a result, the need to address climate change and achieve green emission targets in transportation has become more critical than ever before. Transitioning from gasoline-powered vehicles to electric vehicles (EVs) may be one of the effective strategies to *decarbonize land transportation*.

This chapter discusses the consumption of fossil fuels in road transportation and the subsequent impacts on the CO_2 emissions in the USA, UK, EU countries, China, India, Japan, and South Korea. Moreover, the strategies for decarbonizing road transportation to enhance environmental sustainability are deliberated here. Advancing towards a cleaner transportation system is crucial for reducing emissions in this sector. Nonetheless, the shift to fossil-free transportation must occur at a much accelerated pace. There is an urgent need for long-term strategies to increase the share of green transportation in order to effectively reduce CO_2 emissions from transportation.

2.2 Data Collection

In this chapter, we gathered information on fossil fuel consumption and CO_2 emissions in road transportation across the USA, UK, EU countries, China, India, Japan, and South Korea for the current chapter. We analyzed the collected data on fossil fuel consumption, GHG, and CO_2 emissions, which are imperative for decarbonizing land transportation, together with the switch towards intelligent road transportation systems. Several strategies, including the transition from internal combustion engine (ICE) vehicles to electric vehicles, are expected to effectively address GHG emissions in road transportation.

2.3 Transportation Emissions

Road transportation plays a significant role in global CO_2 emissions across all transport modes [7]. Between 1970 and 2023, global transportation emissions grew nearly threefold, driven by factors such as population growth and expanding economies [8]. In 2023, the biggest source of global CO_2 emissions was electric power generation, which contributed approximately 38%, followed by the transportation sector, which accounted for over 21% of emissions that year (the second-largest share of global CO_2 emissions) [9, 10]. Specifically, in 2023, CO_2 emissions from cars and vans reached 3.8 Gt, accounting for more than 60% of emissions from the road transport sector [11].

2.3.1 Transportation Emissions in the USA

Consumption of fossil fuels in the United States road transportation sector from 2015 is presented in Table 2.1, while Fig. 2.2 provides an overview of total fossil fuel consumption [12]. The significant variations in fossil fuel consumption from 2019 to 2022 were largely influenced by the global coronavirus pandemic, as displayed in Fig. 2.2. In 2023, motor gasoline was the dominant fuel in the US highway transportation sector, consuming approximately 8.5 million barrels per day (equivalent to 357 million gallons per day or 130.305 billion gallons annually), along with diesel fuel usage of less than three million barrels per day (around 45.990 billion gallons per year).

Road travel accounts for the largest source of transportation emissions in the US. As illustrated in Table 2.2, the GHG and CO_2 emissions from the road transportation sector are quite significant. From 1990 to 2005, total emissions in the transportation sector increased by 29% in the USA, driven largely by an increase in vehicle miles traveled (VMT) by cars and trucks [13]. In 2022, emissions from light-duty combustion-engine trucks and heavy-duty trucks (MHDT) were approximately 645 and 400 $MtCO_2e$, respectively, while passenger cars contributed around

Table 2.1 Consumption of fossil fuels in the road transportation sector United States since 2015 (million gallons) [12]

USA (Gasoline, diesel and other fuels)	Light duty vehicle, short wheel base and motorcycle	Light duty vehicle, long wheel base	Single-unit 2-axle 6-tire or more truck	Combination truck	Bus	Total consumption (Gasoline, diesel and other fuels) (million gallons)
2015	90,479	36,437	14,851	28,886	2228	170,653
2016	91,954	37,819	15,338	29,555	2226	174,666
2017	92,171	37,467	15,600	30,364	2350	175,602
2018	92,042	37,189	16,080	30,325	2494	175,636
2019	93,874	38,030	16,657	28,987	2451	177,548
2020	76,936	35,438	15,577	29,186	2047	157,137
2021	85,754	38,494	17,169	30,439	2274	171,856
2022	86,581	37,939	17,181	28,218	2498	169,919
2023						130,305 (motor gasoline) + 45,990 (diesel fuel) = 176,295 (total)

Note: 1 gallon = 3.785412 L; 1 barrel = 42 gallon

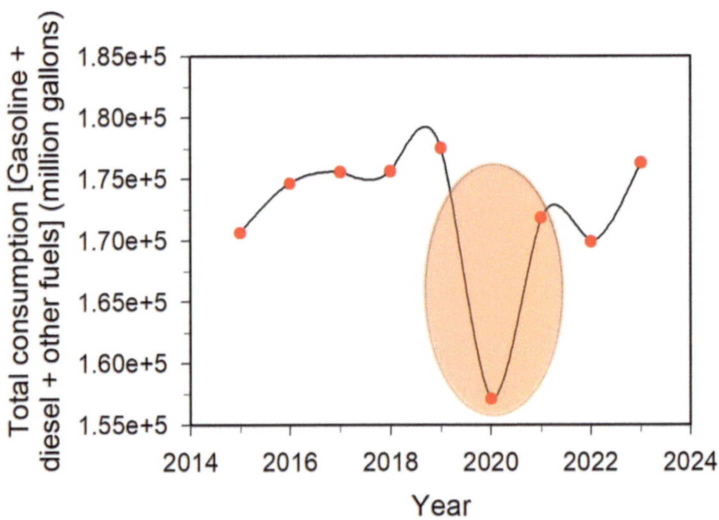

Fig. 2.2 Total consumption of fossil fuels in the road transportation sector of the USA since 2015

360 $MtCO_2e$. The largest sources of transportation-related GHG emissions in 2022 were light-duty trucks, which include sport utility vehicles, pickup trucks, and mini-vans (37%); medium- and heavy-duty trucks (23%); and passenger cars (20%) [14]. Although passenger cars were among the highest emitters in the US, advancements in efficiency and stringent emissions standards have reduced their GHG emissions

Table 2.2 Carbon dioxide emissions from the fossil fuel energy combustion in transportation sector (million metric tons of CO_2)

Year	Total CO_2 emission in transportation sector (On-Road Vehicles, Aircraft, Ships and Boats, Rail, Pipelines, and Pipelines) (million tonnes)	GHG Emissions/ (CO_2 emissions) in road transportation (million tonnes)	CO_2 emissions from usage of motor gasoline in road transportation (million tonnes)
1985	1423		910
1990	1468.9	1202.0	967
1995		1370.4	1027
2000	1757	1572.6	1128
2005	1858.6	1637.6	1177
2010		1541.3	1088
2015		1522.0/(1467.0)	1073
2018	1813.1	1558.2/(1523.3)	1090
2019	1816.6	1549.3/(1512.7)	1086
2020	1572.8	1374.4/(1336.3)	935
2021	1753.5	1496.6/(1459.4)	1025
2022	1751.3	1476.7/(1442.2)	1018
2023	1857		1034
2024	1851		1036

Note: 1 tonne of CO_2 is equal to 12/44 or 0.27 tonnes of CO_2 equivalent

by over 40% since 1990. In contrast, emissions of MHDT, primarily produced by diesel engines, saw an almost 80% increase during the same timeframe. In 2023, per capita transportation emissions in the United States reached an average of 5.05 tonnes of CO_2, marking the highest levels among the world's largest contributors to transportation emissions. During the global pandemic, the GHG and CO_2 emissions were reduced due to the reduced fossil fuel consumption. To gain insight into trends in road transportation emissions, the emissions data across different years was analyzed (see Table 2.2), and a graph depicting the relationship between CHG levels and CO_2 emissions in road vehicle emissions over time was plotted, as displayed in Fig. 2.3 [15–17]. Notably, CO_2 emissions from the usage of motor gasoline in road transportation have almost flattened (see Region 2 in Fig. 2.3). This could indicate a positive shift towards green fuel technologies or an improvement in the efficiency of combustion engines.

2.3.2 *Transportation Emissions in the UK*

Figure 2.4 illustrates the total petroleum fuel energy consumption, which includes petrol, diesel, and liquefied petroleum gas (LPG), as well as the total energy consumption, comprising petroleum fuel energy, renewable fuels, and electricity in the road transport sector of the United Kingdom (UK) from 2000 to 2023 [18]. During this period, the petroleum fuel consumption reduced from 41.1 to 35.2 Mtoe,

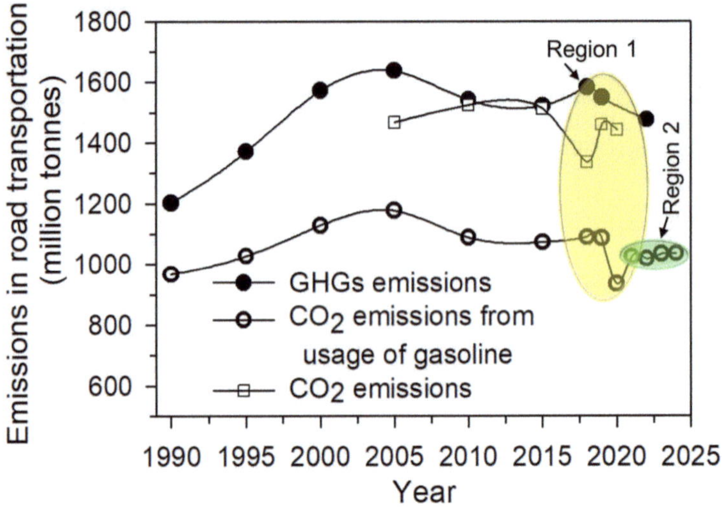

Fig. 2.3 CHGs and CO_2 emissions in road transportation in the USA from 1990

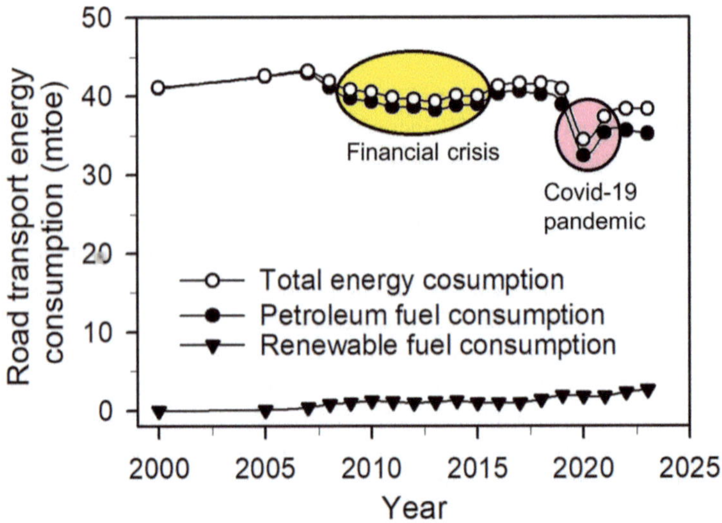

Fig. 2.4 Transport energy (total, petroleum fuel, and renewable fuels) consumption (Mtoe) by road transport mode in the United Kingdom from 2000 to 2023

reflecting a percentage change of −14.35%. Simultaneously, total energy consumption declined from 41.1 to 38.3 Mtoe, a change of −6.81% [19]. The energy consumption in transportation showed a downturn around 2008 due to the financial crisis and again in 2020 during the Covid-19 pandemic (see Fig. 2.4). In 2023, road transport accounted for the largest share of energy consumption, representing about 70% of the sector's total energy consumption (54.4 Mtoe) [20]. Results presented in

Fig. 2.4 suggest a stabilization of energy consumption in road transport following the Covid-19 pandemic. Furthermore, the share of renewable fuels in total road transport energy consumption has shown continuous improvement since 2005, with a significant increase of 136% in renewable fuel consumption from 2013 to 2023.

UK domestic transportation sector emissions increased by 1.6% in 2022, to 113 million tonnes of carbon dioxide equivalent ($MtCO_2e$). While this represents an overall decline of 12% compared with 1990 levels. Emissions from UK road vehicles increased by roughly 1% in 2022, to approximately a value of 101 $MtCO_2e$. This not only accounted for around 90% of domestic transportation-related emissions but also a quarter of total UK GHG emissions. Passenger cars alone were responsible for around 60 $MtCO_2e$ in 2022, or 60% of transportation emissions that year.

Figure 2.5 displays the total domestic GHG emissions, including carbon dioxide, methane, nitrous oxide, hydrofluorocarbons, perfluorocarbons, sulphur hexafluoride, and nitrogen trifluoride (expressed in $MtCO_2e$), beside the direct CO_2 emissions from road transportation, measured in a million tonnes of carbon dioxide ($MtCO_2$), from 1990 to 2022. The transportation sector has consistently produced the highest levels of emissions. According to the Department for Energy Security & Net Zero studies, the United Kingdom has achieved a notable milestone by becoming the world's first major economy to reduce its total GHG emissions by 52.7%, from 810.7 $MtCO_2e$ in 1990 to 384.2 $MtCO_2e$ in 2023, all while growing its economy by 79% [21]. Especially, UK territorial carbon dioxide emissions decreased by 49.8% during the same period. However, emissions from the transportation sector have seen insignificant change over recent decades. Domestic transport, which includes road vehicles, domestic aviation, shipping (including military), fishing vessels, and railways, continues to be the largest source of emissions in the UK,

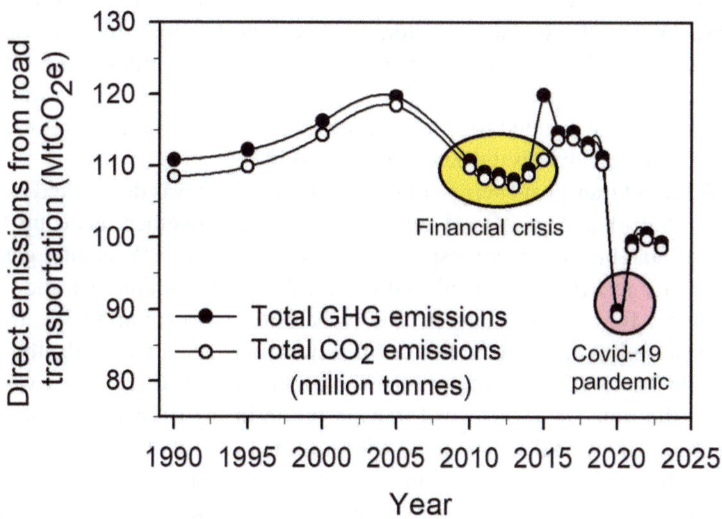

Fig. 2.5 GHG and CO_2 emissions from the road transport sector in the UK 1990–2022

Table 2.3 UK's road transportation GHG emissions per kilometer travelled (gCO_2e) in 2022 [22]

Mode	gCO_2e
Bus (average)	97
Bus (local London)	79
Coach (bus)	27
Diesel car	171
Electric car	47
Motorbike	114
Petrol car	170
Plug-in hybrid	68

accounting for 29.1% of total emissions in 2023. Provisional estimates indicate that GHG emissions from the domestic transport sector and specifically from domestic road transport were approximately 111.6 and 99.4 $MtCO_2e$, respectively, in 2023. Over the period from 1990 to 2023, the domestic road transport GHGs emissions decreased by 10%, while CO_2 emissions from this sector reduced by 9%, from 108.5 $MtCO_2$ in 1990 to 98.6 $MtCO_2$ in 2023. The carbon footprint of travel, measured in grams of carbon dioxide equivalents per passenger kilometer, is displayed in Table 2.3 according to the UK Government, Department for Energy Security and Net Zero (2022) [22].

2.3.3 *Transportation GHG Emissions in the European Union (EU) Countries*

In 2022, the transport sector, comprising road, air, water, rail, and others, accounted for 31.0% of final energy consumption across the European Union (EU)-27 countries [23]. Within the transport sector, road transport, which includes scooters, motorcycles, tricycles, quads, cars, vans, minibuses, buses, trucks, and other on-road vehicles, was responsible for 73.6% of total energy consumed in transportation. Air transport accounted for 11.4%, water transport contributed 13.0%, and rail transport made up 1.4%, as displayed in Fig. 2.6 [23].

In 2022, road transport energy consumption in the EU was dominated by motor gasoline (65.4%) and gas/diesel oil (25.2%), which together made up 90.6%. Renewables and biofuels were responsible for 6.4%, LPG 2.0%, natural gas 0.7%, and electricity just 0.3%, as displayed in Fig. 2.6b [24]. In contrast, the primary energy source for road transport in 1990 was motor gasoline at 54.7%, followed by gas/diesel oil at 43.8% (see Fig. 2.6). Natural gas accounted only for a mere 0.1%, while the contributions from renewables, biofuels, and electricity were negligible. At the time (in 1990), road transportation in the EU was almost entirely dependent on traditional fossil fuels, with minimal contribution from alternative energy sources, likely due to technological and infrastructure limitations. The shift toward biofuels and renewables gained traction much later, particularly during the 2000s.

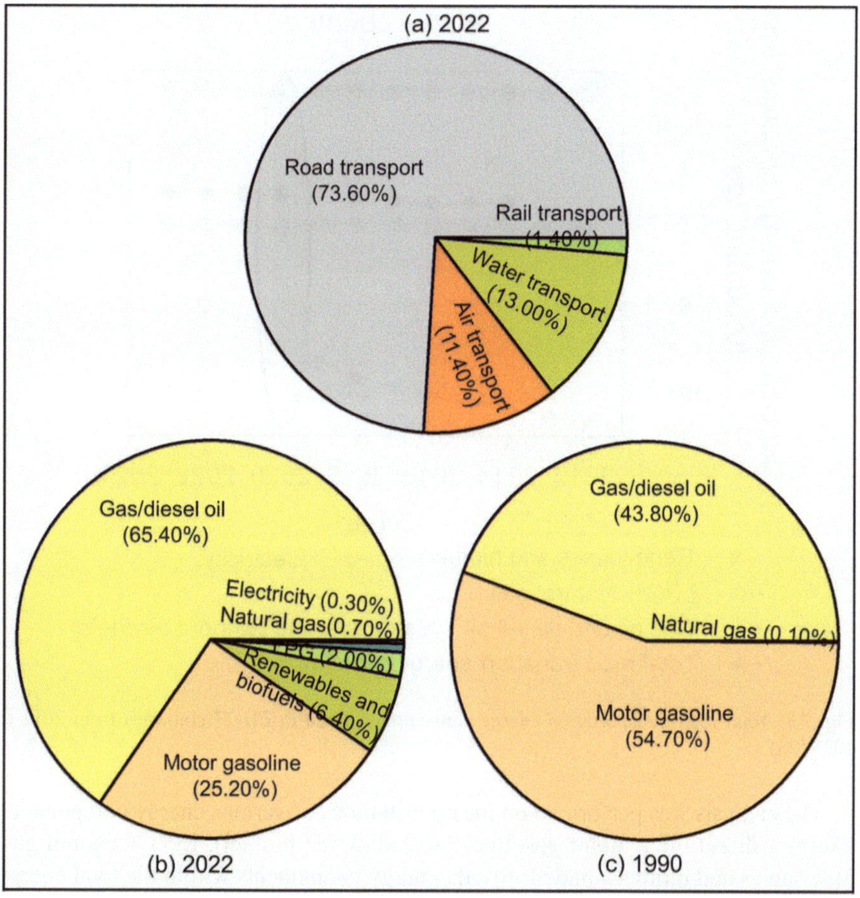

Fig. 2.6 Transport sector energy consumption in the EU 27 countries. (**a**) Road, air, water, rail, and other energy consumption in 2022, (**b**) Road transport energy consumption in 2022, and (**c**) Road transport energy consumption in 1990

Figure 2.7 shows the trend in total road transportation energy consumption across the EU-27 countries [25]. From 2012 to 2023, the total energy consumption remained almost stable. Specifically, the total energy consumption for road transportation experienced a marginal increase of +4.65%, while gas/diesel oil and motor gasoline components (excluding the biofuel portion) saw an increase of +2.26% over the same duration. The depreciation observed around 2020 was linked to the impact of the Covid-19 pandemic. According to the results of Fig. 2.7, gas/diesel oil and motor gasoline continued to be the main energy sources for road transportation during that period in the European nations. Interestingly, there were significant growths in renewables and biofuels, electricity, and LPG + natural gas energy components, which improved by +35.32, +3149.41, and +15.72%, respectively, between 2012 and 2023 (see Fig. 2.7).

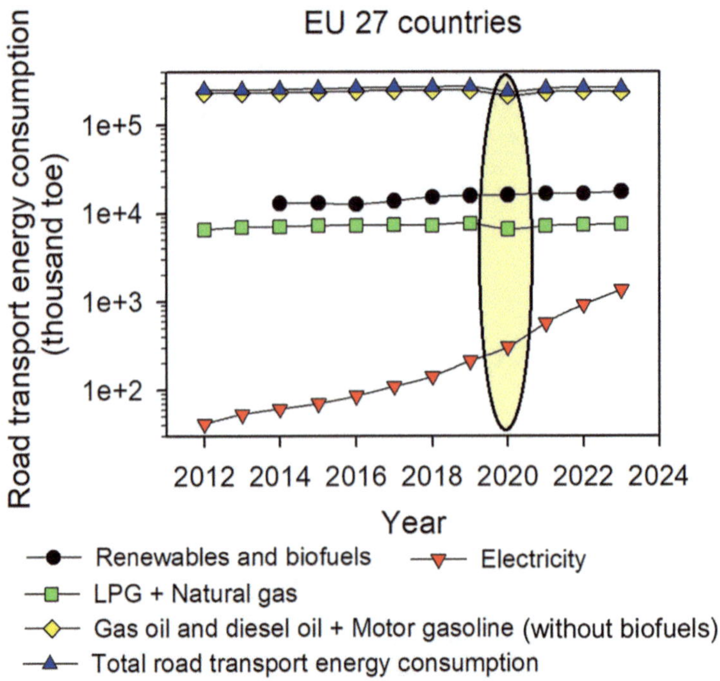

Fig. 2.7 Total road transportation energy consumption trend in EU-27 countries from 2012 to 2023 [25]

The analysis was performed on the contributions of various energy components, like gas/(diesel oil + motor gasoline) (excluding the biofuel), LPG + natural gas, renewables and biofuels, and electricity energy components within the total energy consumption for road transport to have a clear understanding of trends in these energies. Figure 2.8 illustrates the various energy consumption components for the years 2014 and 2023. Table 2.4 details the percentage of each energy constituent in the total energy consumption for road transport across the EU-27 countries. Gas/diesel oil and motor gasoline together remained the predominant energy sources for road transport in both 2014 (92.03%) and 2023 (89.90%). The share of renewables and biofuels in road transportation was negligible, at 5.13% in 2014, increasing slightly to 6.69% in 2023. While LPG and natural gas were used in most EU countries, their collective share in road transport rarely exceeded 3%. Furthermore, the use of electric energy remained minimal, accounting for just 0.024% in 2014 and rising to 0.51% in 2023 (see Fig. 2.8 and Table 2.4). Undoubtedly, substantial reliance on gas/diesel oil and motor gasoline also correspondingly contributed significantly to GHG emissions from road transportation in the EU countries.

It is important to highlight the significant variation in total energy consumption for road transport across different EU countries. Therefore, to highlight this diversity, the study was performed on the total energy consumption and contributions of various energy constituents in road transport for Norway and the Netherlands,

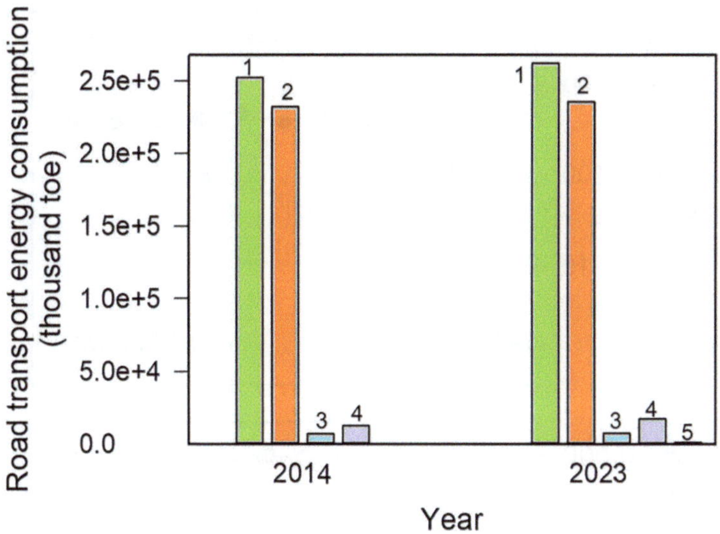

Fig. 2.8 Contribution of various energy constituents in the total energy consumption in road transport in 2014 and 2023 in EU-27 countries

Table 2.4 Percentage portion of each constituent in the total energy consumption in road transport (thousand tonnes of oil equivalent) in EU-27 countries in 2014 and 2023

Year	Total energy consumption (Thousand toe) (1)	Gas oil and diesel oil + motor gasoline (without biofuel portion) (Thousand toe) (2)	LPG + Natural gas (Thousand toe) (3)	Renewables and biofuels (Thousand toe) (4)	Electricity energy (Thousand toe) (5)
2014	252,350.220	232,249.373	7060.72	12,965.468	61.597
	Percentage portion in the total energy consumption (%)	92.03	2.79	5.13	0.024
2023	262,121.717	235,659.685	7558.759	17,545.949	1348.156
	Percentage portion in the total energy consumption (%)	89.90	2.88	6.69	0.51

which serve as representative examples. As displayed in Fig. 2.9, the energy consumption for road transport in both nations from 2012 to 2023 shows that petroleum fuels remained the core energy source during this period [25]. However, visible contrasts exist between the two countries. In Norway, total energy consumption and petroleum fuel usage reduced by 6.37% and 20.66%, respectively, from 2012 to 2023. Meanwhile, the use of renewables, biofuels, and electricity in road transport increased significantly, as seen in Fig. 2.9. Similarly, in the Netherlands, total energy consumption and petroleum fuel reliance decreased by

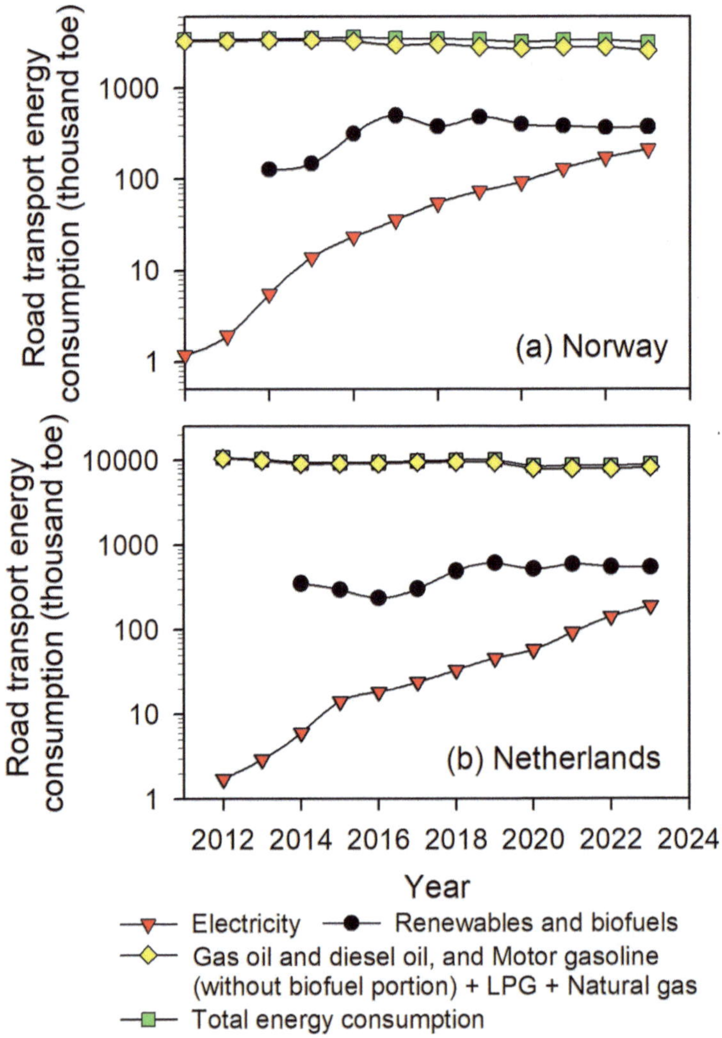

Fig. 2.9 Road transport energy consumption in Norway and the Netherlands from 2012 to 2023 [25]

16.20% and 20.72%, respectively, during the same period, demonstrating a significant rise in the adoption of renewables, biofuels, and electricity energy. Interestingly, the increase in electricity energy usage was vividly manifested in both countries. These trends towards greater confidence in renewables, biofuels, and electricity, alongside decreased petroleum fuel energy consumption and overall energy consumption, signal a positive change towards greener road transportation in both Norway and the Netherlands. These countries serve as exemplary models in managing their road transport energy portfolio for others to follow. In summary, the proportions of gas/diesel oil and motor gasoline (excluding the

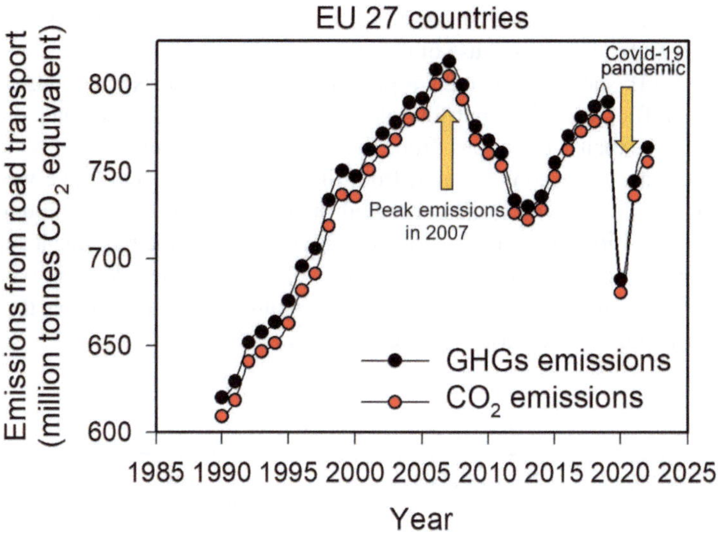

Fig. 2.10 Road transportation GHGs and CO_2 emissions in the EU countries from 1990 to 2022

biofuel) use among the EU-27 countries have increased marginally or, in some instances, lowered.

In recent decades, EU countries have successfully cut emissions across all economic sectors, including transportation to some extent [26]. However, road transportation has still been consistently responsible for a significant share of total transportation emissions, averaging 73.54% from 1990 to 2022. GHG and CO_2 emissions within the EU are displayed in Fig. 2.10 [27]. Since peaking in 2007, the emissions from road transportation have reduced by roughly 6.08% for GHG and 6.11% for CO_2 between 2007 and 2022. This reduction can be attributed to the decarbonization of the vehicle fleet and an increase in electricity use in road transport (refer to Fig. 2.7). In 2022, road vehicles emitted about 763.7 $MtCO_2e$ and 755 $MtCO_2$ of GHG and CO_2 in the EU, respectively. Despite this decline, emissions in 2022 were still considerably higher than the levels recorded in 1990 (base year), displaying increases of 23.15% and 24.01%, respectively.

In 2023, the transport sector accounted for about one-third of Norway's total GHG emissions. GHG and CO_2 emissions in Norway from road transport are illustrated in Fig. 2.11a. Although transport emissions account for over 29% of the country's overall CO_2 output, they have declined sharply since 2012. This reduction can be mostly attributed to the concentrated efforts of all involved stakeholders to gradually transition to electrifying both the vehicle fleet and maritime transport. Despite an increase in transport demand, the decarbonization of the vehicle fleet has been the major cause for this decrease. According to data published by Statistics Norway in November 2024, GHG and CO_2 emissions from road transport in 2023 were approximately 7.9% and 8.6% higher than the base levels of 1990, respectively [28]. However, these emissions from road transport are lower by 21.57% and 28% in

2023 compared to 2014, which can be partly due to the use of biofuels in place of fossil fuels and the increased use of electricity.

Figure 2.11b shows the actual emission into the air in the Netherlands resulting from road traffic, including contributions from foreign vehicles. The data presented encompasses exhaust emissions from the combustion of (motor) fuels, gasoline evaporation, and emissions released from tire, road surface, and brake wear [29]. CO_2 emissions align with trends observed in Norway and other EU countries. In 2023, CO_2 emissions were approximately 12% higher than the baseline levels set in 1990. However, after peaking in 2004, emissions have started declining and were reduced by 10% in 2023. This reduction in emissions is likely attributable to changes in the energy mix of fuels used in road transport.

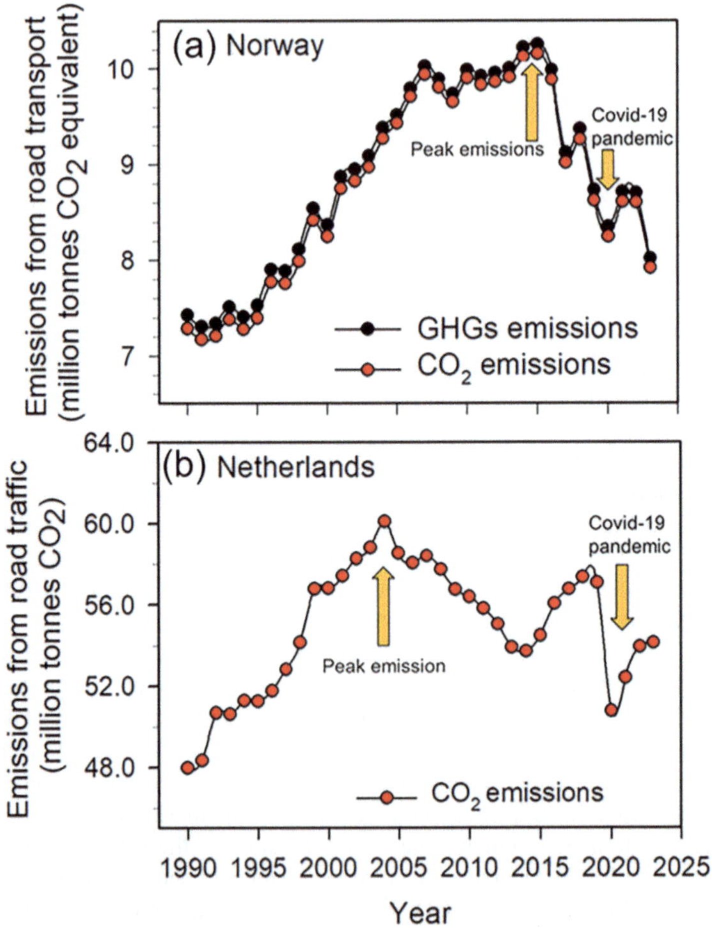

Fig. 2.11 Road transportation GHGs and CO_2 emissions in (**a**) Norway and (**b**) the Netherlands from 1990 to 2023

2.3.4 Transportation Emissions in China

China's oil fuel consumption in the road transportation sector has been undergoing significant changes. Figure 2.12 shows the total fuel consumption (including gasoline, diesel, and gas) and total CO_2 emissions from various transportation modes (light-duty vehicles + heavy-duty vehicles + marine + aviation activities) [30–32]. In recent years, the demand for oil-based fuels like gasoline and diesel has seen a decline. In 2024, the total consumption of these fuels was approximately 470.0835 million tonnes, representing a 19.11% increase compared to 2015 but a 2.5% drop from 2021 levels. The peak consumption recorded was 481.6905 million tonnes in 2021. Projections for 2025 suggest that petroleum fuel consumption in transportation will reduce to 464.28 million tonnes. Factors contributing to the reduction in gasoline consumption in China include increased sales of EVs, sluggish economic growth, and a decline in population [33]. The trend in gasoline consumption is forecasted to decline at an annual compound of 2.7% through 2040 [34]. The transportation sector leads this decline at an annual rate of 2.8%, while still maintaining its share at about 60% of total gasoline demand. Erica Downs and Abhiram Rajendran noted that this decrease in gasoline consumption is largely driven by the increasing adoption of EVs and government policies aimed at promoting energy security and

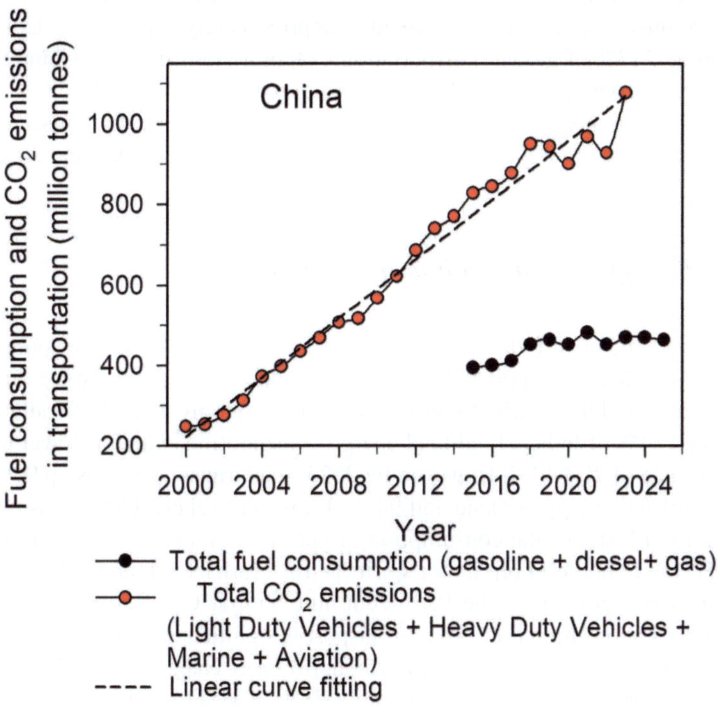

Fig. 2.12 Total fuel consumption and total CO_2 emissions from transportation in China from 2000 to 2023 [30–32]

curbing pollution [35]. Additionally, various substitutes have avoided oil demand growth of around 15% (or 1.2 million barrels per day) in China since 2019, with an expected additional 5% reduction this year, mainly due to the rapid increase in EV penetration [30].

From 2000 to 2021, transportation-related total CO_2 emissions in China showed a steady rise, with a slight decline observed after peaking in 2021 (see Fig. 2.12). In 2021, total CO_2 emission from transportation was 969.447 million tonnes, which shows a 290% increase compared to the emissions in the year 2000. By 2023, these emissions increased further to 1078 million tonnes, indicating a 333% surge from levels documented in 2000. This significant growth was largely attributed to the resumption of economic activity following the end of zero-COVID policies. Furthermore, a study by Teng et al. in 2024 showed that carbon emissions from road transportation in China grew from 11.9 to 33.8 $MtCO_2e$ between 2000 and 2020 [36]. If China implements its stated national policies, such as the Action Plan for Carbon Dioxide Peaking Before 2030 [37], the NEV Industrial Development Plan 2021–2035 [38], and the Technology Roadmap for Energy-Saving and New Energy Vehicles 2.0 [39], it is projected that the country's road transport emissions could peak before 2030, with petroleum consumption potentially peaking before 2027 [40].

The Innovation Center for Energy and Transportation (iCET), a Beijing-based think tank, estimates that transport contributed about 10% of the country's total greenhouse gas emissions in 2019. In 2021, greenhouse gas emissions from the transportation sector in China amounted to approximately 1.1 $GtCO_2e$. China produced around 13% of global transportation CO_2 emissions in 2023. China was the second-largest contributor to global transportation emissions in 2023. The transportation sector accounts for over 27% of China's total greenhouse gas emissions, with projections estimating it will reach its emission peak between 2025 and 2035 [41].

2.3.5 Transportation Emissions in India

The transport sector in India accounts for approximately 50% of the country's oil demand [42]. Road transport is the dominant mode of transport, both in terms of traffic volume and its contribution to the national economy. In 2021, road transport represented 14% of India's total final energy consumption, which is lower than the global average of 20% [43]. However, by 2023, road transport makes up 92% of all transport-related energy demand and 94% of transport-related CO_2 emissions [43]. Figure 2.13a illustrates the consumption of petroleum fuels (including high-speed diesel (HSD), light diesel oil, motor spirit (petrol), furnace oil, and LPG) in India's road transport sector [44]. The Petroleum and Natural Gas Report indicates that reported diesel consumption for road transport is significantly underestimated, as a large portion is categorized under the reseller/retailer category [45]. According to the Energy Statistics report, approximately 78% of HSD reported in the reseller/retailer category is actually used for road transport [46]. Therefore, adding the reported diesel consumption for road transport with the 78% of diesel consumption derived from the reseller/retailer category allows for a more accurate estimate of total diesel consumption in road transport. The results reveal a significant increase in oil fuel consumption. In FY2023–24, oil usage in road transport was 278.1% higher than in FY2000–01. Since the transport sector caters to over a billion people, the rapid urbanization in India continues to drive the addition of new vehicles on the roads each year, leading to rising oil fuel consumption and contributing to environmental pollution.

The GHG emissions from the road transportation sector primarily originate from the exhaust of cars, motorbikes, and commercial vehicles. These emissions result mainly from the burning of fossil fuels used in on-road vehicles. The majority of fuel used in road transportation is petrol or diesel, although some vehicles are now powered by natural gas. In 2014, road transport emerged as the largest contributor to GHG emissions within India's transport sector, accounting for about 90% of total emissions [47]. According to NITI Aayog, India's transport sector ranked as the third most GHG-emitting sector, containing 14% of energy-related CO_2 emissions in 2021 [48]. Data presented in Fig. 2.13b shows CO_2 emissions from 1990 to 2023, highlighting its importance among GHG emissions from transportation (including domestic aviation, road transportation (fuel-based), railways, domestic waterborne navigation, and other forms of transportation, measured in $MtCO_2e$) [49]. In 2023, these emissions had surged to more than five times (427.70%) the levels recorded in 1990, the baseline year under the Kyoto Protocol. With projections indicating that India's urban population will be expected to double by 2050, a further rise in these emissions is anticipated.

The road transport sector, which comprises both passenger and freight transportation, accounts for over 92% of transport-related emissions and is a major contributor to air pollution in India. In 2014, Pandey and Venkataraman reported that on-road vehicles were the major contributors (>96%) to emissions, with super emitters, heavy-duty diesel vehicles, and two-stroke, two- and three-wheelers emerging as the most polluting technologies [50]. The rising demand for vehicles, reliance on conventional fuel, and increased energy consumption indicate a positive correlation with the growth of CO_2 emissions in the country [51, 52]. Recently, Puneet Kamboj et al. reported that emissions from road transport are growing faster than in any other sector and are expected to be a significant challenge to decarbonizing India's energy sector [53].

2.3.6 Transport Emissions in South Korea and Japan

South Korea's energy landscape is heavily dependent on fossil fuels, particularly petroleum and other liquids, accounting for a significant part of its energy consumption. Although specific data on fossil fuel consumption in road transport is not readily available, studies indicate that vehicle emissions contribute notably to the country's air pollution. Recently, Hoseok Nam et al. have shown that road transport accounted for the largest proportion of emissions, growing from 87.1% in 1990 to

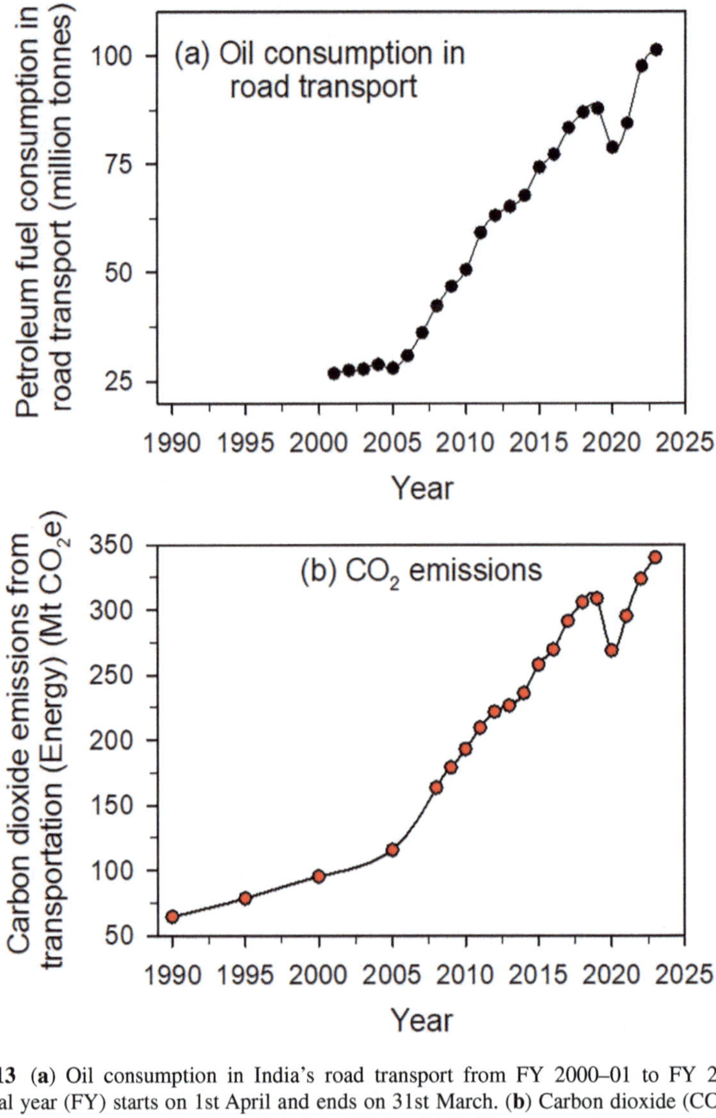

Fig. 2.13 (**a**) Oil consumption in India's road transport from FY 2000–01 to FY 2023–24. Financial year (FY) starts on 1st April and ends on 31st March. (**b**) Carbon dioxide (CO_2) emissions from Transport (energy) (Mt CO_2e) from 1990 to 2023. The measure is standardized to carbon dioxide equivalent values using the Global Warming Potential (GWP) factors of IPCC's 5th Assessment Report (AR5)

96.5% in 2019 [54]. Figure 2.14 displays the total final fuel consumption, comprising city gas, petroleum, biofuel, and electricity, as well as jet oil, fuel oil, and various petroleum products, and the growth in petroleum consumption (gasoline, diesel, and LPG) in South Korea's transport sector [55, 56]. Results indicate a decline in fuel consumption in this sector subsequently by 2022. This reduction in fuel consumption in recent years is expected to positively impact GHG emissions. Since

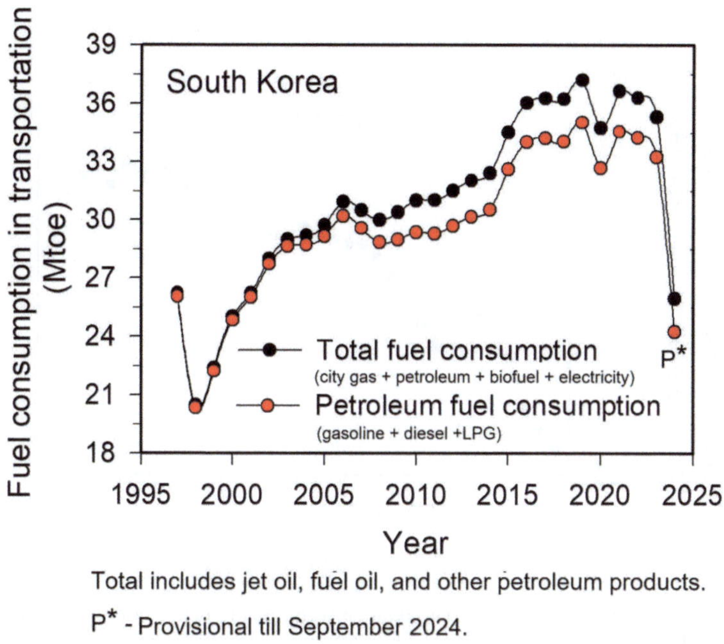

Fig. 2.14 Total final fuel consumption and petroleum consumption by the transport sector in South Korea from 1997 to 2024

2008, South Korea has successfully managed its energy consumption level for transport at reasonable rates. Figure 2.15 indicates GHG emissions from various modes of transportation (maritime, air, rail, road, and others) in the country [57]. As anticipated, the GHG emissions have replicated the trend in fuel consumption, showing a decrease in recent years. However, despite this downward trend, the GHG emission level in 2023 was alarmingly high (+160%) compared to 1990 levels (which were 36.4610 $MtCO_2e$). Consequently, it is imperative to urgently address GHG emissions from road transport to achieve the green transportation goals.

Japan's fossil fuel consumption in road transportation has played a significant role in its overall energy usage. In 2023, transportation total fuel consumption accounted for 68.2517 million tonnes of oil, representing approximately 23% of the country's total fossil energy consumption, which comes to around 296.08 million tonnes (equivalent to 11,476 PJ; here 0.0258 Petajoule (PJ) corresponds to one million tonnes of crude oil). Figure 2.16 shows total energy consumption (including petroleum, electricity, and others) and CO_2 emissions from transportation (such as automobiles) since 1990 [58]. The data show a continuous decline in both fuel consumption and CO_2 emissions. By 2023, the transport-related CO_2 emissions recorded a reduction of 16.30% as compared to the 1990 baseline (217.4 million tonnes), in line with the Kyoto Protocol pledges of Japan. The emissions peaked in 2022 at 267.4 million tonnes but have been declining since, thanks to improved fuel efficiency and the continuing adoption of eco-friendly vehicles. The transportation

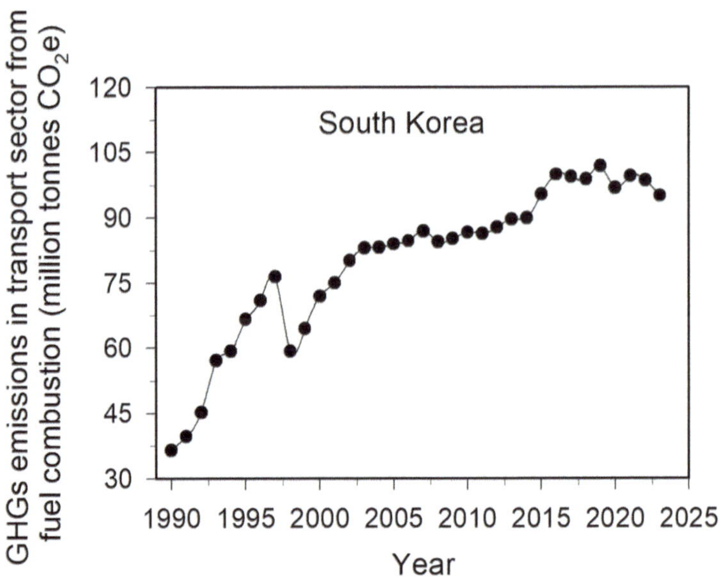

Fig. 2.15 GHG emissions from the transportation sector due to petroleum fuel consumption in Korea from 1990 to 2022

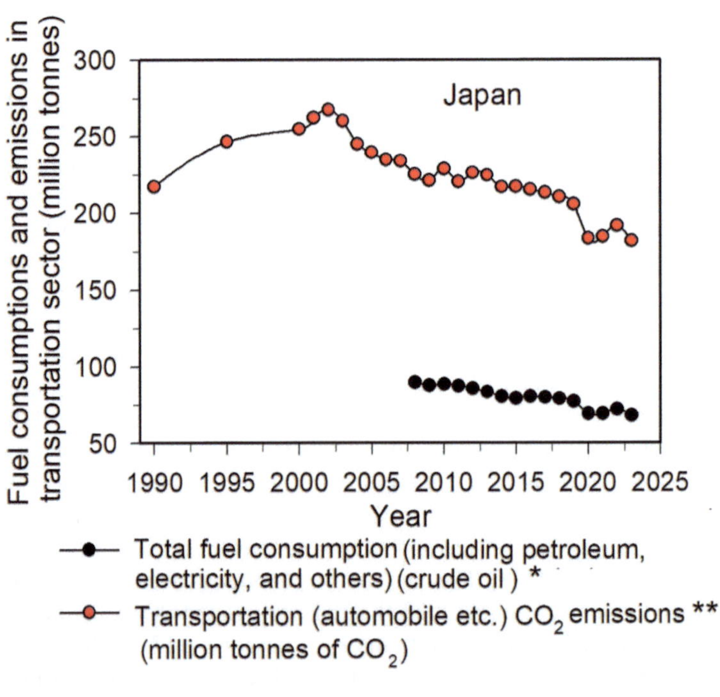

Fig. 2.16 Total energy consumption (including petroleum, electricity, and others) (in million tonnes of crude oil) and the transportation (automobile etc.) CO_2 emissions (in million tonnes of CO_2) in Japan since 1990. (*Source: Director-General's Secretariat, Agency for Natural Resources and Energy, Ministry of Economy, Trade and Industry (METI), Japan. ** Source: Global Environment Bureau, Ministry of the Environment, Japan)

sector, particularly road transport, has been a major source of CO_2 emissions, with efforts underway to reduce reliance on fossil fuels through the development of hybrid, electric, and hydrogen-powered vehicles.

2.4 Decarbonizing Road Transportation

Roughly 99% of the GHGs released from the vehicle's exhaust are carbon dioxide emissions [59]. Burning one gallon of gasoline in a conventional car can produce 8887 g of CO_2, while one gallon of diesel produces roughly 10,180 g of CO_2 [60]. An average passenger car produces 4.6 metric tons of CO_2 annually. This number may change, depending on a vehicle's fuel type, fuel efficiency, and annual mileage. The World Economic Forum estimates that heavy-duty trucks are responsible for 25% of global road emissions, despite making up only 1% of all fleet vehicles [61]. According to a KPMG (UK) analysis, the transportation sector is the largest emitter in the US and the UK, contributing roughly 25% and 30% of the carbon emissions from transportation worldwide, respectively [62]. Nearly a quarter of the global transportation-related emissions originate from China. In India, 90% of CO_2 emissions from transportation come from the road sector. Therefore, setting a precise goal for road transport decarbonization is very imperative.

Road transportation is the primary contributor to GHG emissions in the United States, accounting for 29% of all emissions. Within the transportation sector, heavy-duty vehicles are the second-largest contributor, at 23%. Mitigating GHG emissions is crucial in reducing the impact on climate change. In March 2024, the United States released the National Zero-Emission Freight Corridor Strategy, which outlines a phased plan to establish charging and refueling infrastructure for zero-emission medium- and heavy-duty vehicles [63]. Similarly, around this time, the US Environmental Protection Agency finalized new GHG emissions standards for heavy-duty vehicles [64]. These standards, which aim to reduce greenhouse gas emissions from heavy-duty (HD) vehicles beginning in the 2027 model year, will be applicable to HD vocational vehicles such as delivery trucks, refuse haulers, public utility trucks, transit, shuttle, and school buses, as well as tractors, including day cabs and sleeper cabs on tractor-trailer trucks. Given that transportation is the largest source of GHG emissions in the United States, any initiative to achieve a net-zero GHG economy must aim to bring transportation emissions to near zero. To meet the emissions reduction targets for the transportation sector (19 $GtCO_2$, representing 28% of the projected light-duty vehicle fleet emissions period from 2015–2050), the transition to green vehicles must account for 90% (over 350 million EVs) of new car sales by 2050, or there must be a significant reduction in driving [65].

In 2022, transport (including emissions from international bunkers) accounted for approximately 28.9% of all EU-27 GHG emissions [66]. Road transport is the largest contributor to transport GHG emissions, accounting for 73.2% in 2022 (and 21.1% of all EU-27 GHG emissions in the same year). EU-27 GHG emissions from road transport are almost entirely due to light- and heavy-duty vehicles; powered two-wheelers and other road transport modes are responsible for less than 1% of overall transport emissions [67]. As the EU strives to become the first climate-neutral continent by 2050, it aims for a 90% reduction in transport-related greenhouse gas emissions. Tackling transportation emissions is a main focus point on the EU's road to achieving its climate ambitions. The transition from fossil fuels to clean alternative fuels is set to play a key role in decarbonizing the sector. In May 2024, the EU adopted ambitious CO_2 standards for most new trucks and coaches (90% emissions reduction by 2040) and urban buses (100% zero-emission sales by 2035) [68]. Road transport CO_2 emissions in the EU are expected to peak in 2025 and decline by around 25% by 2035 [69]. On the bright side, EU countries achieved an average decrease of 30.5% between 2005 and 2023. In contrast, the Netherlands surpassed this average with a 32.3% reduction in the same period [70]. Meanwhile, Norway showed an 11.4% decline in transport-related CO_2 emissions between 2012 and 2018, largely due to the transition from petroleum-based fuels to greener alternatives, including the electrification of vehicles and maritime transport [71].

The transportation sector is responsible for around 25% of Canada's annual greenhouse gas emissions, making it the second-largest contributing sector to Canada's overall emissions [72]. Most emissions come from on-road transportation, which includes light-duty vehicles (like cars, sport-utility vehicles, and pickup trucks) and medium- and heavy-duty vehicles (like larger pickup trucks, cargo vans, buses, and freight trucks). As a result, the government foresees that on-road vehicles are the key to achieving net-zero emissions by 2050. In this context, the government proposed its future action to decarbonize on-road transportation under the *Canadian Net Zero Emissions Accountability Act* [73]. As stated in the *Canadian Net-Zero Emissions Accountability Act*, Canada's 2030 nationally determined contribution under the Paris Agreement is to reduce emissions by 40% to 45% below 2005 levels.

Several G20 countries have set goals for reducing GHG emissions in the transportation sector. A policy report by Agora Verkehrswende GIZ (2023) was created for consideration during the G20 meeting in India [74]. The report argues that the G20 nations should and are capable of leading the transition to a carbon-neutral transportation sector, as they are responsible for the majority of global transportation emissions. In 2022, Quanying Lu et al. in China proposed a model for low-carbon transition planning for road transportation, aiming to explore routes toward carbon neutrality by 2060 [75]. Most notably, they observed that the inertia regarding fuel-powered vehicles in the transportation system poses a significant challenge to attaining net zero emissions in road transport by 2060. It is anticipated that the Chinese transportation sector may reach a peak at about 1330.98 million tons of carbon emissions around 2030.

The International Council on Clean Transportation (ICCT), based in Washington, has published a white paper examining India's road transportation industry [76]. The report suggests that under an aggressive policy scenario, India can achieve a cumulative reduction in CO_2 emissions of 45–50% between 2021 and 2050, compared to the business-as-usual (BAU) scenario. Meanwhile, Australia hopes to cut carbon emissions from its passenger road transport sector by 50% by 2030, relative to levels from the year 2000 [74, 77]. Japan also plans to increase diesel fuel efficiency by roughly 13.4% for trucks and heavy vehicles and by about 14.3% for buses by 2025, relative to 2015 fuel efficiency standards [78, 79]. In South Korea, the road transport sector is accountable for approximately 97% of transportation-related GHG emissions [80]. The country aims to improve gasoline fuel efficiency for passenger vehicles to 35 km/L by 2035 and achieve a target of 7.5 km/L for heavy-duty vehicles by 2040 in efforts to reduce GHG emissions [74].

In order to decarbonize UK roads, KPMG proposed three key approaches, such as (i) assisting infrastructure authorities, (ii) financing, and (iii) operators in creating transport decarbonization strategies [62]. Moreover, the UK government has released its "Transport Decarbonisation Plan" and announced a ban on the sale of new petrol and diesel passenger vehicles from 2030 [81]. Under the plan, all new cars and vans must be fully zero-emission at the tailpipe from 2035. The plan calls for all new cars and vans to have completely zero emissions at the exhaust by 2035. All non-zero emission heavy goods motor vehicles (HGVs) would be phased out of production by 2040, and lighter HGVs would start to be sold in 2035. For the Spanish passenger transport system, David Álvarez-Antelo et al. have recently suggested exploring the idea of switching to battery-electric power trains for automobiles, buses, and motorbikes [82]. They demonstrated how, by 2035, passenger transportation will emit 60 Mt of CO_2 annually as opposed to 66 Mt in 2022. Since fossil fuels are Ecuador's primary source of pollution, Atiaja et al. proposed a model focusing on developing a dynamic simulation of energy scenarios for the transportation sector of Ecuador for the year 2035 with the aim of shifting to sustainable mobility [83]. This study demonstrates a large reduction in greenhouse gas emissions. According to the estimate, by 2035, total CO_2 emissions could be reduced from 50,161,432 kilobarrels of oil equivalent (kBOE) to 20,589,720 kBOE if more people drive electric vehicles.

It is important to note that decarbonizing the road transportation system will not come without challenges. However, with coordinated and bold actions across the federal government and with other stakeholders, these challenges could be overcome. One should continue to increase ambition, setting bold targets for improving our transportation systems and transitioning to zero-emission vehicles and fuels on a timeline consistent with achieving economy-wide 2030 and 2050 emissions reduction goals. The blueprint is an exciting first step toward realizing the vision of an improved and sustainable transportation future.

In summary, the decarbonization of road transportation is expected to lead to a significant reduction in GHG emissions. Governments worldwide are taking steps to accelerate the transition towards cleaner transportation. Nonetheless, there is a need for a new strategic vision that addresses the entire transportation ecosystem. To effectively address the climate concern and achieve the net-zero GHG emissions economy-wide goals, nearly all GHG emissions from the transportation sector should be eliminated by 2050. This requires an all-inclusive strategy to develop a future mobility system that is clean, safe, secure, accessible, affordable, and equitable, offering sustainable transportation options for both people and goods. Table 2.5 summarizes the beautiful policy pledges made by some major economies to decarbonize road transportation [40, 41, 70, 81, 84–92].

Table 2.5 Beautiful policy pledges for decarbonizing road transportation by stakeholder countries

Stakeholder country	Policy pledges	References
USA	Net-zero carbon emissions from the transportation sector by 2050. Plan focuses on cutting GHG emissions	[84]
EU	Set ambitious policies to decarbonize road transportation. 90% reduction in transport emissions by 2050	[85, 86]
	Stricter CO_2 Emission Standards:	
	New cars and vans. 55% reduction for cars and 50% for vans by 2030. zero emissions for all new vehicles by 2035	
	Heavy-Duty Vehicle Regulations: 45% reduction by 2030, 65% by 2035, and 90% by 2040	
Norway	Set ambitious goals for decarbonizing road transportation by developing an efficient, environmentally friendly, and safe transport system by 2050	[87]
Netherlands	Set goals for decarbonizing road transportation 2050 for climate neutrality	[70, 88]
	A 60% transport sectoral reduction in emissions by 2050 compared to 1990 levels	
	A 55% reduction in GHG emissions by 2030 compared to 1990 levels	
	A 25 Mt. CO_2 reduction, or 17%, by 2030	
UK	Carbon Neutrality Goal by 2050	[81]
	All new cars and vans must be fully zero emission at the tailpipe from 2035	
	Ending the sale of all non-zero emission HGVs from 2040, with lighter HGVs from 2035	
China	Pledged to achieve carbon neutrality by 2060	[40, 41]
	Mandates for automakers to produce a certain percentage of zero-emission vehicles	
	Road transport emissions could peak before 2030 and decline by 50–95% from 2020 levels by 2060	
India	India aims for net-zero emissions by 2070	[89, 90]
	CO_2 emissions reduction from road transport by 45–50% between 2021 and 2050	
	Promoting fuel efficiency improvements, vehicle electrification, and alternative fuels	
Japan	Outlined a Carbon Neutrality Promotion Strategy for the road sector	[91]
	Aiming to reduce greenhouse gas emissions by 46% by 2030 and achieve full carbon neutrality by 2050	
Korea	Aims to achieve almost zero emissions from the transportation sector by 2050	[92]
	Road transport reduction by more than 92% of GHG emissions by 2050	

2.5 Strategies for Decarbonizing Road Transport

Since the transport sector is a major contributor to climate change, accounting for about 23% of global energy-related CO_2 emissions, advanced technologies, such as intelligent road transport, can help improve this situation. Intelligent road transportation employing new strategies and efficiently using the existing transportation networks plays a crucial role in decarbonization by leveraging smart technologies to reduce emissions and improve efficiency [93]. Smart transport refers to safe, smart, and sustainable mobility by optimizing traffic flow, reducing congestion, and thereby reducing fossil fuel consumption. These types of transport systems have the capability to improve travel safety, revolutionize the concept of mobility, and decrease GHG emissions. Measures (programs, policies, or projects) that will achieve or facilitate the reduction of GHG emissions are broadly grouped into two categories: (i) travel efficiency and (ii) transition to clean vehicles and fuels, as described by the Office of Transportation and Air Quality, United States Environmental Protection Agency [94]. Some key strategies for decarbonization of road transport are illustrated in Table 2.6 [59, 93, 95–109].

Shared mobility has the potential to significantly reduce GHG emissions by optimizing vehicle usage and decreasing reliance on private cars. Studies suggest that widespread adoption of shared mobility services could eliminate, on average, 6.3% of urban passenger transport emissions by 2050 [110]. Recently, Zhao et al. presented findings revealing that an integrated strategy significantly decreases road freight emissions, achieving a reduction of 34% in peak carbon dioxide and 43% in

Table 2.6 Key strategies for decarbonizing road transport

Measure	Description	References
Intelligent Transport Systems (ITS)	These systems optimize traffic flow, reduce congestion, and minimize fuel consumption by using real-time data and automation	[93, 95]
Transition to Electric Vehicles	Transitioning from fossil fuel combustion engine to EVs and integrating charging infrastructure can significantly cut emissions from road transport	[59, 96]
Shift to Public Transport	Encouraging the use of public transportation and reduces reliance on private vehicles	[97]
Smart Traffic Management	AI-driven traffic control systems can reduce idling time and fuel wastage, leading to lower emissions	[98–101]
Digitalization and Shared Mobility	Ride-sharing, carpooling, and mobility-as-a-service (MaaS) platforms help optimize vehicle usage and reduce unnecessary trips	[102–104]
Reduced Vehicle Miles Traveled (VMT)	Decrease the number of vehicles on the road, leading to lower fuel consumption and emissions	[105]
Increased Use of Low-Emission Transport	Integrate electric and hybrid vehicles in transport system for cutting GHGs emissions	[106, 107]
Optimized Urban Transport	Reduce congestion and improve public transport efficiency, leading to lower overall carbon footprints	[108, 109]

nitrogen oxide emissions when compared to scenarios depending solely on road electrification [111]. Ashish Matura et al. proposed the strategies for decarbonizing road transport, highlighting several critical solutions for advancing decarbonization, such as Mobility as a Service (MaaS), sustainable mobility, technological integration with AI/ML techniques, infrastructure upgrades, and the promotion of multimodal transportation [7].

Recently, UN member states made a significant step forward in supporting sustainable low-carbon mobility by adopting recommendations from the Inland Transport Committee's (ITC) Strategy for Reducing GHG Emissions for Inland Transport in 2024 [112]. This strategy emphasizes the importance of public transport in combination with cycling and walking for passenger transport. It also encourages new approaches to urban freight solutions and aims to improve vehicles, infrastructure, and operations, including border crossing operations, to improve efficiency in the inland transport sector. Furthermore, it highlights that achieving decarbonization of mobility will require a transition to electrification and a focus on public transport and active mobility options, as well as substantial societal changes. Kim et al. showed that the strategies for transitioning to battery electric vehicles (BEV), including phasing out diesel vehicles, providing subsidies, and implementing a ban on internal combustion engine vehicles (ICEV), contribute to reducing CO_2 emissions and enhancing gross domestic product in Korea [92]. Moreover, a recent review by Alaa Torkey et al. emphasizes that transportation electrification represents a promising emerging technology for reducing the impacts on climate change [113]. EVs are considered an important player in the fight against global warming.

2.6 Conclusion

In the foreseeable future, roads will remain vital to society, serving a crucial role in freight transport and connecting local and rural areas. The International Energy Agency reports that global CO_2 emissions related to energy source usage hit a new record high of 37.4 Gt in 2023. The World Energy Outlook indicated that road vehicles emitted 5.9 Gt CO_2 worldwide in 2021, with heavy-duty trucks alone accounting for 1.8 Gt out of the total 7.7 Gt CO_2 transport sector emissions, equating to 4% of global CO_2 emissions. Consequently, decarbonizing road transportation is critical for addressing the challenges posed by global warming and climate change.

Undoubtedly, fossil fuel consumption and its impact on CO_2 emissions in road transportation have been declining in countries such as the USA, UK, EU, Japan, and South Korea. In contrast, developing countries like China and India have experienced an increase in these figures. In response, many countries worldwide have already formulated strategies that utilize smart technologies to reduce transportation emissions and improve vehicle efficiency. The concept of smart transport envisions safe, intelligent, and sustainable mobility by optimizing traffic flow and reducing congestion, ultimately lowering fossil fuel consumption. Transitioning from gasoline-powered vehicles to EVs is widely considered one of the most practical ways for decarbonizing land transportation. In summary, these types of transport systems, incorporating advanced technology, are expected to increase travel safety, transform mobility, and help in reducing GHG emissions.

References

1. International Energy Agency (IEA), France (2024) CO_2 Emissions in 2023. https://iea.blob.core.windows.net/assets/33e2badc-b839-4c18-84ce-f6387b3c008f/CO2Emissionsin2023.pdf
2. International Energy Agency (IEA), France (2023) CO_2 emissions in 2022. https://iea.blob.core.windows.net/assets/3c8fa115-35c4-4474-b237-1b00424c8844/CO2Emissionsin2022.pdf
3. Ritchie H (2020) Cars, planes, trains: where do CO_2 emissions from transport come from? OurWorldinData.org. https://ourworldindata.org/co2-emissions-from-transport
4. IEA (2022) World Energy Outlook 2022. IEA, Paris. https://www.iea.org/reports/world-energy-outlook-2022
5. Jaramillo PS, Kahn RP, Dhar NS, Diemuodeke OE, Kajino T, Lee DS, Nugroho SB, Ou X, Hammer Strømman A, Whitehead J (2022) Transport. In IPCC, 2022: Climate Change 2022: mitigation of climate change. Contribution of Working Group III to the Sixth Assessment Report of the Intergovernmental Panel on Climate Change. Shukla PR, Skea J, Slade R, Al Khourdajie A, van Diemen R, McCollum D, Pathak M, Some S, Vyas P, Fradera R, Belkacemi M, Hasija A, Lisboa G, Luz S, Malley J (eds.). Cambridge University Press, Cambridge. https://doi.org/10.1017/9781009157926.012
6. Alam M, Rakib AKM, Monjurul Hasan ASM, Siddique MNI, Kabir MA, Trianni A (2025) Decarbonizing road transportation: barriers and drivers in an emerging economy context. Transp Res Part D: Transp Environ 143:104723. https://doi.org/10.1016/j.trd.2025.104723
7. Matura A, Singh RK, Kumar R (2025) Decarbonizing road transport: a systematic literature review based on use case analysis. Case Stud Transport Policy 20:101416. https://doi.org/10.1016/j.cstp.2025.101416
8. https://www.statista.com/statistics/1291615/carbon-dioxide-emissions-transport-sector-worldwide/. Accessed 22 Feb 2025
9. Greenhouse Gas (GHG) Emissions (n.d.) Climate Watch. https://www.climatewatchdata.org/ghg-emissions?end_year=2022&start_year=1990. Accessed 22 Feb 2025
10. Climate Watch (2024) World Resources Institute (WRI), Washington, DC. https://www.climatewatchdata.org and https://www.climatewatchdata.org/data-explorer/historical-emissions
11. IEA (n.d.) Car and vans. https://www.iea.org/energy-system/transport/cars-and-vans#tracking. Accessed 22 Feb 2025
12. Fuel consumption by mode of transportation in physical units. Bureau of Transportation Statistics, United States Department of Transportation. https://www.bts.gov/content/fuel-consumption-mode-transportation. Accessed 19 Mar 2025
13. Lawson A, Ahmad FM (2018) Decarbonizing U.S. transportation, climate innovation 2050. Center for Climate and Energy Solutions. https://www.c2es.org/wp-content/uploads/2018/07/innovation-transportation-background-brief-07-18.pdf. Accessed 15 Apr 2025
14. Transportation Sector Emissions (n.d.) United States Environmental Protection Agency. https://www.epa.gov/ghgemissions/transportation-sector-emissions#sources. Accessed 15 Apr 2025
15. EPA (2024) Inventory of U.S. Greenhouse Gas Emissions and Sinks: 1990-2022. U.S. Environmental Protection Agency, EPA 430-R-24-004. https://www.epa.gov/system/

files/documents/2024-04/us-ghg-inventory-2024-main-text_04-18-2024.pdf and https://www.epa.gov/ghgemissions/inventory-us-greenhouse-gas-emissions-and-sinks-1990-2022
16. Monthly Energy Review (MER) (2025) U.S. Energy Information Administration's (EIA). https://www.eia.gov/totalenergy/data/monthly/pdf/mer.pdf
17. Fast Facts U.S. Transportation Sector Greenhouse Gas Emissions 1990–2022. https://nepis.epa.gov/Exe/ZyPDF.cgi?Dockey=P101AKR0.pdf
18. Energy consumption in the UK. Department for Energy Security and Net Zero. Statistical Release. 26 Sept 2024. https://www.gov.uk/government/collections/energy-consumption-in-the-uk
19. ENV0102a: transport energy consumption (million tonnes of oil equivalent) by transport mode and source of energy, United Kingdom, 1998 to 2023. Updated 12 Dec 2024. https://www.gov.uk/government/statistical-data-sets/energy-and-environment-data-tables-env. Accessed 24 Apr 2025
20. Department for Energy Security and Net Zero (Energy Consumption in the UK) and Department for Energy Security and Net Zero (Digest of UK Energy Statistics). https://assets.publishing.service.gov.uk/media/66f13235f188a93404379f94/Energy_Consumption_in_the_UK_2024.pdf
21. 2023 UK greenhouse gas emissions, provisional figures (2024) Department for Energy Security & Net Zero. https://assets.publishing.service.gov.uk/media/6604460f91a320001a82b0fd/uk-greenhouse-gas-emissions-provisional-figures-statistical-release-2023.pdf
22. Ritchie H (2023) Which form of transport has the smallest carbon footprint? OurWorldinData.org. https://ourworldindata.org/travel-carbon-footprint
23. Eurostat (2024) Final energy consumption in transport—detailed statistics. https://ec.europa.eu/eurostat/statistics-explained/index.php?oldid=652642#cite_note-1. Accessed 20 Mar 2025
24. Eurostat (2024) Final energy consumption in transport—detailed statistics. https://ec.europa.eu/eurostat/statistics-explained/index.php?title=Final_energy_consumption_in_transport_-_detailed_statistics#Road_transport
25. (a) Final energy consumption in road transport by type of fuel (2025) https://ec.europa.eu/eurostat/databrowser/view/ten00127/default/table?lang=en. (b) Data Source Eurostat https://doi.org/10.2908/NRG_BAL_C
26. Domagała J, Kadłubek M (2023) Economic, energy and environmental efficiency of road freight transportation sector in the EU. Energies 16(1):461. https://doi.org/10.3390/en16010461
27. EU transport in figures—statistical pocketbook (2024) European Union Publications, Luxembourg. https://transport.ec.europa.eu/document/download/ee264fc5-ec49-4751-9d92-08c038856ce1_en?filename=MI-AA-24-001-EN-N.pdf
28. Emissions to Air. Statistics Norway. https://www.ssb.no/en/statbank/table/13931/tableViewLayout1/. Accessed 5 May 2025
29. Emissions to Air on Dutch Territory; Road Traffic. Statistics Netherlands (CBS). https://www.cbs.nl/en-gb/figures/detail/85347ENG#shortTableDescription. Accessed 6 Feb 2025
30. IEA (2025) Fuel demand and substitution in China, 2009–2025. IEA, Paris. https://www.iea.org/data-and-statistics/charts/fuel-demand-and-substitution-in-china-2009-2025
31. China Statistical Yearbooks 2022, 2021, 2019, 2018, 2016, 2015, 2014. https://www.stats.gov.cn/sj/ndsj/2024/indexeh.htm
32. China National Bureau of Statistics (n.d.) China General Administration of Customs, and Bloomberg L.P. https://www.stats.gov.cn/english/
33. Today in Energy (2024) What's driving decreasing gasoline consumption in China? U.S. Energy Information Administration, Washington DC. https://www.eia.gov/todayinenergy/detail.php?id=63764
34. Binici E (2024) China's gasoil demand predictions for 2025 and beyond. Kpler, Paris. https://www.kpler.com/blog/chinas-gasoil-demand-is-set-to-peak-in-2026-as-sweeping-changes-in-the-transportation-sector-take-a-toll. Accessed 23 May 2025

35. Downs E, Rajendran A (2024) China's slowing oil demand growth is likely to persist and could impact markets. Center for Global Energy Policy Columbia SIPA, New York. https://www.energypolicy.columbia.edu/chinas-slowing-oil-demand-growth-is-likely-to-persist-and-could-impact-markets/
36. Teng W, Zhang Q, Guo Z, Ying G, Zhao J (2024) Carbon emissions from road transportation in China: from past to the future. Environ Sci Pollut Res 31:48048–48061. https://doi.org/10.1007/s11356-024-34344-0
37. Action plan for carbon dioxide peaking before 2030 (2021) State Information Center (SIC), National Development and Reform Commission (NDRC), Beijing, People's Republic of China. https://en.ndrc.gov.cn/policies/202110/t20211027_1301020.html
38. NEV Industrial Development Plan 2021–2035, China's New Energy Vehicle Industrial Development Plan for 2021 to 2035. Policy update (June 2021) International Council on Clean Transportation. China's New Energy Vehicle Industrial Development Plan for 2021 to 2035
39. The "Energy-saving and new energy vehicle technology roadmap 2.0" was officially released (2020) Automobile Industry, Departments and Bureaus, Ministry of Industry and Information Technology, People's Republic of China. https://www.miit.gov.cn/jgsj/zbys/qcgy/art/2020/art_7eea943abda746339d899bd5fd520c92.html
40. Xue L, Liu D (2022) Decarbonizing China's road transport sector: strategies toward carbon neutrality. https://doi.org/10.46830/wrirpt.21.001455
41. Huang S (2024) China's focus on sustainable transportation heats up. S&P Global Mobility. https://www.spglobal.com/mobility/en/research-analysis/china-sustainability-transportation-regulations.html
42. IEA (2021) India energy outlook 2021, vol 32. International Energy Agency, OECD Publishing, Paris. https://doi.org/10.1787/ec2fd78d-en
43. IEA (2023) Transitioning India's road transport sector. IEA, Paris. https://www.iea.org/reports/transitioning-indias-road-transport-sector
44. NITI Aayog Dash Board, Government of India. https://iced.niti.gov.in/energy/end-use/transport/road-transport
45. Indian Petroleum and Natural Gas Statistics 2023–2024 (31 Jan 2025) Economic and Statistics Division, Ministry of Petroleum and Natural Gas, Government of India. https://mopng.gov.in/files/TableManagements/IPNG-Statistics-Report_2023-24_Final.pdf
46. Energy Statistics India (2025) Ministry of Statistics & Programme Implementation, Government of India. https://www.mospi.gov.in/sites/default/files/publication_reports/Energy_Statistics_2025/Energy%20Statistics%20India%202025_27032025.pdf
47. IEA (2020) Energy consumption in road transport in selected IEA countries, 2000–2018. IEA, Paris. https://www.iea.org/data-and-statistics/charts/energy-consumption-in-road-transport-in-selected-iea-countries-2000-2018
48. Decarbonising transport: redefining mobility policies in India (24 June 2021) National Portal of India, NITI Aayog. https://niti.gov.in/decarbonising-transport-redefining-mobility-policies-india
49. Data from database: World Development Indicators. https://data.worldbank.org/indicator/EN.GHG.CO2.TR.MT.CE.AR5?end=2023&locations=IN&start=1970&view=chart
50. Pandey A, Venkataraman C (2014) Estimating emissions from the Indian transport sector with on-road fleet composition and traffic volume. Heliyon 9(9):e19871. https://doi.org/10.1016/j.atmosenv.2014.08.039
51. Jain S, Rankavat S (2023) Analysing driving factors of India's transportation sector CO_2 emissions: based on LMDI decomposition method. Heliyon 9(9):e19871. https://doi.org/10.1016/j.heliyon.2023.e19871
52. Singh N, Mishra T, Banerjee R (2019) Greenhouse gas emissions in India's road transport sector. In: Venkataraman C, Mishra T, Ghosh S, Karmakar S (eds) Climate change signals and response. Springer, Singapore. https://doi.org/10.1007/978-981-13-0280-0_12
53. Kamboj P, Malyan A, Kaur H, Jain H, Chaturvedi V (n.d.) India transport energy outlook. Council on Energy, Environment and Water, ISID Campus, New Delhi. https://www.ceew.

in/sites/default/files/ceew-research-transport-energy-use-carbon-emissions-decarbonisation.pdf. Accessed 23 February 2025
54. Nam HS, Byun JH, Nam HG (2025) Prediction greenhouse gas emissions from road freight flow in South Korea for sustainable transportation planning. Heliyon 11(2):e41937. https://doi.org/10.1016/j.heliyon.2025.e41937
55. Korea energy economics institute, monthly energy statistics, March 2022, p. 12. Monthly Energy Statistics, February 2025. Table 2.3.2 Transport, p. 33. https://www.keei.re.kr/pdfOpen.es?bid=0033&list_no=124606&seq. Accessed 20 May 2025
56. Total final consumption by sector: transport. Korea Energy Economics Institute, Energy Supply and Demand Statistics. https://kosis.kr/statHtml/statHtml.do?orgId=339&tblId=DT_339001_009&conn_path=I2&language=en. Accessed 8 May 2025
57. The Trends of the National GHG Inventories. Korea GHGs emissions: greenhouse gas inventory and research center (Ministry of Environment). https://kosis.kr/statHtml/statHtml.do?orgId=106&tblId=DT_106N_99_2800020&conn_path=I2&language=en. Accessed 8 May 2025
58. Statistics Bureau, Ministry of Internal Affairs and Communications, Japan (2025) Japan statistical yearbook. https://www.stat.go.jp/english/data/nenkan/74nenkan/1431-17.html
59. Patil G, Pode G, Diouf B, Pode R (2024) Sustainable decarbonization of road transport: policies, current status, and challenges of electric vehicles. Sustainability 16(18):8058. https://doi.org/10.3390/su16188058
60. Tailpipe Greenhouse Gas Emissions from a Typical Passenger Vehicle. Office of Transportation and Air Quality EPA-420-F-23-014 June 2023. U.S. Environmental Protection Agency. https://nepis.epa.gov/Exe/ZyPDF.cgi?Dockey=P1017FP5.pdf. Accessed 23 May 2025
61. Supply Chain and Transport, World Economic Forum. (16 Aug 2021) How to decarbonize heavy-duty transport and make it affordable. https://www.weforum.org/agenda/2021/08/how-to-decarbonize-heavy-duty-transport-affordable/. Accessed 10 July 2024
62. Bose, M (2023) Decarbonisation of transport and infrastructure: translating strategy to action. KPMG. https://kpmg.com/uk/en/home/insights/2023/07/decarbonisation-of-transport-and-infrastructure.html. Accessed 1 July 2024
63. Joint Office of Energy and Transportation (2024) National Zero-Emission Freight Corridor Strategy. https://driveelectric.gov/files/zef-corridor-strategy.pdf. Accessed 23 May 2025
64. Final rule: greenhouse gas emissions standards for heavy-duty vehicles—phase 3 (27 Apr 2023) Environmental protection agency. Federal Register 88 (81): Proposed Rules. https://www.govinfo.gov/content/pkg/FR-2023-04-27/pdf/2023-07955.pdf. Accessed 23 May 2025
65. Milovanoff A, Posen ID, MacLean HL (2020) Electrification of light-duty vehicle fleet alone will not meet mitigation targets. Nat Clim Chang 10:1102–1107. https://doi.org/10.1038/s41558-020-00921-7
66. Arrigoni A, Bravo DL (2022) Hydrogen emissions from a hydrogen economy and their potential global warming impact. JRC. https://publications.jrc.ec.europa.eu/repository/handle/JRC130362. Accessed 31 July 2024
67. Koellinger C (2024) Sustainability of Europe's mobility systems. Report no. 01/2024. 12 Nov 2024. EN HTML: TH-01-24-007-EN-Q. https://doi.org/10.2800/8560026
68. The CO_2 Emission Performance Standards for New Heavy-Duty Vehicles (PE-29-2024-REV-1). (14 May 2024) The European Union Parliament, Brussels. https://data.consilium.europa.eu/doc/document/PE-29-2024-REV-1/en/pdf. Accessed 23 May 2025
69. Sen A, Teter J, Miller J (2025) Vision 2050: update on the global zero-emission vehicle transition in 2024. The International Council on Clean Transportation. Washington, DC. https://theicct.org/wp-content/uploads/2025/01/ID-260-%E2%80%93-Vision-2050-update_report_final.pdf
70. Erbach G, Dewulf N (2024) The Netherlands' climate action strategy. Roadmap to EU climate neutrality—Scrutiny of Member States, PE 767.176—December 2024, European Parliamentary Research Service. https://www.europarl.europa.eu/RegData/etudes/BRIE/2024/767176/EPRS_BRI%282024%29767176_EN.pdf

71. Simonet G (n.d.) Transport: the progressive electrification of land and maritime transport, Case study Norway 2019. Climate Change—Global Observatory on Non-State Climate Action. https://www.climate-chance.org/wp-content/uploads/2019/11/cp4-2019_transport-norway-vf-en_20191126_complet.pdf
72. Canada's Action Plan for Clean On-Road Transportation (Catalogue No. T42-30/2022E-PDF) (2022) Minister of Transport, Government of Canada. https://tc.canada.ca/sites/default/files/2023-03/ROAD-04-ON_ROAD_ACTION_PLAN_REPORT_EN_V09.pdf
73. Canadian Net-Zero Emissions Accountability Act (S.C. 2021, c. 22) (Last amended on March 31, 2023) (2025). The Minister of Justice, Government of Canada. https://www.laws-lois.justice.gc.ca/PDF/C-19.3.pdf
74. Riehle E, Le Claire E, Rublack E, Schmidt T, Sinha SJ, Teja J, Sharma G (2023) Towards decarbonising transport. A Stocktake on Sectoral Ambition in the G20; Report on behalf of Agora Verkehrswende and Deutsche Gesellschaft für Internationale Zusammenarbeit, GIZ 2023. Agora Transport Transformation gGmbH, Berlin. https://www.agora-verkehrswende.de/fileadmin/user_upload/98_Towards_Decarbonising_Transport_2023.pdf. Accessed 24 July 2024
75. Lu Q, Duan H, Shi H, Peng B, Liu Y, Wu T, Du H, Wang S (2022) Decarbonization scenarios and carbon reduction potential for China's road transportation by 2060. Urban Sustain 2:34. https://doi.org/10.1038/s42949-022-00079-5
76. Kumar M, Shao Z, Braun C, Bandivadekar A (2022) ICCT white paper. Decarbonizing India's road transport: a meta-analysis of road transport emissions models. https://theicct.org/wp-content/uploads/2022/05/Meta-study-India-transport_final.pdf. Accessed 24 July 2024
77. Smit R, Khan T, Yang Z (2024) How Australian light-duty vehicle CO_2 emissions compare with the rest of the World. International Council on Clean Transportation. https://theicct.org/wp-content/uploads/2024/02/ID-114-%E2%80%93-Australia-CO2_final.pdf. Accessed 5 Apr 2024
78. Japan: heavy-duty: fuel economy. https://www.transportpolicy.net/standard/japan-heavy-duty-fuel-economy/. Accessed 1 Sept 2024
79. Second-phase fuel economy standards for on-road heavy-duty vehicles in Japan. International Council on Clean Transportation. Policy Update January 2019. https://theicct.org/sites/default/files/publications/Japan_HDV_FE_Phase_2_20190129.pdf. Accessed 1 Sep 2024
80. Oh IY, Jang MJ, Yeo YJ, Ahn YH (2023) 2050 Carbon neutrality measures and costs in the road transport sector of South Korea. J Clim Change Res 14(6-1):651–671. https://doi.org/10.15531/KSCCR.2023.14.6.651
81. Department of Transport (2021) Decarbonising transport—a better greener Britain. https://assets.publishing. service.gov.uk/media/610d63ffe90e0706d92fa282/decarbonising-transport-a-better-greener-britain.pdf. Accessed 4 Aug 2024
82. Álvarez-Antelo D, Lauer A, Capellán-Pérez I (2024) Exploring the potential of a novel passenger transport model to study the decarbonization of the transport sector. Energy 305:132313. https://doi.org/10.1016/j.energy.2024.132313
83. Atiaja J, Arroyo F, Hidalgo V, Erazo J, Remache A, Bravo D (2024) Dynamic simulation of energy scenarios in the transition to sustainable mobility in the Ecuadorian transport sector. Sustainability 16(15):6640. https://doi.org/10.3390/su16156640
84. The U.S. National Blueprint for Transportation Decarbonization (2023) DOE/EE-2674. https://eelp.law.harvard.edu/wp-content/uploads/2025/01/the-us-national-blueprint-for-transportation-decarbonization-compressed_1.pdf
85. Road transport. Climate Action, EU. https://climate.ec.europa.eu/eu-action/transport-decarbonisation/road-transport_en
86. Clean and sustainable mobility. European Council. https://www.consilium.europa.eu/en/policies/clean-and-sustainable-mobility/#Road
87. Chapter 5: Initiatives relating to the five policy objectives for an efficient, environmental-friendly and safe transport system in 2050 in National Transport Plan 2022–2033 (Meld. St. 20 (2020–2021) Report to the Storting (white paper) English summary). Norwegian

Ministry of Transport. https://www.regjeringen.no/contentassets/117831ad96524b9b9eaadf72d88d3704/en-gb/pdfs/stm202020210020000engpdfs.pdf
88. Turk J (2017) Decarbonising transportation in The Netherlands—a policy case study. ClimateXChange, Scottish Government. https://www.climatexchange.org.uk/wp-content/uploads/2023/09/eu_case_study_dutch_transportation.pdf
89. Gupta R, Malik D, Sankhe S, Unni N (2022) Decarbonising India: charting a pathway for sustainable growth. McKinsey India. https://www.mckinsey.com/capabilities/sustainability/our-insights/decarbonising-india-charting-a-pathway-for-sustainable-growth#/
90. Kumar M, Shao Z, Braun C, Bandivadekar A (2022) Decarbonizing India's Road transport: a meta-analysis of road transport emissions models, ICCT white paper. International Council on Clean Transportation, Washington, DC. https://theicct.org/wp-content/uploads/2022/05/Meta-study-India-transport_final.pdf
91. Nagao S (n.d.) Low-carbon technologies & initiatives for quality road infrastructure in Japan. Ministry of Land, Infrastructure, Transport and Tourism (MLIT). https://uncrd.un.org/sites/uncrd.un.org/files/15th-est_ps3-p1.pdf
92. Kim YG, Lim HW, Lee JU (2024) Decarbonizing road transport in Korea: role of electric vehicle transition policies. Transp Res Part D: Transp Environ 128:104084. https://doi.org/10.1016/j.trd.2024.104084
93. Intelligent transport systems enable the decarbonization of road transportation. William Sales Partnership (WSP) Global Inc., Quebec.
94. Environment Protection Agency (2024) Reducing Greenhouse Gas Emissions from the Transportation Sector through Climate Planning (EPA-420-F-24-042). Climate Pollution Reduction Grants (CPRG) program. https://www.epa.gov/system/files/documents/2024-12/420f24042.pdf
95. Zakharova E, Minashina I, Pashchenko F (2023) Smart transport as an enhancement of the urban infrastructure. In: Vershinin YA, Pashchenko F, Olaverri-Monreal C (eds) Technologies for smart cities. Springer, Cham. https://doi.org/10.1007/978-3-031-05516-4_7
96. Veza I, Asy'ari MZ, Idris M, Epin V, Rizwanul Fattah IM, Spraggon M (2023) Electric Vehicle (EV) and driving towards sustainability: comparison between EV, HEV, PHEV, and ICE vehicles to achieve net zero emissions by 2050 from EV. Alexandria Eng J 82:459–467. https://doi.org/10.1016/j.aej.2023.10.020
97. EASAC (2019) Decarbonization of transport: options and challenges. The European Academies' Science Advisory Council. https://easac.eu/fileadmin/PDF_s/reports_statements/Decarbonisation_of_Tansport
98. Pillai AS (2024) Traffic management: implementing AI to optimize traffic flow and reduce congestion. J Emerg Technol Innov Res 11(7):c272–c278. https://doi.org/10.2139/ssrn.4916398
99. Mohanty A, Mohapatra AG, Mohanty SK (2025) Real-time traffic monitoring with AI in smart cities. In: Abraham A, Prasad S, Alhammadi A, Lestable T, Chaabane F (eds) Internet of vehicles and computer vision solutions for smart city transformations, Lecture notes in intelligent transportation and infrastructure. Springer, Cham. https://doi.org/10.1007/978-3-031-72959-1_7
100. Dimitrov D (2025) The future of transportation: AI, connectivity and safety in 2025. Forbes Media LLC, Jersey City, NJ. https://www.forbes.com/councils/forbestechcouncil/2025/02/10/the-future-of-transportation-ai-connectivity-and-safety-in-2025/
101. Terrence Gatsby (2025) AI-powered traffic management: reducing congestion and emissions in smart cities. https://beyondthehype.terrencegatsby.com/ai/ai-powered-traffic-management-reducing-congestion-and-emissions-in-smart-cities/
102. Shepard S (2023) Guide to governments for successful regional and open MaaS implementation. In: Vershinin YA, Pashchenko F, Olaverri-Monreal C (eds) Technologies for smart cities. Springer, Cham, pp 43–59. https://doi.org/10.1007/978-3-031-05516-4_4
103. Labee P, Rasouli S, Liao F (2022) The implications of mobility as a service for urban emissions. Transp Res Part D: Transp Environ 102:103128. https://doi.org/10.1016/j.trd.2021.103128

104. Mitropoulos L, Kortsari A, Mizaras V, Ayfantopoulou G (2023) Mobility as a Service (MaaS) planning and implementation: challenges and lessons learned. Future Transp 3(2):498–518. https://doi.org/10.3390/futuretransp3020029
105. Millar R (2023) Vehicle Miles Traveled (VMT) targets—final report. Washington State Department of Transportation. https://wsdot.wa.gov/sites/default/files/2023-06/VMT-Targets-Final-Report-June2023.pdf
106. Delso-Vicente AT, Camperos MC, Almonacid-Durán M (2025) The evolution of electric and hybrid vehicles and their influence on sustainable transport: a review and future research lines. Sustain Technol Entrep 4(2):100100. https://doi.org/10.1016/j.stae.2025.100100
107. Pata SK, Erdogan S, Pata UK, Meo MS (2025) Greening road transport: comparison of technologies in conventional, hybrid, and electric vehicles. J Environ Manag 380:124908. https://doi.org/10.1016/j.jenvman.2025.124908
108. Li J, Huang Y, Zhang C, Yao D (2025) How does public transport development contribute to carbon emission reduction? Transp Res Part A Policy Pract 191:104327. https://doi.org/10.1016/j.tra.2024.104327
109. Lyu S, Huang Y, Sun T (2025) Urban sprawl, public transportation efficiency and carbon emissions. J Clean Prod 489:144652. https://doi.org/10.1016/j.jclepro.2025.144652
110. Tikoudis I, Martinez L, Farrow K, Bouyssou CG, Petrik O, Oueslati W (2021) Ridesharing services and urban transport CO_2 emissions: simulation-based evidence from 247 cities. Transp Res Part D: Transp Environ 97:102923. https://doi.org/10.1016/j.trd.2021.102923
111. Zhao P, Li Z, He Z, Chen Y, Xiao Z (2025) Reducing the road freight emissions through integrated strategy in the port cities. Nat Commun 16:2563. https://doi.org/10.1038/s41467-025-57861-z
112. Draft Inland Transport Committee Strategy on Reducing Greenhouse Gas Emissions from Inland Transport (ECE/TRANS/2024/3). Economic and Social Council, United Nations Economic Commission for Europe, Geneva, 20–23 February 2024. https://unece.org/sites/default/files/2024-02/ECE-TRANS-2024-3e.pdf
113. Torkey A, Zaki MH, El Damatty AA (2024) Transportation electrification: a critical review of EVs mobility during disruptive events. Transp Res Part D Transp Environ 128:104103. https://doi.org/10.1016/j.trd.2024.104103

Chapter 3
The Electric Vehicle Transition

Contents

3.1 Introduction

Over 100 years ago, road transport was revolutionized by the replacement of horses with automobiles, resulting in the rise of oil. The next revolution is currently taking place with oil being progressively replaced with electricity. In fact, electric cars were very popular in the early 1900s (Fig. 3.1); it was around 1910 that the electric car was gradually replaced due to major developments in ICE vehicles. As an illustration, by 1912 a Ford Model T (Fig. 3.2) was sold for barely 40% of the price of an electric car. The progress in ICE vehicles was combined with the development of road networks beyond urban areas; the ease of installing gasoline stations and the discovery of new oil sources made fuel relatively affordable. This caused electric automobiles to be archaic around 1935.

Presently, about 1.5 billion ICE cars operate around the world [1], corresponding to almost 45% of the global fuel consumption [2] as represented in (Fig. 3.3). Diesel and gasoline engines have dominated the road transport market since the first automobile revolution without any credible alternative. Crude oil is affordable, and engines are more and more performant. Road transport holds an important contribution to air pollution, greenhouse gas emissions, and consequently global warming.

R. Pode, B. Diouf, *Decarbonization of Road Transportation*, Green Energy and Technology, https://doi.org/10.1007/978-3-032-11388-7_3

Fig. 3.1 Fritchle electric car, 1910

Fig. 3.2 Ford Model T, 1921

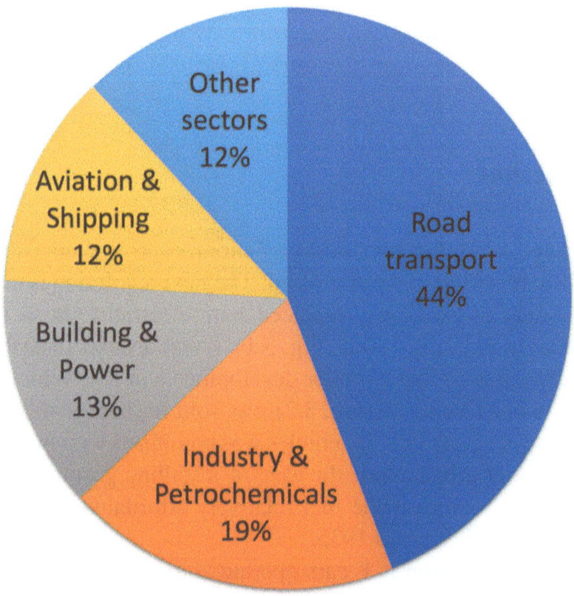

Fig. 3.3 Breakdown of oil usage in different sectors [2]

There is a need for alternatives for economic and environmental considerations; there are different offers with different levels of integration in industrial production [3, 4]. In the early 2010s, EV witnessed an important second revolution with the development of Li-ion batteries with improved power and energy densities, allowing better performance and range [5–7]. Since 2010, there has been a steady progression of electric car sales around the world, reaching over 10% of new car registrations globally in 2022 [8] out of a total of 66.2 million new vehicles on the market [8]. On the other hand, since the 1970s, the price of crude oil has known immense periodic variations directly affecting the price of fuel. The variation of crude oil price was primarily the decision of oil-producing countries (OPEC) that control the production, as in the oil shock of 1973 (Fig. 3.4) [9].

EVs are supposed to take the lead in the global vehicle market by 2040 (Fig. 3.5) [10]. For various motivations, such as environmental, economic, and political, governments establish different electric mobility policies.

In this chapter, we present our research related mainly to the non-subsidy drivers of the EV market. We compare ICE vehicles to EVs, with the performance of the latter essentially driven by the present battery technology and its future requirements, challenges, and the necessity of more research with a top-down approach. In second place we will present EV market penetration criteria with leading countries and the different variables that show how ready or prepared they may be to guarantee 100% EV transition: cost of ICE vehicle ownership, cost of electricity, density of population, Gross Domestic Product (GDP) per capita, rate of urbanization, and percentage of renewable energy in their energy sources. Here, we have presented an upgraded diffusion model of EV technology based on the traditional S-curve equation, the challenges of a potential secondhand EV market, and the future of oil and lithium resources as well.

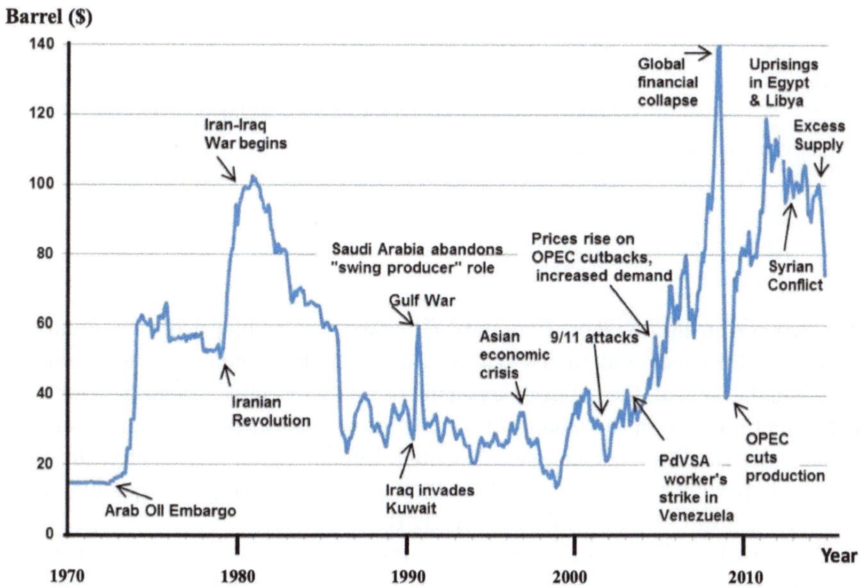

Fig. 3.4 Crude oil prices react to many types of geopolitical events, from weather disasters to wars, revolutions, and economic growth or recessions [9]

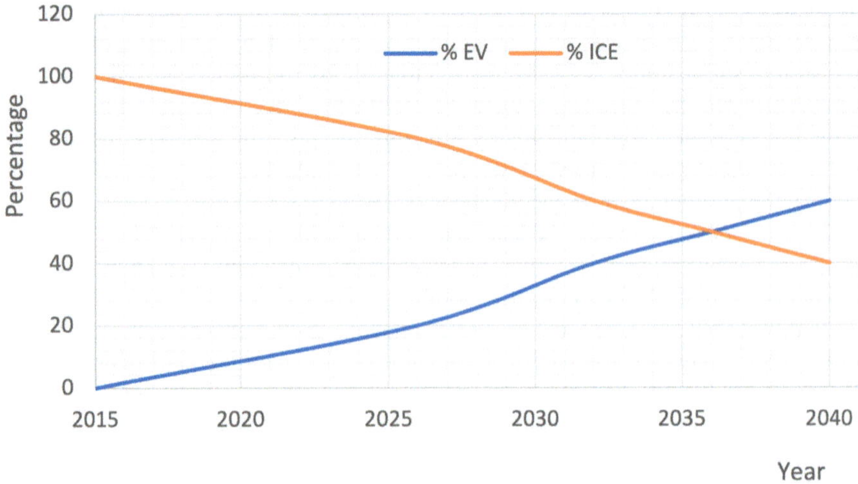

Fig. 3.5 Global share of total annual passenger vehicle sales by drivetrain [10]

Table 3.1 EV vs. ICE Kia Ray [12]

Kia Ray	EV	ICE
Weight	1185 kg	998 kg
Capacity	14.6 kWh/330 V	38 L
Range	138 km	400 km < r < 500 km
Max speed	130 km/h	200 km/h
Consumption	1.06 km/kWh–10.5 kWh/100 km	7.6 L/100 km–9.4 L/100 km
Charging time	6 h/220 V–25 min fast charge	
Cost[a]	US $21,000	US $13,000

[a] At the current exchange rate $1 = 1300 KRW

3.2 EV Vs. ICE Vehicles: Lithium the Driver of EV Technology

The first non-institutional driver of the EV market, along with price, is the combination of performance and safety that should match the standards of ICE vehicles [5–7]. In that regard, the battery plays a major role. Presently, around the globe, for equivalent models on the catalog, EVs cost more than their ICE counterparts; in some cases the difference can be substantial. Depending on the country, subsidies play an important role in setting off the total cost of ownership or acquisition [11]. Table 3.1 below compares EV and ICE Kia Ray [12].

A conventional ICE vehicle can generally last between 200,000 and 300,000 miles. It corresponds to a total of 10–15 years. On the other hand, EV batteries last 10–20 years in theory. The average cost of an EV battery is between 30% and 60% of the vehicle's total value. Replacing the battery will be from free of charge to $20,000 on average, depending on the warranty. The demand for Li-ion batteries

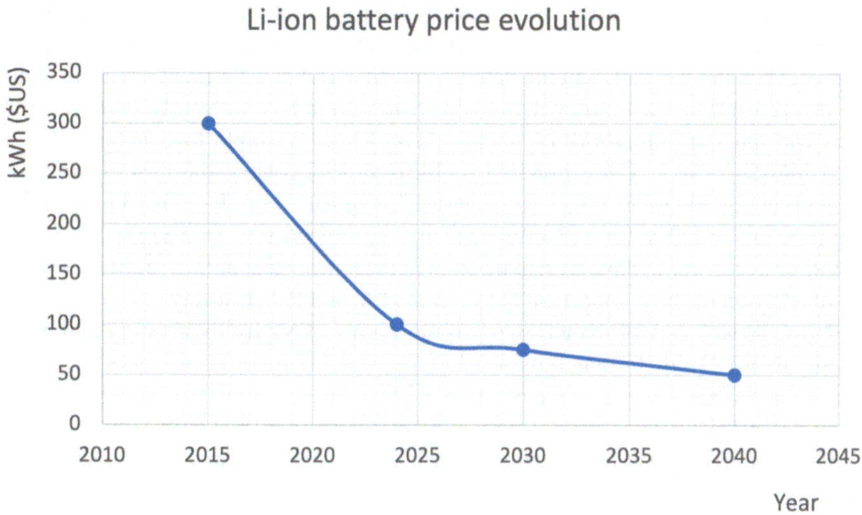

Fig. 3.6 Li-ion battery price evolution [14]

driven by the electric automobile industry will continue growing to reach 4.7 TWh by 2030 [13]. The continuous drop of the battery price (Fig. 3.6) will pull the EV total acquisition cost down, making it more affordable. There is a decrease from $300 in 2015 to approximately $100/kWh currently, and the cost of the kWh may be as low as $50 by 2040, as presented in Fig. 3.6 below [14–17].

EV Li-ion battery demand jumped from 330 GWh in 2021 to 550 GWh in 2022, a 65% increment, corresponding to a 55% increase in new car registrations made of EVs from 2021 to 2022 [18].

Obviously, the battery demand drives the basic material needs to critical limits. In 2022 lithium demand surpassed the supply [19]. As a reference year so far, in 2022, 60% of the available lithium on the market was for the EV industry, 30% of cobalt, and 10% of nickel. In the near future, lithium mining may not keep up with the market demand to guarantee the progress in EV transition [19].

Presently, the main cathode materials for the EV battery industry are lithium nickel manganese cobalt oxide (NMC), 60% of the market, followed by lithium iron phosphate (LFP), about 30%, and nickel cobalt aluminum oxide (NCA) for approximately 8% market share. Commercial LFP batteries have a density of energy below 200 Wh/kg while for ternary batteries (NMC), the energy density is between 200 and 300 Wh/kg [20].

Sodium-ion (Na-ion) batteries are a credible alternative to Li-ion technology. Among others, they present the advantage of using lower-cost materials even if their energy density (75–160 Wh/kg) is lower than for Li-ion (150–170 Wh/kg) [20] they may make a good match with urban transit, which requires shorter range and therefore smaller energy storage.

If we compare EV and ICE vehicles, filling a gas tank takes 10 min at the most for an average vehicle, compared to at least about 6 h of home charging for an EV or about 25 min while using fast public charging [21, 22]. One of the main drivers of Li-ion technology in electric transport is the range, corresponding to the comparison between the gas tank and the battery capacity. The development of Li-ion technology [23, 24] for the future will include greater autonomy. Currently, fuel-powered cars can have an autonomy of up to 731 miles on a single tank (Jaguar X20d 731 miles) for 56 L, corresponding to 14.62 miles/L, while the best electric car range is about 500 miles on a single charge (Lucid Air 531 miles) for 13 h of charging time under 240 V for 118 kWh capacity, 4.5 miles/kWh. To bridge such a gap, it will be necessary to have a breakthrough in the battery technology. Table 3.1 gives an illustration of the comparison between the EV and ICE Kia Ray.

NMC as well as LFP batteries are used to power EVs for their safety and good energy density [20]. To match the world record of 731 miles range for a passenger car, it will be necessary to have batteries with an energy density of about 400 Wh/kg, with obviously comparable safety guarantees, lifespan, power density… Such a density of energy is yet to be achieved for batteries that can satisfy safety and lifespan in EV application requirements. This shows the necessity of more investment and research in battery technology to make them more affordable and with higher performance. The electrolyte is a good direction for more research.

In fact, the electrolyte plays an important role in the battery capacity, power density, life cycle, and safety. Modern Li-ion batteries use liquid electrolyte. Liquid electrolytes present issues as low boiling points and flashing points. These batteries are prone to catch fire or explode in case the liquid leaks or heats up above a certain level. As well, liquid electrolytes do not allow the use of high-voltage cathode material and consequently limit the battery energy density.

Solid-state lithium batteries are said to be the next-generation batteries with a higher energy density and improved safety. They use solid electrolytes and no separator (Fig. 3.7) [25, 26].

It is very likely that lithium will continue to play a central role in the EV industry. Car manufacturers have been working on a rapidly emerging new battery

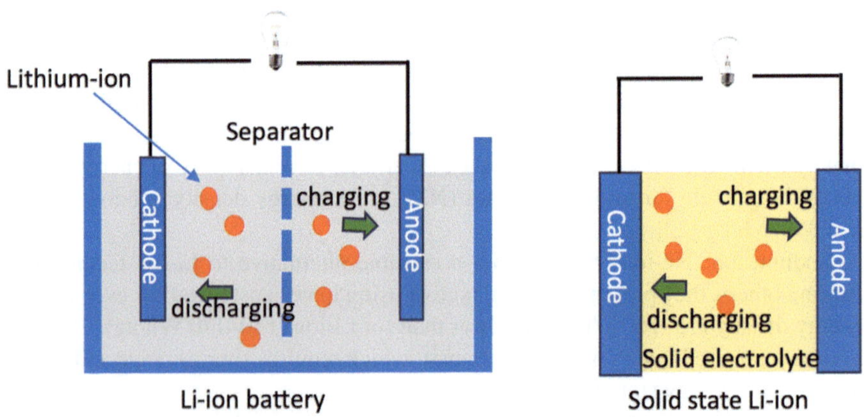

Fig. 3.7 Liquid vs. solid electrolytes [26]

technology known as solid-state batteries. Solid-state lithium batteries use a solid material rather than a potentially combustible gel. These materials are neither volatile nor combustible. The higher energy density would increase the range of EVs by between one-third and 50%. Toyota has a solid-state battery-powered car prototype with production expected in 2024 [27].

Energy storage for electric transport still needs a longer lifespan, a lower rate of degradation, higher energy density, less sensitivity to high temperatures and temperature variations, and less sensitivity to humidity as well. There is an important amount of research being conducted to improve battery quality and performance based on experiment, theory, or simulation [28–33]. The main orientations to reach such a goal are design and heat management, material composition, and structure as well [28–33].

3.3 Electric Car Market Drivers

3.3.1 Market Drivers

It is a complex task to discuss the difference in incentives between countries unless it is direct cash received by EV buyers. Incentives can apply differently depending on countries; as an illustration, in the UK buyers receive direct flat subsidies, in the US they are based on the battery capacity, in France they evaluate carbon dioxide emission, and in China subsidies are range-based. In Norway, the objective of the government is to arrange subsidies so that the acquisition of an EV will cost at most the same as its ICE vehicle counterpart. Norway has the substantial tax incentives for EVs, with all taxes waived, even if such an approach is being scaled down.

What is the secret to leading the world on the EV transition? The leaders, Norway and Iceland, (Table 3.2) have approached EV incentives differently from many other countries, and such an approach started decades ago. Rather than important direct subsidies to individuals, they created a context favorable to the adoption of EVs, including considerable tax breaks and infrastructure such as charging stations [34, 35]. Norway has a high taxation on ICE cars, which makes it easier to incentivize EVs through taxation benefits. All this combined with high fossil fuel prices compared to electricity and high urbanization makes it an ideal market for EVs. Furthermore, Iceland and Norway have the capacity to provide virtually 100% of their electricity from renewable sources.

Table 3.2 shows the rate of penetration of EVs in different markets with Norway and Iceland. In 2022, they come far ahead of any other nation, as shown in Table 3.2 with 86% of new car registrations in Norway and 72% in Iceland. The third country in the top 20, Sweden, comes far behind with "only" 43%. Out of the 20 countries that make the list, the 11 countries that present the lowest rates are all under 20%; 7 of those are under 10% (Table 3.2).

Table 3.2 Non-subsidy drivers of the EV market in the top 20 markets

	[38]	[39]	[40]		[41]	Energy sources [42–44]			Density [45]	[46]	Cars/1000 people [47]
Country	% of e-cars (%)	1 L Gasoline ($US)	1 kWh ($US)	1 L/1 kWh	GDP/ Capita	Renewable	Fossil	Nuclear	pop/km^2	%Urban	
Norway	86	2.091	0.129	16.209	89,154.30	96.7	3.3	0	43	84	634.33
Iceland	72	2.242	0.146	15.356	68,727.60	100	0	0	4	94	695.18
Sweden	43	1.853	0.361	5.132	61,028.70	55.9	2.4	40.2	60	88	544.53
Denmark	35	2.066	0.576	3.586	68,007.80	40.2	57.5	0	352	88	546.03
Finland	31	2.209	0.454	4.865	53,654.80	32.8	27.5	31.6	43	86	577.57
Netherlands	30	1.945	0.489	3.977	57,767.90	11	83.3	3.7	520	93	589.33
Germany	26	1.94	0.554	3.501	51,203.60	20.5	60.1	17.9	605	78	633.46
Switzerland	22	2.014	0.234	8.606	91,991.60	66	1	34	545	74	603.77
Portugal	20	1.778	0.293	6.068	24,567.50	57	44	0	289	67	648.40
United Kingdom	19	1.806	0.412	4.383	46,510.30	41.9	39	19.1	715	84	639.14
France	19	1.979	0.217	9.119	43,659.00	11.6	8.5	79.4	303	82	673.75
Belgium	18	1.793	0.521	3.441	51,247.00	20.5	60.1	17.9	985	98	591.19
China	16	1.196	0.08	14.95	12,556.30	17	81.2	1.8	385	64	225.22
Italy	9	1.959	0.576	3.401	35,657.50	27.6	71.4	0	509	72	765.75
Spain	8	1.702	0.369	4.612	30,103.50	29.8	50.2	20	243	81	629.01
Greece	7	1.984	0.272	7.294	20,192.60	29	71	0	205	80	631.57
Canada	7	1.3	0.119	10.924	51,987.90	62.3	22.8	14.7	11	82	704.82
South Korea	6	1.233	0.101	12.207	34,997.80	1.5	68.7	29.8	518	81	457.80
United States	5	1.011	0.18	5.616	70,248.60	12.2	68.6	19.2	91	83	872.55
New Zealand	4	1.622	0.193	8.404	48,781.00	76	24	0	49	87	864.92

The electric car sale is up to 14% of global sales in 2022, and some studies project electric car sales to reach 54% of new cars on the market by 2040 [8, 36] (Fig. 3.8), with projections to reach about 70 million EVs to be sold in 2040.

A survey [37] (Fig. 3.9) shows that ahead of any other criteria, affordability is the first motivation for car buyers to acquire an EV, followed by range and availability of public charging stations. Most of the other criteria can be translated in terms of affordability (Fig. 3.9). The EV market is certainly impacted by governments' policies visible in acquisition and operation costs, particularly compared to ICE vehicles. Acquisition subsidies can certainly guarantee affordability to get the industry to pick up, but they are not sustainable in the longer run. GDP per capita, as a translation of average income, would rather be more determinant.

Among the countries with the highest percentage of EV sales, Norway and Iceland, the first and second market shares in percentage in the list of the top 20

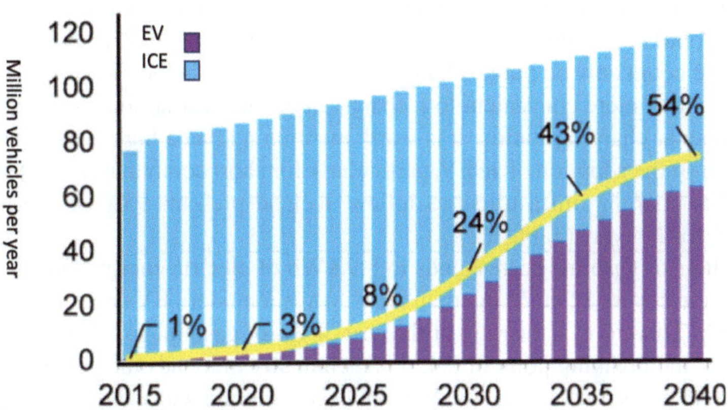

Fig. 3.8 Annual global light-duty vehicle sales [36]

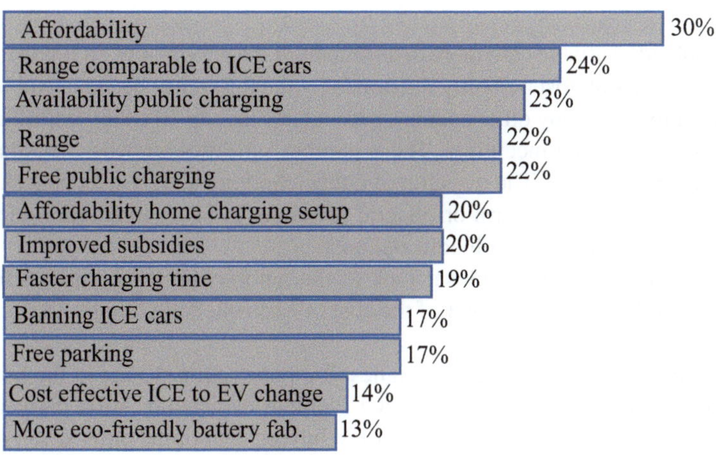

Fig. 3.9 Motivations for EV adoption [37]

countries with the highest proportion of electric cars sold in 2022, rank second and fourth highest in GDP per capita in the list presented in Table 3.2. Government policies can be seen as well with the percentage of renewable energy in their energy sources, with Norway and Iceland coming in second and first positions with 96.7% of renewable sources for Norway and almost 100% for Iceland. This shows that these governments' policies as well as the potential they have in renewable energy make a good match with the transition to 100% EV.

These countries present as well other remarkable specificities: the highest cost of gasoline and one of the highest ratios of cost of gasoline to electricity. Norway is the country that presents the highest ratio, followed by Iceland. This shows that without subsidies, the operation cost of an EV would be much more beneficial in these countries compared to all the other countries in the list.

Iceland presents the lowest density of population, and Norway comes in third position; they both have a high rate of urbanization. This gives a good indication about populations having more chance to live in individual houses compared to densely populated countries where people are more likely to live in apartment complexes and count only on public charging stations. Living in individual flats or villas makes it much more convenient to recharge an electric car at home or even have a personal solar charging station. It is worth mentioning that a high urbanization rate is in favor of the adoption of an EV, particularly for short commutes, provided, obviously, that charging infrastructures follow. Iceland presents 94% of urbanization and 84% for Norway (Table 3.2).

One liter of gasoline corresponds to 8.9 kWh of electric energy. On average an ICE vehicle covers 12 km/L in urban areas, while a modern EV can run about 5 km on 1 kWh. Therefore, if the ratio of the cost of 1 L of gasoline to 1 kWh of electricity is over the breaking point of 2.4, it is technically more advantageous to operate an EV than a gasoline car. In Saudi Arabia, for example, currently 1 L of gasoline costs $0.621 while 1 kWh of electricity is sold for $0.18 for households; the ratio is only 3.45 versus 16.21 in Norway and 15.36 in Iceland. A Saudi driver needs a different motivation to buy an EV compared to someone who lives in Norway or Iceland.

Non-subsidy variables are more determinant in the adoption of an EV than direct subsidies. This is almost perfectly illustrated by the difference between Norway and Iceland compared to any other country, particularly the top 20 countries, which look to propose very strong incentives. These indirect subsidies are what will guarantee the sustainability of EV. The question of public charging stations can be technically solved by either using on-grid or off-grid solutions. A better integration of renewable energies looks to be an obvious option.

It is necessary to have a closer look at the automobile reality in the different markets to understand the weight of the cost of car ownership (Table 3.3) in market drivers.

Out of the 20 countries with the highest EV penetration, 13 are in the list of the 20 markets with the highest cost of ICE vehicle ownership. Such markets have much more room to offer tax rebates for EV buyers. Out of the other seven remaining countries, Iceland, Switzerland, Belgium, Italy, China, South Korea, and New

Table 3.3 Most expensive countries for car ownership (average cost in USD) [48]

Ranking	Country	Annual depreciation	Annual insurance	annual inspection	Annual tax & registration	Annual maintenance cost	Annual fuel	Annual parking	Annual toll fees	Annual washing	Fuel (L)	Licensing
1	Denmark	$1809	$651	$125	$355	$794	$2892	$120	$62	$82	$2.07	$36
2	Germany	$1662	$717	$236	$217	$2045	$1717	$136	$41	$96	$1.89	$43
3	Finland	$1481	$720	$75	$351	$2143	$1711	$149	$21	$109	$2.06	$27
4	UK	$1336	$692	$110	$174	$2068	$1670	$167	$38	$67	$1.89	$39
5	Netherlands	$966	$660	$177	$521	$1765	$1366	$107	$52	$55	$2.03	$41
6	Sweden	$1517	$615	$33	$259	$2232	$1816	$221	$77	$116	$1.95	$25
7	Ireland	$1261	$970	$81	$335	$632	$2042	$155	$186	$82	$1.82	$54
8	Norway	$1699	$848	$249	$226	$608	$1519	$203	$417	$110	$2.16	$19
9	Canada	$1455	$1316	$92	$88	$711	$1835	$223	$71	$85	$1.37	$66
10	USA	$1613	$1018	$85	$151	$1931	$1525	$171	$139	$104	$1.06	$47
11	Australia	$1250	$704	$147	$235	$2048	$1616	$132	$137	$87	$1.09	$40
12	Spain	$1541	$519	$47	$123	$1835	$1505	$215	105	$57	$1.75	$24
13	Greece	$934	$366	$56	$249	$1821	$1527	$81	$149	$73	$2.02	$18
14	Guatemala	$853	$502	$160	$88	$1713	$1316	$237	$117	$109	$1.21	$65
15	Portugal	$1280	$346	$33	$111	$1493	$1237	$107	$217	$44	$1.83	$29
16	Ecuador	$932	$581	$157	$278	$1476	$1202	$151	$189	$86	$0.63	$142
17	France	$1416	$672	$66	$78	$1461	$1173	$144	$157	$57	$1.63	$29
18	Czech Republic	$1218	$425	$78	$42	$2379	$1908	$84	$77	$67	$1.75	$2
19	Uruguay	$419	$543	$71	$371	$2076	$1701	$170	$150	$94	$1.82	$7
20	Poland	$782	$366	$74	$25	$1591	$1392	$73	$64	$68	$1.45	$36

Zealand, five are part of the list of top countries with the highest direct incentives for the acquisition of an EV [48].

The cases of Switzerland and Iceland, the only two countries that are in none of the lists, deserve attention. Switzerland has applied, at least initially, an exclusively sanction policy based on cars' greenhouse gas emission level before including subsidies and a heavy investment in public charging infrastructures [49].

Iceland presents other specificities: a high urbanization rate (94%), a small population and small country, a high density of charging stations, and a low cost of electricity [50, 51].

The bottom line is that markets with the highest EV penetration are those where ownership of ICE vehicles is more expensive and countries where subsidies are higher.

Most of the other variables are directly related to batteries, as the cost of the vehicle, the performance of the EV, safety, or lifespan. Based on market requirements, more research is still needed even if EV technology has reached a competitive maturity. In the longer run, more research is needed to develop higher-performance batteries to be more competitive in cost, energy density, lifespan, and complexity of materials for easier recycling.

The full adoption of EVs will need to overcome a certain number of challenges, such as public charging stations. In the immediate practical aspect, access to public charging stations is particularly important for densely populated countries such as South Korea or Japan. Grid infrastructures are not designed to support such an overload in production, transmission, or distribution. It is almost obvious that off-grid renewable energy solutions will be needed. This may include more consideration of photovoltaic chargers in EVs, in the form of foldable high-efficiency solar panels or even high-efficiency photovoltaic paint that would cover vehicles. Gas stations may be associated with an easier integration of EV charging stations, as Shell Company is already doing (Fig. 3.10). Such an initiative will help in the transition from ICE cars to EVs in terms of infrastructure or employment.

Fig. 3.10 Shell offering EV charging stations next to standard gasoline stations [52]

3.3.2 EV Diffusion Model

The success of the EV transition can somehow be measured by the will of governments to fight pollution or limit dependence imported oil, particularly in the geopolitical context of the last 50 years. Table 3.4 shows the evolution of the EV transition in the 20 most dynamic markets between 2013 and 2022.

EVs represent a new technology, and they are still more expensive than their ICE counterparts. Under the light of statistics, as an innovation, EV market growth follows an S-curve (Fig. 3.11) [54].

The slope of the curve, representing how fast the technology is adopted, will depend on the context of each market before 100% transition is eventually achieved. The pickup is slow before an exponential growth, and a slowdown before saturation is eventually reached for a full transition, as in any S-curve [54] characterized market penetration.

The major increase in new registrations for all countries occurs at an inflection point as if there was a form of confidence from buyers after pioneers opened the game. There is then an exponential growth till a certain level before a slowdown to convince the most reticent buyers. Once efficiency is proven and a solid base of trust built, the majority of buyers will invest in what is supposed to be a new product. The last category of buyers steps in at a slow pace to complete the full EV transition.

Table 3.4 Passenger plug-in (battery electric and plug-in hybrid) market share of total new car sales for selected countries and selected regional markets since 2013 [53]

Year	2013	2014	2015	2016	2017	2018	2019	2020	2021	2022
Country										
Norway	6.1	13.8	22.4	29.1	39.2	49.1	55.9	74.7	86.2	87.8
Iceland	1.3	2.1	3.6	5.6	11.7	17.2	22.6	45.0	54.7	56.1
Sweden	0.71	1.53	2.62	3.5	5.2	8.2	11.4	32.2	43.3	56.1
Denmark	0.29	0.88	2.29	0.6	0.6	2	4.2	16.4	35.2	38.6
Finland	–	–	–	1.2	2.5	4.7	6.9	18.1	30.8	37.5
Netherlands	5.55	3.87	9.9	6.7	2.2	6.2	14.9	24.6	29.8	34.9
Germany	0.2	0.43	0.73	1.1	1.58	1.9	3.0	13.5	26.0	31.4
Switzerland	0.44	0.75	1.98	1.8	2.55	3.2	5.5	14.3	22.5	25.9
Portugal	–	–	–	–	1.9	3.6	5.7	13.5	20.0	21.7
United Kingdom	0.16	0.59	1.07	1.37	1.86	2.53	2.9	10.7	18.6	22.9
France	0.83	0.70	1.19	1.4	1.98	2.11	2.8	11.2	18.3	21.6
Belgium	–	–	–	1.8	2.7	2.5	3.2	10.7	18.4	26.5
China	0.08	0.23	0.84	1.31	2.1	4.2	4.9	5.4	15	30
Italy	0.07	0.08	0.09	0.08	0.1	0.26	0.6	4.3	9.3	8.8
Spain	0.07	0.08	0.09	0.08	0.1	0.26	0.6	4.3	9.3	8.8
Greece	–	–	–				0.2	0.8	6.2	7
Canada	0.18	0.28	0.35	0.58	0.92	2.2	3.0	3.5	6.6	9.4
South Korea	–	0.01	–	–	1	2.5	2.8	3.4	–	6
United States	0.60	0.72	0.66	0.90	1.1	2.1	1.9	2.2	4.0	7.7
New Zealand	–	0.1	0.2	0.6	1.2	1.9	2.8	2.8	3.6	13.0
Global	0.27	0.43	0.7	0.89	1.4	2.3	2.6	4.2	8.7	14

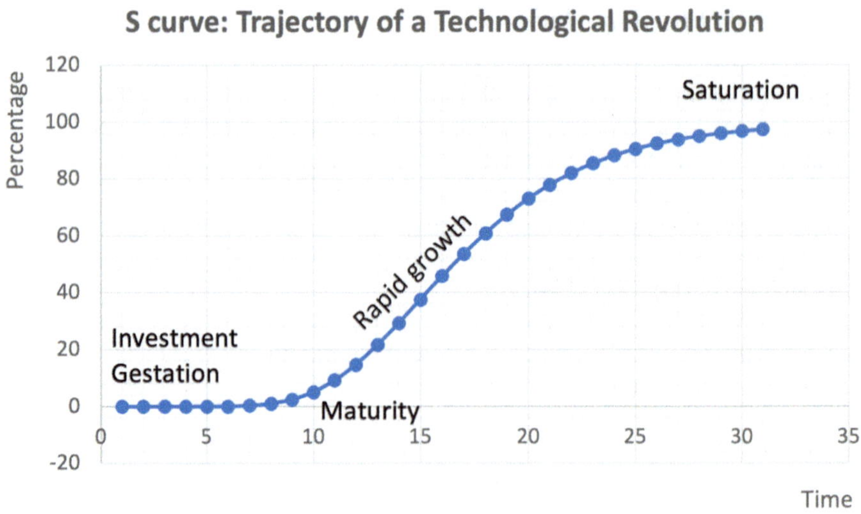

Fig. 3.11 S-curve as a typical trajectory of a technological transition

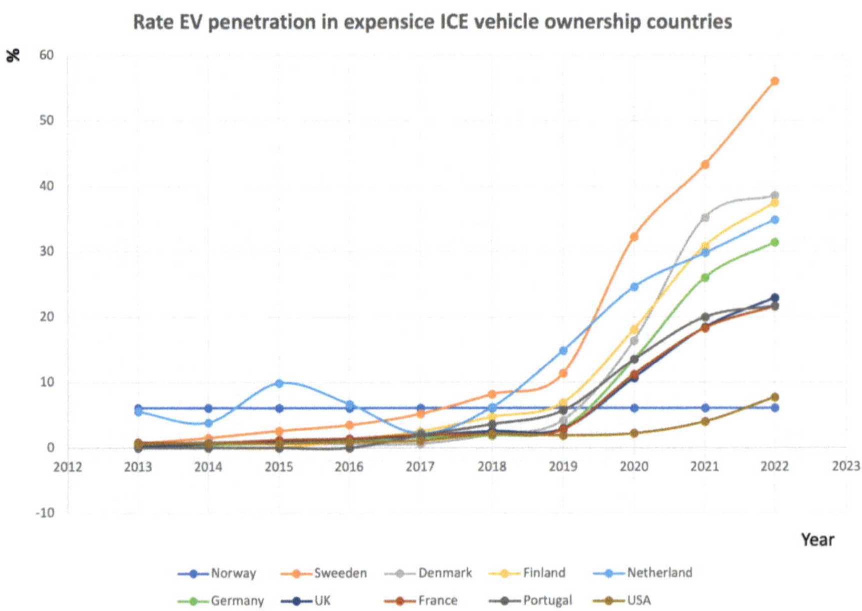

Fig. 3.12 Rate of EV penetration in expensive ICE vehicle ownership countries from 2013 to 2022

Figure 3.12 shows the market evolution of some countries with a meaningful EV penetration. These markets are those with the highest cost of ICE vehicle ownership. This Fig. 3.12 displays the curve of growth for some markets with a meaningful EV penetration from 2013 to 2022.

The way an innovation is adopted follows a non-linear path; the first stage is slow, and the second is fast before the last stage that corresponds to saturation. These curves are known as S-curves [44] due to their shape. The shape of the S-curve can be attributed to system feedback: economies of scale, social diffusion, and technological reinforcement. The S-curve model [44] can provide a description of these markets. The curve gives the rate R(t) of market penetration of the EV against time by the Eq. (3.1) below, where each market can be well represented by two output values, α and k, to be found by adjusting the curve R(t) with the actual data. So α and k are the outputs to be found by scanning through a range of values until the graph R(t) matches the graph obtained from actual data.

$$R(t) = \frac{100}{\left[1+\exp(-\alpha t)\right]^{k}} \quad \textit{where}\, \alpha > 0\, \textit{and}\, k > 0\, \textit{and where}\, t\, \textit{is the time}, \tag{3.1}$$

Here are some significant markets and their characteristics:

Norway (Fig. 3.13) α = 0.29 and k = 11
Sweden (Fig. 3.14) α = 0.25 and k = 30
Global (Fig. 3.15) α = 0.12 and k = 11

Based on these curves, Norway could be expected to have witnessed a full EV transition for new registrations around 2025 (Fig. 3.13) and 2030 for Sweden (Fig. 3.14), while globally (Fig. 3.15) a full transition will not occur before 2050 when considering the most common types of electric vehicles: battery electric vehicles (BEV), hybrid electric vehicles (HEV), and plug-in hybrid electric vehicles (PHEV).

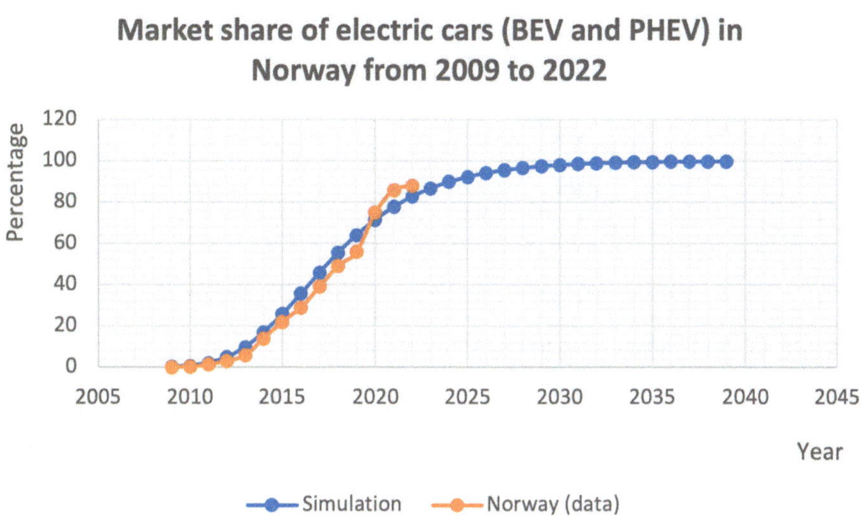

Fig. 3.13 Norway EV transition statistics and simulation

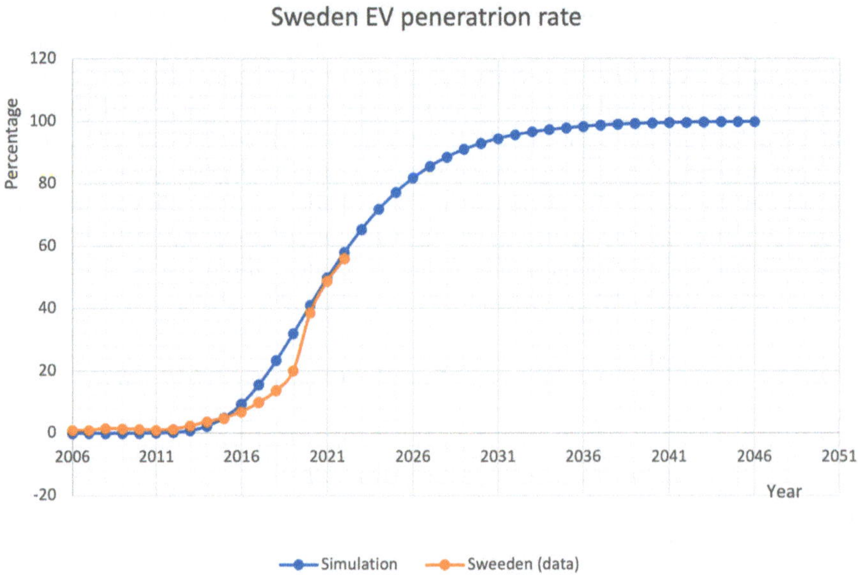

Fig. 3.14 Sweden EV transition statistics and simulation

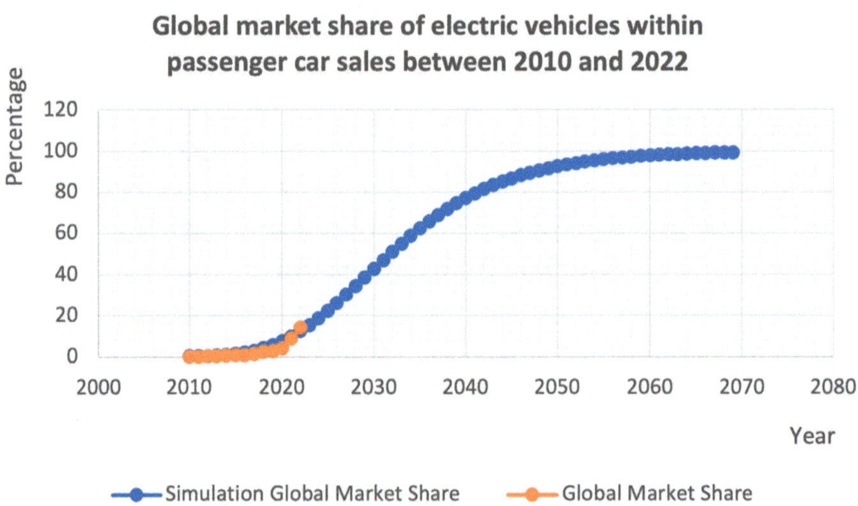

Fig. 3.15 Global EV transition statistics and simulation [55]

Each market is described by a couple (α, k) that determines the time for the technology to take off and the slope of the S-curve that corresponds to the rapidity of growth.

3.3.3 *Would the Future of Oil Affect Global EV Market Penetration?*

Crude oil is beyond a mere source of energy; it has become a real strategic tool in the hands of oil-producing countries (OPEC) to play major political roles on the international scene.

From ICE to EV, the dependence is switching from oil to lithium. Countries around the world develop different strategies to prepare for the EV transition. Presently, lithium Earth reserves are evaluated at 88 million tons [56]; with 8 kg of lithium on average per car, there is potentially enough material to guarantee a 100% EV global transition even if the demand is growing at a very high rate (Fig. 3.16).

The main questions that arise with the higher penetration of EVs and their future market domination, despite the potential of reserves, are the availability of enough lithium to build batteries on the one hand, as stated above, and on the other hand, what would be the impact of a high percentage of EVs on the oil demand and price, knowing that the global transport sector accounts for almost 50% of the current oil demand?

It is predicted that by 2025, 75% of all mined lithium will go into EVs, as presented in Fig. 3.16. Lithium reserves may not present any potential shortage, but there are difficulties related to the extraction of raw materials. Lithium supply shortage may be one of the challenges that the manufacturing of EVs may face in the near term, as shown in Fig. 3.17, unless there is a technological breakthrough.

The development of EV will undoubtedly lead to a lower fuel demand; oil supply will be higher than demand. In such a case, with the steady increase of EV sales,

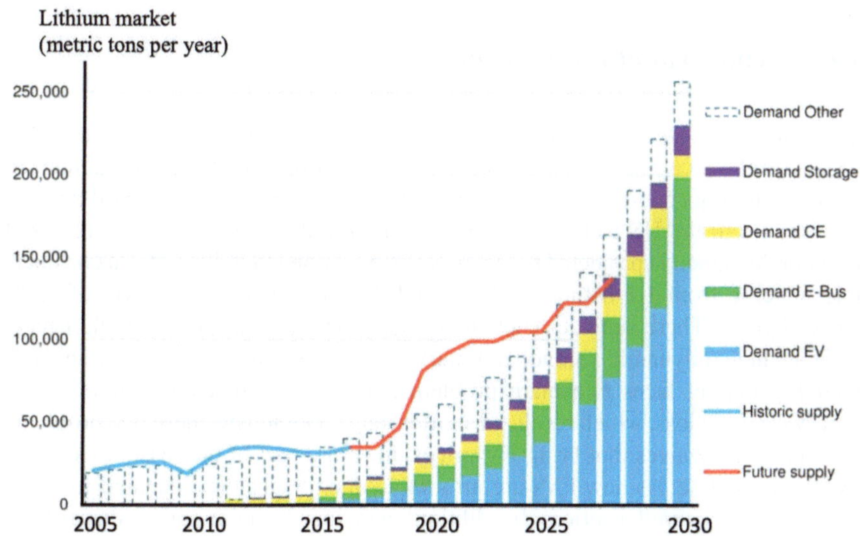

Fig. 3.16 Demand for lithium [57]

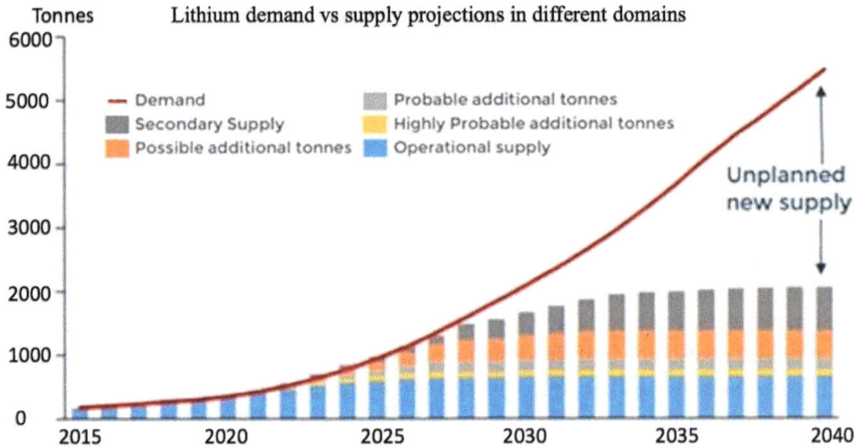

Fig. 3.17 Lithium demand vs. supply forecast [58]

predicted to reach 54% of new registrations by 2040 [58], it could be expected to have a future oil price drop due to lower demand.

Surprisingly, analysts predict prices will increase for oil, due primarily to more difficulties of oil extraction, less accessible sources that would require more important investments, and the growing demand from India and China. The direct consequence is a price increase for fuels.

It can easily be predicted then that there will be an increasing demand in EVs due to the higher price of oil.

3.4 Second-Hand EV Market

According to research released by AA Cars, in total, almost 75% of drivers in the US, as a reference market, purchased a used car as their most recent vehicle. In most developing countries this rate goes up to almost 100%. Sales of used passenger vehicles in the US were around 38.6 million units in 2022. In 2027 the used automobile market is projected to reach over 1.5 trillion US dollars; it is a considerable increase from 1.2 trillion in 2020, and it corresponds to a growth of 3.2% across 7 years. These statistics show that a successful EV transition should necessarily include a dynamic secondhand market. Nevertheless, presently, the secondhand EV market faces two main problems: the rapid depreciation of EVs and battery capacity loss, which results in shorter range. Obviously, these two problems are related. Incentives that bring EV prices down are part of the reason why used EV prices decline. As new EVs are affordable due to subsidies, demand for used EVs decreases as well as prices. In addition, there may be an important disparity in interest rates that can be higher for used cars than for new vehicles. The height of the coronavirus 19 disease (COVID-19) was an unusual period, due to factories' lower activity there was a shortage of new cars, demand on the used market drove prices up.

Based on a Cox Automotive study, EV owners perceive that EV batteries last about 7 years on average; after that period there is a sensitive range drop. The odometer only is not enough to understand the state of health (SOH) [59] of the battery; a certain number of other parameters come into play in the battery life of an EV: manufacturing variances, operation conditions such as the cooling system, and charging conditions such as fast or slow charging. The challenge is working out how much charge is still available after a certain period of operation. The best guarantee of a good knowledge of the battery health is by getting a diagnostic from the onboard diagnostics port of the EV. While 90–95% of the original capacity is acceptable, below 80% should raise some worries. Tesla claims about 90% of capacity will be left after 200,000 miles of use on the Model S, and X; that may not be accurate based on drivers' opinions. No other manufacturer has made a report of such good battery and management qualities. Such a claim is beyond expectations. The reality for most manufacturers is 2–3% of capacity loss per year, and it is even worse when fast charging is used, one of the worst stresses on a battery.

The secondhand market is presently indeed a major issue in EV technology, in contrast to ICE vehicles. Almost all EV drivers buy their vehicles new, and generally incentives apply exclusively to new EVs, as a replacement of an ICE vehicle or as their first vehicle. Despite the remarkable performance of EVs and their competitiveness, they show some limits compared to ICE vehicles, as presented above. It is reported that the depreciation of EVs is over 50% of their original value within the first 3 years of ownership compared to less than 40% for ICE vehicles. Due to the high price of premium EVs, such depreciation corresponds to a loss of thousands of dollars for the original owner. EV prices fall further down than their ICE counterparts, while the EV premium can be over $10,000 more expensive; this gives more importance to their value loss. The depreciation of EVs is also about a question of demand and supply. It was common to be on a waiting list to purchase an EV, while presently EV makers can keep up with the demand. A third point is related to the battery lifespan as stated above. In fact, due to the battery technology limitations, buyers wonder if it is worth it to buy a secondhand EV, as the range on a single charge may be much shorter than for a new car [60].

In the current state of technology, the battery does not outlast the vehicle. Satisfying such criteria would have been a turning point in the development of the second-hand EV market.

Figure 3.18 below shows the average capacity loss of an EV battery with time of use, with the average range dropping to about 80% after about 6 years of operation while considering 3% of capacity loss per year and after about 10 years when 2% capacity loss per year is considered [61] for the most common chemistries used in EVs: NMC or LFP.

An ecological EV transition will necessarily go with a sustainable secondhand vehicle market; the conditions for such a market are so far missing. It should also be noted that the EV transition in developing countries will most likely strongly depend on a secondhand vehicle market, as is already the case with ICE vehicles.

Fig. 3.18 Average EV battery degradation with time [61]

It is very likely that the concept of secondhand cars as known with ICE vehicles will be replaced with that of refurbished vehicles with a serviced battery and updated software.

The secondhand EV market is a good argument to illustrate the need to build batteries with better performance, longer lifespans, slower degradation rates, and less susceptibility to fast charging.

3.5 Conclusion

The development of Li-ion batteries has been the main tool to achieve the successful penetration of EVs in our societies, but they also remain the main threat to possibly limit the expansion of EVs. Li-ion batteries will need to have higher energy density to match current standard ICE cars; they should allow faster charging without compromising their lifespan, and they should be lower cost to make EVs more affordable and competitive compared to their ICE counterparts. More research and development will be needed to produce the battery for a global and sustainable EV transition for acquisition and operation as well as the recycling of batteries.

In the present situation, some countries seem to be in a more favorable situation for a successful EV transition. It is the case of Norway and Iceland, due to their geography, energy policies, and obviously their economies, that subsidies can guarantee affordability to their populations. In some other countries, opting for EV is still a luxury, and the transition looks to be not realistic in the near future.

The approach of subsidies has certainly been proven efficient to get the market started, but in the longer run, local policies should be able to prepare a more favorable context without subsidies.

The main driver of the EV market is primarily cost before environmental convictions. Strong-willed governments consequently have policies designed to deliver technically performant EVs to buyers at a lower cost or a cost comparable to ICE vehicles and make available a dense network of public charging infrastructures. Such a situation is favorable when the cost of ICE vehicle ownership is high and when subsidies are consistent; in both cases the transition to EV corresponds to saving money. The cost of EVs and range are the other main drivers attracting car owners to EVs when charging infrastructures are available. The cost of an EV as well as its range mainly depends on the cost and performance of the battery. It can then be expected that when Li-ion battery prices considerably drop, the adoption of EVs will be wider. Governments' policies may include an important battery development approach to reduce its cost. The development of sodium-ion batteries (Na-ion) is promising to substantially drop battery cost without losing performance, as it helps for the development of solid-state batteries.

The diffusion of EV obeys the S-curve that allowed us to extract parameters α and k specific to each country and to predict the period to reach 100% EV transition.

Oil dependence is changing to lithium dependence, with the demand already surpassing the supply due to the EV industry.

The present state of EV battery technology and subsidies are the main barriers hampering the development of a meaningful secondhand EV market.

References

1. The Car Drive. https://www.thedrive.com/guides-and-gear/how-many-cars-are-there-in-the-world
2. International Institute for Sustainable Development. https://www.iisd.org/articles/deep-dive/canada-can-expect-declining-oil-demand
3. Verma J, Goel S (2023) State-of-the-art in bioresources for sustainable transportation. Int J Hydrogen Energy 48:3768–3790. https://doi.org/10.1016/j.ijhydene.2022.10.206
4. Verma J, Kumar D (2021) Metal-ion batteries for electric vehicles: current state of the technology, issues and future perspectives. Nanoscale Adv 3:3384. https://doi.org/10.1039/d1na00214g
5. Cooper H, Brauer S, Kalowb DA (2020) Battery energy storage in stationary applications, special edition-battery energy storage. https://www.aiche.org/. Accessed 31 Mar 2022
6. Cheng K, Zou Y, Xin X, Gong S (2020) Optimal Lane expansion model for a battery electric vehicle transportation network considering range anxiety and demand uncertainty. J Clean Prod 276:124198. https://doi.org/10.1016/j.jclepro.2020.124198
7. Diouf B, Pode R (2015) The potential of lithium-ion batteries in renewable energy. Renew Energy 76:375–380. https://doi.org/10.1016/j.renene.2014.11.058
8. International Energy Agency. https://www.iea.org/news/demand-for-electric-cars-is-booming-with-sales-expected-to-leap-35-this-year-after-a-record-breaking-2022#
9. US Department of Energy. https://theconversation.com/oil-price-shocks-have-a-long-history-but-todays-situation-may-be-the-most-complex-ever-178861

10. https://about.bnef.com/blog/electric-vehicle-sales-to-fall-18-in-2020-but-long-term-prospects-remain-undimmed/
11. Data Force https://www.dataforce.de/en/news/incentives-electric-cars-2023/
12. Kia Seoul, South Korea (2024) (Department of Information Display, Kyung Hee University, Seoul, Republic of Korea). Personal communication
13. McKinsey. https://www.mckinsey.com/industries/automotive-and-assembly/our-insights/battery-2030-resilient-sustainable-and-circular
14. Rizopoulos D, Esztergár-Kiss D (2020) A method for the optimization of daily activity chains including electric vehicles. Energies 13:906. https://doi.org/10.3390/en13040906
15. Greenwood D (2018) Automotive batteries 101. Advanced Propulsion Systems. https://warwick.ac.uk/fac/sci/wmg/. Accessed 13 Apr 2021
16. Scot L (2019) Behind the scenes take on lithium-ion battery prices. Bloomberg NEF. https://about.bnef.com. Accessed 21 Mar 2022
17. Bajolle H, Lagadic M, Louvet N et al (2022) The future of lithium-ion batteries: exploring expert conceptions, market trends, and price scenarios. Energy Res Soc Sci 93:102850. https://doi.org/10.1016/j.erss.2022.102850
18. https://www.iea.org/reports/global-ev-outlook-2023/trends-in-batteries
19. https://www.afr.com/wealth/investing/lithium-prices-to-keep-rising-as-demand-outpaces-supply-20230508-p5d6m3
20. Lindsay R (n.d.) 2018 why lithium-ion batteries are the future of UPS energy storage for data centers and colocation facilities. Data Center Frontier Special Report Series. https://datacenterfrontier.com/white-paper. Accessed 1 Nov 2021
21. Gupta V, Kumar R, Panigrahi BK (2020) User-willingness-based decentralized EV charging management in multiaggregator scheduling. IEEE Trans Ind Appl 56:5704–5715. https://doi.org/10.1109/TIA.2020.2993988
22. Abdullah HM, Gastli A, Ben-Brahim L (2021) Reinforcement learning based EV charging management systems—a review. IEEE Access 9:41506–41531. https://doi.org/10.1109/ACCESS.2021.3064354
23. Manzetti S, Mariasiu F (2015) Electric vehicle battery technologies: from present state to future systems. Renew Sust Energ Rev 51:1004–1012. https://doi.org/10.1016/j.rser.2015.07.010
24. Fotouhi A, Auger DJ, Propp K, Longo S, Wild M (2016) A review on electric vehicle battery modelling: from lithium-ion toward lithium-sulphur. Renew Sust Energ Rev 56:1008–1021. https://doi.org/10.1016/j.rser.2015.12.009
25. https://www.tycorun.com/blogs/news/solid-state-lithium-battery
26. Yang R, Xie Y, Li K, Tran MK, Fowler M, Panchal S, Deng Z, Zhang Y (2023) Comparative study on the thermal characteristics of solid-state lithium-ion batteries. IEEE Trans Transport Electr. https://doi.org/10.1109/TTE.2023.3289997
27. The Nikkei in December (2020) Available online: https://asia.nikkei.com/Spotlight/Most-read-in-2020/Toyotas-game-changing-solid-state-battery-en-route-for-2021-debut (accessed on 1 June 2025)
28. Koorata PK, Panchal S (2023) Thermal management of large-sized LiFePO4 pouch cell using simplified mini-channel cold plates. Appl Thermal Eng 234:121286. https://doi.org/10.1016/j.applthermaleng.2023.121286
29. Joshi AK, Kakati P, Dandotiya D, Prashanth SP, Patil NG, Panchal S (2023) Computational analysis of preheating cylindrical lithium-ion batteries with fin-assisted phase change material (PCM). Int J Modern Phys C. https://doi.org/10.1142/S0129183124500475
30. Subhedar D, Chauhan KV, Panchal S, Bais A (2023) Numerical investigation of performance for liquid-cooled cylindrical electrical vehicle battery pack using Al_2O_3/EG-water nano coolant. Mater Today Proc. https://www.sciencedirect.com/science/article/pii/S2214785323043079
31. Bais A, Subhedar D, Panchal S (2023) Experimental investigation of longevity and temperature of a lithium-ion battery cell using phase change material-based battery thermal management system. Mater Today Proc. https://www.sciencedirect.com/science/article/pii/S2214785323043614

32. Talele V, Moralı U, Patil M-S, Panchal S, Fraser R, Fowler M, Thorat P, Gokhale Y-P (2023) Computational modelling and statistical evaluation of thermal runaway safety regime response on lithium-ion battery with different cathodic chemistry and varying ambient condition. Int Commun Heat Mass Transfer 146:106907. https://doi.org/10.1016/j.icheatmasstransfer.2023.106907
33. Myatezh AV, Smirnov MA, Malozyomov BV, Shtang AA (2017) Performance increase of lithium-ion batteries. Adv Eng Res 133:519–524. https://doi.org/10.2991/aime-17.2017.84
34. Sun Z, Gao W, Li B, Wang L (2020) Locating charging stations for electric vehicles. Transp Policy 98:48–54. https://doi.org/10.1016/j.tranpol.2018.07.009
35. Kong Q, Fowler M, Entchev E, Ribberink H, McCallum R (2018) The role of charging infrastructure in electric vehicle implementation within smart grids. Energies 11:3362. https://doi.org/10.3390/en11123362
36. https://about.bnef.com/blog/electric-vehicles-accelerate-54-new-car-sales-2040/
37. https://www.insiderintelligence.com/chart/235386/factorsfeatures-uk-drivers-would-consider-towards-purchasing-electric-vehicle-jan-2020-of-respondents
38. IEA. Global EV Outlook (2022) Securing Supplies for An Electric Future. 2022. Available online: https://www.iea.org/reports/global-ev-outlook-2022
39. Global Petrol Prices. https://www.globalpetrolprices.com/gasoline_prices/
40. Global Petrol Prices. https://www.globalpetrolprices.com/electricity_prices/
41. World Bank. https://data.worldbank.org/indicator/NY.GDP.PCAP.CD
42. https://wisevoter.com/country-rankings/renewable-energy-by-country/
43. https://en.wikipedia.org/wiki/List_of_countries_by_renewable_electricity_production
44. https://yearbook.enerdata.net/renewables/renewable-in-electricity-production-share.html
45. World Bank. https://data.worldbank.org/indicator/EN.POP.DNST?locations=OE
46. World Bank. https://data.worldbank.org/indicator/SP.URB.TOTL.IN.ZS
47. https://countryeconomy.com/business/motor-vehicle-in-use
48. https://www.comparethemarket.com.au/car-insurance/features/the-most-expensive-countries-for-car-ownership/
49. https://www.hivepower.tech/blog/electric-vehicles-evs-in-switzerland-what-you-should-know
50. https://www.reykjavikcars.com/blog/driving-iceland/electric-car-iceland
51. https://en.wikipedia.org/wiki/Iceland
52. https://www.goodnewsnetwork.org/giant-oil-companies-install-ev-chargers-at-gas-stations/
53. https://en.wikipedia.org/wiki/Electric_car_use_by_country
54. Sawaguchi M (2011) Innovation activities based on s-curve analysis and patterns of technical evolution "From the standpoint of engineers, what is innovation?". Procedia Eng 9:596–610. https://doi.org/10.1016/j.proeng.2011.03.145
55. https://www.statista.com/statistics/1371599/global-ev-market-share/
56. Washington Post. https://www.washingtonpost.com/climate-environment/2023/02/02/critical-minerals-run-out-shortage/
57. https://about.bnef.com/blog/end-sight-near-term-lithium-supply-shortages/
58. Institutional Investor. https://www.institutionalinvestor.com/article/2bswqvt5gr58738z0fwu8/innovation/can-lithium-supply-keep-up-with-strong-ev-demand
59. Chen J, Manivanan M, Duque J, Kollmeyer P, Panchal S, Gross O, Emadi A (2023) A convolutional neural network for estimation of lithium-ion battery state-of-health during constant current operation. In: Conference: 2023 IEEE transportation electrification conference & expo (ITEC). IEEE, pp 1–6
60. https://www.standard.co.uk/tech/why-electric-car-prices-dropping-b1085646.html
61. https://batteryuniversity.com/article/bu-1003a-battery-aging-in-an-electric-vehicle-ev

Chapter 4
Electric Vehicle Transition Policies and Strategies

Contents

4.1 Introduction

At present, carbon dioxide (CO_2) and other greenhouse gas (GHG) emissions from the transport sector account for one/fourth of the global energy-related emissions. These emissions are one of the primary driving factors driving the current increase in global temperatures and climate change. According to the International Energy Agency report, global energy-related CO_2 emissions hit a new record high of 37.4 billion tonnes (or gigatonnes, Gt) in 2023 [1]. In 2022, transportation emissions surged by 254 tonnes, compared to 7.73 Gt in 2021. The World Energy Outlook reported that road vehicles worldwide emitted 5.9 Gt CO_2 in 2021, with heavy-duty trucks responsible for 1.8 Gt, which represents 4% of global CO_2 emissions [2]. Clearly, road transportation is the leading source of GHG emissions. The use of petroleum fuels in this sector is primarily responsible for tailpipe emissions, with

R. Pode, B. Diouf, *Decarbonization of Road Transportation*, Green Energy and Technology, https://doi.org/10.1007/978-3-032-11388-7_4

Table 4.1 Sales of battery EVs in 2024 and projections for 2025 in some major countries

Region	Total sales in 2024 (million units)	2025 predictions (million units)/ (% increase)
Global	Over 20	Over 23.6/(+18%)
China	12.9	15.1/(+17%)
EU & EFTA & UK	3.5	4.0/(+15%)
USA & Canada	2.1	2.4/(+16%)
Rest of World	0.6	0.8/(+36%)

approximately 99% of the GHGs released from the vehicles coming in the form of carbon dioxide [3]. Particularly in developing economies and non-OECD countries, emissions from road transport are sharply rising due to the increase in the number of vehicles powered by internal combustion engines. As a result, there is an urgent need to significantly reduce emissions from road transportation. Setting a precise goal for decarbonizing this sector and transitioning to zero-emission vehicles is very imperative. The adoption of battery electric vehicles (BEVs) seems to be the most feasible alternative for decarbonizing road transport. Several countries around the globe have formulated policies to reduce the transportation emissions by encouraging EV adoption and offered various attractive incentives to customers.

Recently, Rho Motion, the leading electric vehicle (EV) research house, predicted that the global EV market will record a growth of +18% in 2025, with over 20 million electric vehicles sold in 2024. Table 4.1 shows the sales of battery EVs in 2024 and projections for 2025 in some major countries [4]. Around 1.6 million units of EVs are sold globally.

In this chapter, the policies of some countries, such as the USA, China, India, and EU countries, to enhance the share of EVs in the transport sector are presented here. Accordingly, these countries have developed and implemented policies to promote electrification of road transport. The EVs have a high potential for adoption as rapid public transportation options and taxis in urban areas, and they are likely to become the main mode of transport in the future.

4.2 Strategies for Decarbonization of Transportation

Policies and strategies are essential for decarbonizing road transport because they provide a systematic approach for reducing GHG emissions, setting clear goals, and ensuring advancements. Emissions reduction efforts would be fragmented and inconsistent in the absence of policies. Well-crafted policies and strategies play a pivotal role in driving this transformation forward. Earlier, Noussan et al. presented a wide range of policies in detail that are available and that can be implemented to ensure effective decarbonization of transport sectors [5].

Initiatives (including programs, policies, or projects) directed at reducing GHG emissions are generally grouped into two main categories: (i) improving travel efficiency and (ii) transitioning to clean vehicles and fuels, as described by the Office of Transportation and Air Quality, United States Environmental Protection Agency [6]. In 2015, Matthew Barth et al. from the National Center for Sustainable Transportation at the University of California published a white paper demonstrating that enhancing travel efficiency through measures such as traffic monitoring systems, traffic incident management, integrated corridor management, and travel demand management can lead to emissions reductions ranging from 5% to 15% [7]. Recently, Zhao et al. showed that a 30% reduction in CO_2 emission can be achieved by improving fuel efficiency by 35%, a crucial step for limiting emissions from road freight [8]. The energy efficiency of road freight contributed to emissions trends. It has significantly helped to limit emission increases during periods of growing demand. Enhancing the energy and emissions efficiency of vehicles is expected to be even more vital in the future [9].

Figure 4.1 shows the various policies/strategies for decarbonizing road transportation. These are categorized in six different groups, each addressing different aspects of the transition. Infrastructure development consists of (1) expanding the fast-charging network along highways, urban areas, and workplaces; (2) augmenting power grid resilience to handle large EV demand; and (3) encouraging renewable energy integration to make EV charging more sustainable [10]. Policy and regulatory support comprise (1) providing tax breaks, purchase incentives, and reduced registration fees for EV buyers; (2) ensuring stricter emission standards to

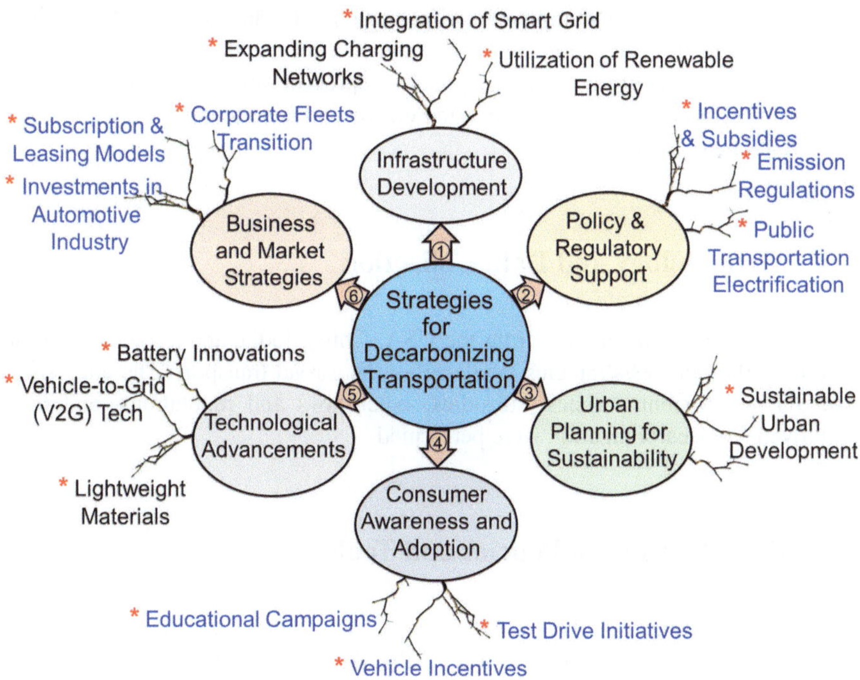

Fig. 4.1 Strategies and tools for decarbonizing road transportation

encourage a shift away from fossil-fuel vehicles; and (3) motivating for electric buses and government fleet conversions [11, 12]. Under sustainable urban planning, cities are designed with mixed-use developments and accessible public transport options, lowering the dependence on private vehicles and hence reducing emissions [13, 14]. Consumer awareness and adoption campaigns (1) create awareness among the public about the benefits and cost savings of EVs, (2) provide lower insurance premiums and reduced tolls for EV owners, and (3) allow buyers to have firsthand experience of EVs to address range anxiety and performance concerns [15]. The technological advancements strategy includes (1) improving battery life, charging speed, and recyclability to make EVs more appealing; (2) allowing EVs to feed power back into the grid, enhancing energy storage and efficiency; and (3) reducing weight and improving aerodynamics to enhance vehicle efficiency [16, 17]. Lastly, business and market strategies include (1) encouraging businesses to adopt EVs for their transportation needs, (2) providing more flexible ownership options to make EVs accessible, and (3) stimulating R&D for better EV models with improved range and affordability [18, 19]. Each of these strategies plays a vital role in accelerating EV adoption and transforming road transportation into a more sustainable system.

The policy measures, such as primarily purchase subsidies and/or vehicle purchase and registration tax rebates, were implemented as early as the 1990s in Norway, in the United States in 2008, and in China in 2014 [12]. Since 1990, Norway has exempted battery EVs from registration tax and from value-added tax since 2001. Individual countries will have particular and specific advantages/disadvantages with regard to the implementation of each individual measure. While there is no single ideal policy framework to achieve effective decarbonization of the transport sector, the aforementioned policies represent key areas where countries have made efforts or are currently making efforts and need to persist in striving for positive climate impact.

4.3 Methodology and Data Collection

We have gathered information from the USA, China, India, and a few EU nations regarding the policies/strategies for decarbonizing road transport. The analysis of data on government policies, subsidies, other laws and regulations, and tools employed for decarbonization were performed.

4.4 Decarbonization Policies and Tools

There are several successful case studies of electric vehicle (EV) adoption that highlight effective strategies and impactful results. As of now, Norway (1990), the US (2008), and China (2014) have established the most supportive policy frameworks for the adoption of EVs [12]. Since 1990, battery EVs (BEV) in Norway have benefited from exemptions from registration tax, and they have also been free from value-added tax since 2001. To provide a new tax credit for eligible plug-in electric drive motor cars, the US Senate approved bill H.R. 6049 during the 110th Congress (2007–2008) in 2008 [20]. Meanwhile, China launched a series of policies regarding EVs, laying out a preliminary roadmap that included energy, science, and technology planning from 2004 to 2008 [21]. This policy placed a strong emphasis on the development of hybrid vehicles (HVs) and the examination of EV-related policies. In Europe, Norway was the first country to enter the EV market in 2009 and promoted EVs for public usage from 2009 to 2012. Subsequently, the focus of the demonstration projects transitioned from the public to the private domain (2013–present), offering policy incentives to overcome challenges such as price, driving range, and charging time, thereby enhancing their competitiveness [22].

The European Automobile Manufacturers' Association (ACEA) predicts that the adoption of battery electric vehicles will exceed 70% in EU + EFTA + UK member countries, 44.8% in the United States, and 58.5% in China by 2030 [23]. According to Reuters, global EV sales jumped 69% in January 2024 as compared to January 2023 but were down 26% from December 2023 [24]. Subsidy cuts or tighter rules in Germany and France and seasonally weaker sales in China are influencing global EV sales, as per market research firm Rho Motion (London) [25]. In January 2024, sales of fully electric cars, or battery EVs, and plug-in hybrids reached 1.1 million, a significant increase from 660,000 in January 2023, even with the conclusion of 11 years of subsidies in China.

The EV policies, laws, and incentives for battery EVs in the USA, China, India, and the EU are dealt with in the following sections. These deliberations are essential to raising the EV adoption rate.

4.5 EU Member Countries Scenario

In 2019, the European Parliament proposed EU-level measures for the adoption of electric mobility that encourage the use of renewable electricity and smart charging, help develop and standardize charging infrastructure, and support research on batteries [26]. Additionally, incentives at the local, regional, and national levels, such as reduced taxes or complimentary public parking for EVs, have been suggested to encourage electric mobility. In 2020, Sandra Wappelhorst et al. analyzed the market trends and key national and local electric vehicle promotion policies across Europe, concluding that the EU encourages EV deployment [27]. Further, they showed that the important policies of union member countries (incentives, charging infrastructure, availability of electric vehicle models, etc.) are critical to overcoming key barriers to electric vehicle affordability, convenience, and awareness. Timeline of EU's electric vehicle policy is shown in Table 4.2 [28–41]. This policy timeline is marked with ambitious goals for reducing emissions from transportation, with a strong focus on phasing out ICE vehicles. The EU aims to achieve carbon neutrality across various sectors, including transportation.

Table 4.2 Timeline of EU's electric vehicle policy

Early stage & foundation (pre–2020)	Expansion & planning (2020–2024)	Implementation & review (2024–2025)	Looking ahead (beyond 2025 ~)
Early stages **(2008–2014)** * **2008**: Launch of the *European Green Cars Initiative* under the EU Economic Recovery Plan, investing €5 billion to support sustainable transport technologies [29] * **2014**: Adoption of the *Alternative Fuels Infrastructure Directive (AFID)*, setting targets for public charging infrastructure and promoting harmonization across Member States [31]	* **2021**: EAFO policy paper outlines the need for one million public recharging points by 2025 to support 13 million zero- and low-emission vehicles. [31] * 2022: The European Parliament voted to ban the sale of new ICE vehicles from 2035 [35] * **2023**: Germany withdraws its €6750 EV grant, signaling a shift from subsidies to regulatory pressure [33] * **2024**: The EU introduced definitive countervailing duties on imports of battery EVs from china [36] * **2024**: The EU set targets to cut CO_2 emissions from cars by 55% and vans by 50% by 2030 [37]	* **2025**: EU CO_2 targets tighten again, expected to significantly boost EV sales. France plans €1 billion in EV subsidies, including a €4000 eco-bonus and social leasing schemes [33] * **2025**: The EU is sticking to its clean mobility targets, including emission reduction goals for new cars and vans [38] * **2025**: Carmakers are given until 2027 to meet emission standard [39] * **2025:** The EU's 60 km rule. It mandates charging stations every 60 km along trans-European transport network [40] * **2025**: Many EU countries are implementing national programs to incentivize EV adoption and charging structure development. As of April 1, 2025, BEVs are no longer exempt from the motor-related insurance tax and are taxed similarly to internal combustion engine (ICE) vehicles [41]	* **2030**: The EU aims for significant increase in the uptake of EVs, with targets to cut CO_2 emissions for cars by 55% and vans by 50% by 2030 [42] * By **2035:** The EU aims to phase out sales of petrol and diesel cars entirely, aligning with its climate neutrality goals [33] Ongoing efforts include scaling battery production, improving affordability, and expanding charging infrastructure * **2050**: Aims to achieve carbon neutrality across various sectors, including transportation [35]
(2015–2020) * **2017**: EU begins standardizing charging infrastructure and supporting battery research [32] * **2019**: Regulation (EU) 2019/631 sets binding CO_2 emission targets for new passenger cars and vans, pushing automakers toward cleaner fleets [33] The European green deal was formalized, setting the long-term climate neutrality by 2050 [34] * **2020**: EV market share jumps to 11% across Europe, driven by tighter CO_2 standards and national incentives [31]			

Table 4.3 Highlights of key policy drivers to achieve carbon neutrality in transportation by 2050

Drivers	Key policies and initiatives
European Green Deal [43]	Sets the overarching framework for climate action, including transport decarbonization • Aims to reduce *transport emissions by 90% by 2050* • Sets the overarching climate-neutrality goal for the EU • Promotes cleaner fuels, zero-emission vehicles, and sustainable infrastructure
Sustainable and Smart Mobility Strategy [44]	• Introduced with an *Action Plan of 82 initiatives* • Focuses on making transport *green, digital, and resilient* • Targets include: – 30 million zero-emission cars by 2030 – Doubling high-speed rail traffic – Climate-neutral urban mobility in 100 cities
Fit for 55 Package [41, 45]	• Legislative bundle to cut *net greenhouse gas emissions by 55% by 2030* • Includes revised CO_2 standards for cars, vans, and heavy-duty vehicles • Supports the deployment of *alternative fuels and clean energy*
Energy Performance of Buildings Directive [46]	Address building renovation and integration of EV charging infrastructure • Provisions on *recharging points* for electric vehicles and bicycle parking spaces • Recharging points will have to enable *smart charging* and, where appropriate, bi-directional charging
Alternative Fuels Infrastructure Regulation (AFIR) [47]	• Mandates public charging infrastructure expansion • Sets targets like *one public charger per 10 EVs* • Promotes hydrogen and renewable fuels deployment
Corporate Sustainability Reporting Directive (CSRD) [48]	• Requires companies to *disclose transport emissions*, including supply chains • Drives corporate investment in low-carbon mobility solutions

Key policy drivers to realize sustainable and smart mobility in EU countries are highlighted in Table 4.3 [41–47]. These strategies lay the foundation for how the EU transport system can achieve its green and zero-carbon transportation. As outlined in the European Green Deal, the result will be a 90% reduction in emissions by 2050, delivered by a smart, competitive, safe, accessible, and affordable transport system. There is growing evidence that EU green mobility policies are making a measurable impact. Cities adopting the Urban Mobility Framework have seen growth in cycling infrastructure, electrified public transport, and low-emission zones [48].

Separately, EU member states offer some form of incentive to stimulate the market uptake of EVs, but the tax benefits and purchase incentives vary greatly across the European Union countries [49]. Among the 27 EU member states, 21 provide incentives for purchasing EVs, while six countries do not offer direct purchase incentives but instead provide tax reductions or exemptions for EVs. Figure 4.2 displays the EV purchase incentives in some EU countries [3, 49]. Earlier, the EU Parliament and the Washington Post reported that the member countries of the

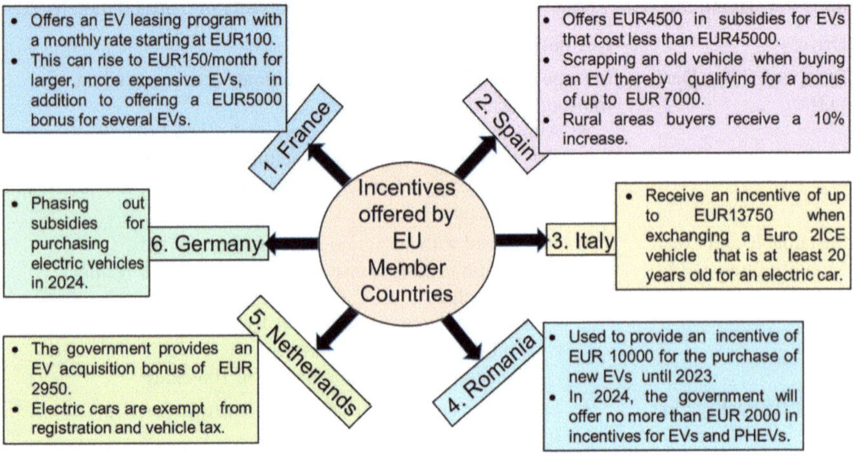

Fig. 4.2 EV purchasing incentives in some European Union countries [3, 49]

European Union reached an agreement to ban the sale of new non-electric cars starting in 2035, aiming for a complete carbon reduction for new cars and vans by that year [50, 51]. This agreement sets the stage for a modern and competitive automotive industry within the EU. The zero-emission target will help slow down climate change, which can cause severe disruptions in many sectors of our society, including the environment, migration, food security, and the economy. To significantly reduce passenger car and truck CO_2 emissions, it was proposed to target the need for a significant share of new passenger car (PC) and commercial vehicle (CV) sales to be EVs by 2030. EU member states necessitate the development of an EV-charging infrastructure across Europe. According to a 2020 McKinsey EV Consumer Survey, charging infrastructure remains a crucial bottleneck for consumer adoption; faster and earlier uptake of public charging infrastructure is fundamental [52].

As a result, the European Automobile Manufacturers' Association created the EV Charging Masterplan in 2022 [53]. According to the European Automobile Manufacturers' Association (ACEA), the new electric car market was about 47.5% (battery-electric cars 10.9% + hybrid-electric cars 28.8% + plug-in hybrid electric cars 7.8%) of the total car sales in the EU in January 2024 [54]. In January 2024, the share of gasoline and diesel vehicles reached nearly 50%, a decrease from 54% the previous year. The European Union has planned to ban all new sales of ICE vehicles by 2035 [55], whereas the UK is acting even more aggressively, with a 2030 ICE ban [56].

The EU policies and incentives for decarbonizing transport are received positively by all stakeholders in the EU and have made a positive impact on the expansion of zero-emission vehicles. This is well reflected in the lowering of CO_2 emissions. The growth of zero-emission cars (BEV + PHEV) in Europe and the rest of the world (excluding Europe, China, and the US) is shown in Fig. 4.3 over a period from 2014 to 2025. In 2025, zero-emission electric car sales are projected to

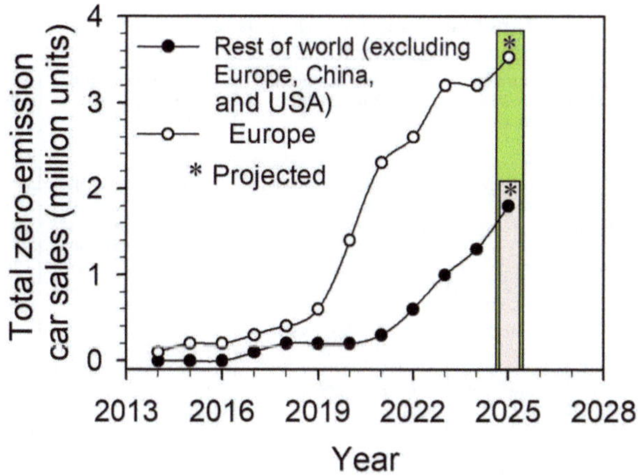

Fig. 4.3 Sales of zero-emission cars (BEV + PHEV) in Europe and the rest of the world (excluding Europe, China, and the US) over a period from 2014 to 2025

increase by 10% and 38% in Europe and the rest of the world, compared to 2020. EVs are expected to represent 71% of the European market by 2032, partly due to these actions and interventions by the EU [57].

In summary, all stakeholders across the continent, including policymakers and industry leaders, are making every effort to implement ambitious regulatory frameworks to accelerate the shift toward zero-emission vehicles, renewable energy integration, and sustainable transportation modes. Citizens across the EU have shown a mix of support, skepticism, and adaptation toward green mobility initiatives, shaped by local contexts, perceived fairness, and access to alternatives.

4.6 US's EV Policies

The US EV policies are characterized by changes in focus and support, shaped by technological advancements, economic situations, and political priorities. Sherilyn Hayashida et al. analyzed the changes in US EV policies from 2010 to 2018 and concluded that the political economy of EV policy is a complex interaction of state politics, citizen ideology, structural factors, and environmental costs generated by fossil fuel use in each state [58]. They observed that environmental ideology and state politics are the main factors supporting state EV subsidies. From 1990 to 1992, new federal and state regulations renewed interest in EVs in the US. As a result, automakers started modifying popular vehicle models into EVs. The timeline of key EV policy developments is displayed in Fig. 4.4, showing how the country has evolved from early incentives to full-scale infrastructure and manufacturing support [59–62].

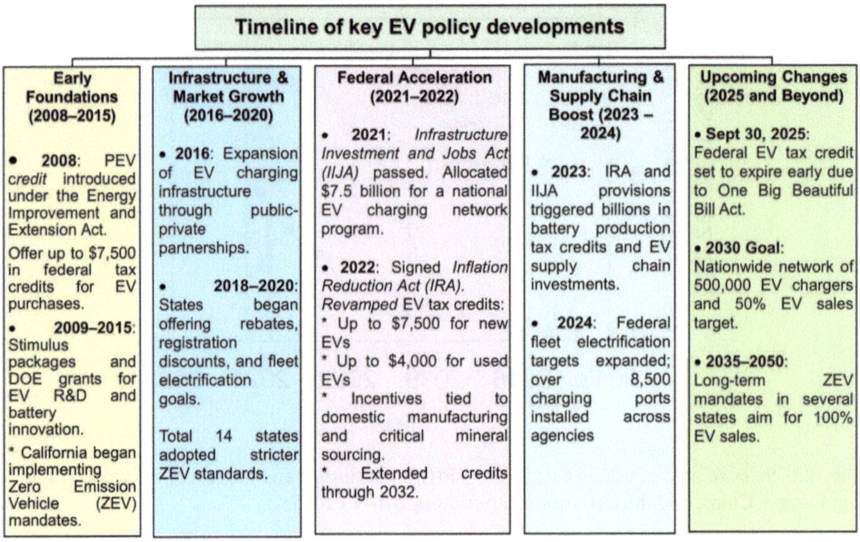

Fig. 4.4 Timeline of key EV policy developments in the USA

The Inflation Reduction Act of 2022 (Public Law 117–169) has amended the Qualified Plug-in Electric Drive Motor Vehicle Credit (IRC 30D), now referred to as the Clean Vehicle Credit. Under this law, US taxpayers who buy an eligible EV may qualify for a tax credit of up to USD 7500 for vehicles acquired between 2023 and 2032 [63]. This federal EV tax credit aims to encourage the adoption of EVs. Specifically, vehicles that satisfy the critical mineral requirements can earn a tax credit of USD 3750, while those that meet the battery component criteria are also eligible for USD 3750. Vehicles fulfilling both the key mineral and battery component specifications qualify for the full USD 7500 tax credit. Beyond federal incentives, various state programs offer additional benefits, such as discounts on residential charging stations, access to High Occupancy Vehicle (HOV) lanes regardless of the number of occupants, rebates for zero-emission motorcycle purchases, rebates of up to USD 12,000 for new EV purchases, and special time-of-use rates for charging during off-peak hours, among other incentives. Additionally, buyers could save from USD 800 to USD 1000 annually by charging EVs at home rather than filling up on gasoline [64]. As a result, nearly half of US car buyers intend to purchase an electric vehicle, according to the EY research report 2023 [65]. However, the majority (57%) of US car buyers cite the dangers of home charging and safety as key deterrents. According to the La Grada Spanish agency, the EV strategy of the US was flawed from the beginning [66]. The emphasis has been on utilizing EVs merely as replacements for gasoline cars, which are known for their poor fuel efficiency, instead of considering them as part of a broader strategy for sustainable transportation. As a result, larger transportation challenges and emissions targets may be overlooked, leaving various other issues unaddressed.

California has implemented aggressive policies to promote EVs, including (i) *zero-emission vehicle mandates*, requiring automakers to sell a certain percentage of EVs; (ii) *financial incentives*, such as rebates and tax credits; and (iii) *investment in charging networks*, ensuring accessibility across the state. Some major private companies have also played a critical role in promoting EV adoption. Companies like Amazon and UPS have committed to electrifying their delivery fleets, reducing emissions and operational costs. Strategies include (i) *bulk purchasing agreements* with EV manufacturers, (ii) *charging infrastructure development* at distribution centers, and (iii) *integration of AI for route optimization*, maximizing efficiency.

Both automotive dealers and consumers believe that battery range and cost are the primary obstacles to purchasing EVs [67]. Additionally, the challenge of installing a home charging station, particularly in rental units where tenants may not reside long-term and may be unsure about the landlord's policies or the feasibility of fitting charging options, also complicates the situation. Recently, Reuters News reported a rise in sales of gas–electric hybrid vehicles and a decline in EV sales in the United States [68]. Consequently, automakers and suppliers are betting that consumers will seek a middle ground between combustion and fully electric as a durable trend. In conclusion, financial incentives alone are not sufficient to enhance the presence of EVs on the road.

Despite all these consumer-friendly policies, merely 7.6% of new car sales were reported to be electric in the U.S. in 2022, according to Kelley Blue Book (information services, e-commerce software, Automotive, Irvine, California, U.S.) [86]. Total electric (BEV + PHEV) car sales from 2010 to 2025 (till June 30) in the US are displayed in Fig. 4.5 [69]. By June 2025, electric car sales in the United States reached 1.6 million, representing a significant increase of over 400% compared to 2020. However, the growth rate for EV sales slowed in 2024, rising by only 10%, a sharp decline from the 40% increase observed in 2023 [69].

Transitioning to electric transportation will reduce the economic, national security, and public health issues that are related to the United States' reliance on oil. By

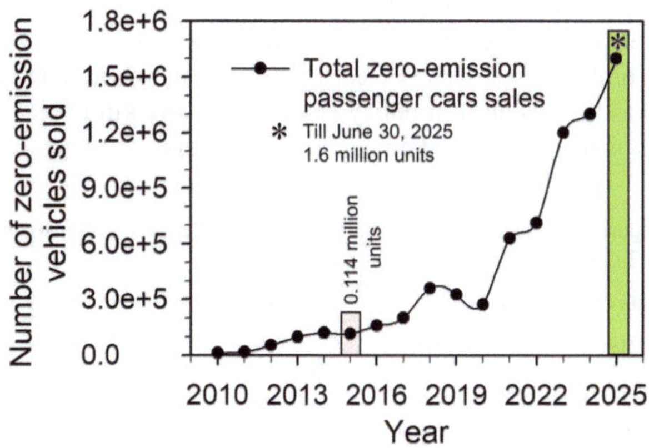

Fig. 4.5 Total electric (BEV + PHEV) car sales from 2010 to 2025 (till June 30) in US

investing in domestic electric transportation infrastructure, we can generate new jobs, rejuvenate the American manufacturing industry, provide monetary savings for consumers and fleet operators, and fortify supply chains for essential military and industrial technologies.

4.6.1 *Future of the US's EV Policies*

The future of EV policies in the U.S. is entering a turbulent phase, with major shifts happening that could reshape the landscape for years to come. Drastic changes and reversion in EV policies due to the new legislation (One Big Beautiful Bill Act) are highlighted in Table 4.4 [70]. The current administration has rolled back EV mandates from previous administrations, including eliminating the goal of achieving 50% EV sales by 2030 and pausing federal funding for charging infrastructure. Furthermore, this new bill phases out tax credits for wind, solar, and electric cars while making it cheaper to drill and mine for fossil fuels on federal lands.

Short-term implications of EV policy uncertainties are manifold. Some individuals from industries have reacted negatively to the reversion of EV policies. Automakers like Ford and GM may scale back EV production targets due to rising costs and reduced incentives. Despite policy headwinds, experts say EV demand remains strong, especially among higher-income buyers [61]. On the other side, international players like the EU and China continue pushing aggressive EV adoption, forcing U.S. automakers to stay competitive abroad. At the same time, US tariffs on Chinese EVs and battery components are raising domestic production costs, complicating supply chains, and adding to the cost of EVs.

Furthermore, projections for U.S. EV adoption have taken a sharp turn. According to BloombergNEF's 2025 outlook, it is projected that EV market share is likely to make just 27% by 2030 under the new EV policy outlook, a significant drop from last year's (2024) forecast of 48% [71]. Some analysts are optimistic and still expect 37% EV sales by 2030, citing tech improvements and consumer interest. According to Princeton University's Zero Lab Analysis, the loss of the EV tax credit could lead to roughly 8.3 million fewer battery electric and plug-in hybrid light vehicles on US roads in 2030 [72]. This shift in policy could also likely kill off the nascent clean hydrogen, CO_2 management, and nuclear power sectors.

Table 4.4 Changes and reversion in EV policies due the new legislation due to One Big Beautiful Bill Act (July 9, 2025)

End of federal tax credits support	Infrastructure & climate funding rolled back
Repeals core provisions of the *Inflation Reduction Act* * *EV tax credits* of up to $7500 for new vehicles and $4000 *for used EVs* will *end on September 30, 2025* * New *annual registration fees* for EVs ($250) and hybrids ($100) will be introduced	* Federal grants for *EV manufacturing, charging stations*, and clean energy upgrades are being *phased out* * Enforcement of *fuel-efficiency standards* and dealership transparency rules has been weakened

In summary, the recent EV policy shifts are considerably affecting EV buyers, automakers, and climate goals. The repeal of federal tax credits and rollback of climate funding are sending shockwaves through the entire electric vehicle ecosystem.

4.6.2 Domestic Impact of Reversal in EV Policy

The domestic impact of U.S. EV policy shifts on the EV buyers, automakers, or climate goals is lucidly illustrated in Table 4.5.

The uncertainty surrounding U.S. EV policies is shaking up the auto industry, but not everyone is losing sleep over it. Surely, some sectors are benefiting from the EV policy reversals, as displayed in Table 4.6.

4.6.3 Long-Term Effects of Uncertain EV Policies

The long-term effects of *uncertain EV policies* in the U.S. ripple far beyond just car dealerships; they touch everything from climate goals to global competitiveness, as illustrated in Table 4.7. These changes in EV policies pose a direct threat to the economic competitiveness of US automakers.

Table 4.5 Domestic impact of new U.S. EV policy shift bill on the EV buyers, automakers, and climate goals

EV buyers	Automakers	Climate goals
* **Higher Costs**: EV buyers face significantly higher upfront prices without the federal tax credit * **Reduced Incentives**: While certain states might still provide rebates, the loss of federal support makes EVs less financially appealing for middle-income consumers * **New Fees**: Starting October 2025, EV owners will incur an annual registration fee of $250, and hybrid owners will pay $100, further discouraging adoption * **Market Confusion**: Uncertainty around future incentives is causing many buyers to delay purchases in anticipation of new programs to emerge	* **Investment Risk**: Automakers like Ford, GM, and Tesla have invested billions into EV production. The abrupt shift in policy jeopardizes those investments * **Production Cuts**: Companies might their EV manufacturing goals in response to lower demand and escalating costs * **Supply Chain Disruption**: Rising tariffs on Chinese EV components and batteries are increasing expenses and complicating supply chain logistics	* **Slower Decarbonization**: Transportation is the largest source of U.S. emissions. Without strong EV adoption, achieving climate targets may be at risk * **Missed Milestones**: The U.S. is falling behind its efforts to achieve net-zero transportation goals by 2050. Even the most optimistic forecasts indicate that the emission will only decrease by 27–37% by 2030 * **Global Pressure**: With the EU and China advancing robust EV mandates, the U.S. may risk its position in climate leadership and clean tech innovation

Table 4.6 Sectors benefiting by the EV policy reversals

Traditional automakers	Oil & gas industry	Chinese EV manufacturers	Used EV market & leasing companies
* Companies such as Toyota and Honda, which have lagged in EV development, now have the opportunity to focus on hybrids without facing penalties * Stellantis may reduce its EV plans and prioritize more profitable combustion models, capitalizing on relaxed fuel economy standards	* Slower EV adoption means continued demand for gasoline, particularly as fewer consumers switch to electric * Reduced pressure on automakers to electrify fleets aids in sustaining the production of internal combustion engines (ICE)	* While tariffs negatively impact direct exports, uncertainty in the U.S. market could boost Chin's global dominance, as U.S. automakers scale back EV ambitions * BYD, having surpassed Tesla in global EV sales in Q4 2024, may benefit from weakened U.S. competition	* With new EV tax credits ending, used EVs and lease options may become more attractive and appealing * Leasing companies can continue to realize residual value and offer competitive pricing, especially if they build their inventory before the September 30 deadline

In brief, policy uncertainty doesn't just slow down EVs, but it also risks derailing an entire ecosystem built around clean mobility.

4.6.4 Impact on Global Clean Mobility Ecosystem

The *uncertainty in U.S. EV policy* is sending ripples through the global clean mobility ecosystem [73]. This will impact the entire global ecosystem built around clean mobility, which is depicted in Table 4.8.

Reversing EV policies in the US is likely slow down the global transition towards clean mobility and could undermine its position as a leader in clean energy. Some analysts believe that this is a disaster in the making, gutting the US industrial base and creating a self-inflicted economic, national security, and jobs disaster [74]. In short, the U.S. isn't just a big player, but it's also a bellwether. When it pulls back, the entire clean mobility movement feels the tremor. The rollback of electric vehicle incentives under the *One Big Beautiful Bill Act* is creating a vacuum in global clean mobility leadership, which is rapidly being filled by other players like China [75, 76].

Table 4.7 Long-term effects of uncertain EV policies in the U.S.

Stalled EV adoption	Industry disruption	Climate setbacks	Global competitiveness risks	Economic ripple effects
* **Consumer hesitation**: Absence of stable incentives could lead buyers to delay switching to EVs, hindering market expansion * **Reduced affordability**: Ending federal tax credits reduces the accessibility of EVs, especially for middle-income households * **State-level incentives**: State-level programs might fill some gaps, but patchy benefits across different regions could create confusion for buyers and hinder widespread adoption	* **Automaker strategy shifts**: Companies like GM and Ford already decided to delay launching of EV and shifted their focus toward hybrids and ICE vehicles * **Supplier uncertainty**: Tier 1 suppliers that made significant investments in EV components face delayed returns and may shift their focus back to ICE technologies * **Innovation slowdown**: R&D budgets may decrease as automakers adopt a cautious approach, potentially impeding breakthroughs in battery technology and charging infrastructure	* **Missed emissions targets**: Transportation accounts for the highest share of U.S. emissions. Without strong EV uptake, achieving net-zero goals could become unattainable * **Grid strain and inefficiency**: Slower EV adoption could delay smart grid integration and energy storage solutions that enhance the resilience of the grid * **Lifecycle emissions**: EVs offer substantial reductions in GHG emissions over their lifetimes, even when charged from fossil fuel intensive grids. Fewer EVs result in higher overall emissions	* **China and EU surge ahead**: While the U.S. pulls back, China and Europe are doubling down their investment in EVs, threatening U.S. leadership in clean technology * **Trade tensions**: Tariffs and fragmented supply chains may increase costs and hinder U.S. automakers' ability to compete globally * **Talent drain**: Uncertainty may drive engineers and innovators to pursue markets with clearer EV roadmaps	* **Investment volatility**: Policy reversals pose risks for investors, which could potentially deter capital investments in EV startups and infrastructure * **Job market instability**: EV-related manufacturing and infrastructure jobs may shrink or shift, particularly in areas relying on green growth * **Consumer spending**: Higher EV prices and fees could negatively impact overall auto sales, which could, in turn, affect broader economic indicators

Table 4.8 Impact of U.S. EV policy reversal on the global clean mobility ecosystem

Global leadership vacuum	Supply chain disruption	Investment volatility	Market fragmentation	Climate consequences
* The U.S. was a major driver of EV innovation and investment. Rolling back incentives and support for infrastructure undermine its role as a clean technology leader, consequently creating a space for China and the EU to dominate EV markets * Countries that are aligned with U.S. EV policy, like Canada and Mexico, may reconsider their own strategies, especially if cross-border supply chains are disrupted	* Changes in U.S. policy affect battery sourcing, rare earth minerals, and EV component manufacturing, which are deeply globalized * Tariffs and trade restrictions on Chinese EVs and components are compelling companies to reconfigure supply chains, often at higher expenses	* Global investors are cautious about policy reversals that could stall or reroute billions in funding for clean mobility initiatives * Since 2021, the U.S. had attracted over $200 billion in EV manufacturing investments; however, future commitments may shift to more stable regions	* Lacking a stable and reliable U.S. policy, automakers may opt on markets with clearer roadmaps, such as the EU's Fit for 55 or China's NEV mandates * This could lead to disparities in EV availability, pricing, and innovation across different regions	* Transportation is a major contributor to global emissions. A slowdown in U.S. EV adoption could hinder global decarbonization goals, especially if other countries do the same * The Electric Vehicles Initiative (EVI), co-led by the U.S., may lose momentum or credibility if domestic policies are inconsistent with its goals

4.7 China's EV Policies

China holds the position as the world's largest EV market for EVs and currently leads the manufacturing sector for these vehicles. As the Chinese market approaches maturity, particularly following the culmination of government subsidies in 2022 and the growing alignment in cost between EVs and ICE vehicles, significant changes have occurred. Around the year 2000, the Chinese auto industry was in turmoil and primarily focused on producing traditional ICE vehicles for only foreign brands. In response to this situation, the Chinese government shifted its focus from conventional internal combustion engine technology to pioneering a new frontier: battery-powered vehicles. This transition has led China to aggressively promote the adoption of EVs since 2009. To accelerate market penetration, five pilot cities were selected in 2010 to offer subsidies for private EVs and to create a charging infrastructure for fast EV charging. China introduced various policies and provided incentives, including: (i) exempting energy-efficient and alternative-energy vehicles from vehicle license auctions and lottery-based vehicle purchases, as well as imposing traffic restrictions; (ii) implementing supportive measures for parking charges, electricity rates, and tolls; (iii) promoting purchase of energy-efficient and alternative-energy vehicles among government bodies, enterprises, institutions, and individuals; (iv) enhancing infrastructure, such as the charging station network; and (v) establishing a rapid-charging grid [21, 77]. China's journey in the EV sector has been rapid, with significant developments occurring almost every year. Figure 4.6 illustrates important milestones that have shaped the EV policy.

According to the KPMG China Report 2021, China has committed to achieving carbon neutrality by 2060, with a significant emphasis on the electrification of transportation [78]. This commitment, driven by demand and policy changes, has prompted stakeholders in the automotive industry to ramp up their electrification initiatives. To expedite the transition towards EVs, China has implemented a range of incentives and policies, which are summarized in Table 4.9 [79, 80].

China has established itself as the leader in the EV industry, thanks to significant advancements in battery manufacturing capabilities, charging infrastructure, and the development of innovative EV models. Its vast manufacturing capabilities enable it to produce EVs at lower costs. At present, China stands as the largest car market globally and has solidified its position as the dominant force in the EV sector. According to a report from the International Energy Agency, more than half of the EVs on roads worldwide were located in China in 2022 [81]. Additionally, the country accounted for 35% of global EV exports during the year. The Chinese government has invested billions to support citizens in purchasing electric cars, introducing tax incentives beginning in 2014 that have been renewed several times. In 2023, the country launched a 4-year tax break program for EVs and other environmentally friendly vehicles, with a total worth of CNY 520 billion (approximately EUR 68 billion or USD 72.3 billion) [82, 83]. Furthermore, EVs will not be subject

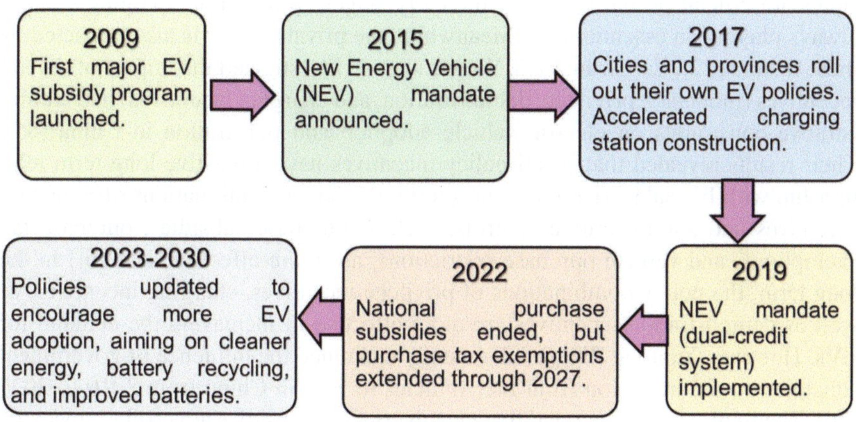

Fig. 4.6 Timeline of China's EV Policy

Table 4.9 China's incentives and policies for EVs transition

Category	Tools	References
Policy and regulatory	* **Tax Exemptions & Reductions**: *NEV purchase tax exemption* until *2027*, providing financial relief for buyers From 2024 to 2025, buyers can receive up to RMB 30,000 (US$4170) in tax exemptions per vehicle From 2026 to 2027, the exemption will be halved to RMB 15,000 (US$2078) * **Government Procurement**: EVs are prioritized for government fleet purchases	[79, 80]
Business and market strategies	* **Government Targets**: Set ambitious goals, including a 20% NEV sales share by 2025 and EV quotas for manufacturers * **Manufacturing Subsidies**: provides financial support to EV manufacturers to boost production All-electric plug-in cars with a range over 400 km are eligible for subsidies of RMB 12,600 (approximately $2000) All-electric plug-in cars with a range of 300–400 km are eligible for subsidies of RMB 9100 (approximately $1400) Plug-in hybrid cars are now eligible for subsidies of RMB 4800 (approximately $750)	[80]
Infrastructure development	* **Charging Infrastructure**: encourage the construction of EV charging stations, making EV adoption more convenient	[80]

to a purchase tax for 2024 and 2025, up to CNY 30,000 (EUR 3900 or USD 4170) [84]. The exemption will be reduced to half and capped at CNY 15,000 for purchases made in 2026 and 2027, according to a statement from the Ministry of Finance [83]. Certainly, government subsidies significantly influence EV adoption across various sectors. In the public sector, while purchase subsidies and the construction of charging stations show negligible effects, in the commercial sector, subsidies are crucial primarily during the early stages, with public charging stations always playing an essential role. Meanwhile, the private sector is also impacted by these factors [22]. Earlier in 2023, Yiran Liu et al. investigated the impact of several incentives (monetary, privilege, demonstration, and charging incentives) and administrative constraints on electric vehicle adoption and penetration in China [85]. Their results revealed that (i) all policy incentives have a positive long-term relationship with EV sales; (ii) in the short term, the policy combinations of monetary incentives and administrative controls, such as purchase subsidies, purchase tax exemptions, and vehicle purchase restrictions, are more effective; and (iii) in the long term, the policy combinations of privilege incentives, charging incentives, as well as demonstration incentives are more effective in increasing the demand for EVs. However, Yuntong Zhao et al. recently examined the influence of government subsidies on the transition from fuel vehicles to EVs in China from 2010 to 2020 [86]. The findings demonstrated that (1) government subsidies adversely impact the degree of enterprise transformation towards EVs; (2) this adverse effect can be alleviated or even reversed by implementing stricter technical requirements and reducing the intensity of subsidies.

Additionally, the Chinese government has urged domestic EV companies to partner with international manufacturers for technology development and to establish supply chains that are beneficial for both parties. Furthermore, in response to global trade restrictions and to facilitate the creation of overseas supply chains, China has promised to support its EV manufacturers as part of a major initiative to expand the industry globally, as detailed in a document released by the Ministry of Commerce of China [87].

Recently, Liyin Shen et al. reported technical (driving range, charging time, battery cost, bulk, and weight) and infrastructural (insufficient charging infrastructures, improper distribution of charging stations, and lack of sharing between charging piles) barriers to widespread EV promotion in China [88]. With such intriguing knowledge about the EV sector, the government is now shifting its focus from consumer-based incentives to technology development and innovation-based incentives [89]. It should be mentioned that Chinese automakers realized long before, in 2022, that battery costs would account for 30–40% of the entire manufacturing cost of a complete EV vehicle. Emphasizing on technology and innovation paves the way for newcomers in the industry. As a result, BYD Auto Company, headquartered in Shenzhen, China and recognized as the largest EV manufacturer globally, collaborated with Daimler and Toyota to exchange insights on EV manufacturing while sharing its expertise in battery manufacturing technology. Additionally, BYD collaborates with Foshan Plastics Group to develop optoelectronics systems for electronic sensors that manage lighting systems. Geely Auto, a prominent automobile manufacturer based in Hangzhou, China, has created an ecosystem that spans low-orbit satellites and smart hardware, aimed at gathering and analyzing data to potentially enhance EV battery performance. Jidu Auto, a joint venture of Geely Auto and Baidu, focuses on the production of intelligent EVs. Furthermore, Geely Auto acquired Drivetrain Systems International, an Australian company specializing in automatic transmission.

In summary, autonomous vehicles and new mobility options are now well appreciated by Chinese customers [90]. In summary, autonomous vehicles and new mobility options are now well appreciated by Chinese customers [90]. For Chinese EV companies, international expansion is the logical next step. They are equipped with all the essential elements, like advanced battery technology, a solid grip on the battery supply chain, and various operational advantages that position them well to achieve this objective, helping them to maintain their leadership in the EV market. China is eager to exploit the potential of EVs and is proactively addressing the challenge of limited model variety. With this advantageous landscape, they hope to highlight their commitment to sustainability, appealing to consumers seeking eco-friendly vehicles. Furthermore, they are collaborating with conventional dealerships to establish robust electric charging networks.

4.8 EV Policies of India

The history of EVs in India began in the 1990s and has since evolved with the help of technological innovations, government policies, and the growing demand for cleaner mobility solutions. Its first wave of EVs began in the 1990s, primarily through small electric scooters and three-wheelers designed to combat urban pollution. At that time, the country lacked the infrastructure necessary to support the widespread adoption of EVs. Limited charging stations, underdeveloped battery technology, and minimal government support made it difficult for these vehicles to gain popularity. The slow adoption of EVs in India during the initial years changed with the launch of favorable government policies and incentives.

The Indian government launched a number of regulatory initiatives in an effort to increase the number of EVs and meet India's goal of net zero emissions by 2070. The timeline of India's EV policies and incentives is illustrated in Table 4.10 [91–95]. This demonstrates an increasing dedication to sustainable zero-emission transportation, beginning with early incentives and evolving into a comprehensive strategy for both manufacturing and adoption. Notable milestones include the launch of the National Electric Mobility Mission Plan (NEMMP) in 2013, the introduction of the

Table 4.10 Key milestones and incentives of India's EV policy

Early stages (Pre–2013)	Expansion (2015–2021)	Recent developments (2024–Present)
* **2010**: The Ministry of New and Renewable Energy (MNRE) introduced an early subsidy scheme to encourage EV production [92] * **2013**: Launch of the National Electric Mobility Mission Plan NEMMP 2020 arget: six to seven million EV sales annually by 2020 [93]	* **2015**: FAME I (*Faster Adoption and Manufacturing of Hybrid and Electric Vehicles*) scheme was launched, providing incentives for EV purchases. Focused on R&D, pilot projects, and demand creation [93] * **2017**: The FAME-I scheme was extended for another 2 years, with a total allocation of INR 5.29 billion utilized over 4 years * **2019**: FAME II was introduced with larger budget (INR 10,000 crore). Supported e-buses, e-3 wheelers, e-cars, and e-2 wheelers. Also reduced GST rates on EVs and chargers * **2021**: Two major *Production Linked Incentive* (PLI Schemes) schemes launched: INR 25,938 crore for EVs and auto components INR 18,100 crores for Advanced Chemistry Cell (ACC) battery storage [94]	* **2024**: Electric Mobility Promotion Scheme (EMPS 2024) launched with INR 500 crore for e-2 W and e-3 W adoption, valid from April to July 2024 [94] * **2024**: Scheme to Promote Manufacturing of Electric Passenger Cars (SPMEPCI) Aimed at attracting global EV manufacturers with minimum INR 4150 crore investment and 50% domestic value addition within 5 years. To promote manufacturing of electric passenger cars in India [95] * **2025**: New EV Policy Focused on attracting global manufacturers, reducing import dependency, and boosting job creation through localized production. Targeting 30% EV penetration in private cars, 70% in commercial vehicles, 40% in buses, and 80% in two and three wheelers by 2030 [93]

FAME India scheme in 2015, FAME II in 2019, and several state-level policies designed to encourage EV adoption and local production. Initiatives aimed at promoting manufacturing of electric passenger cars and the Electric Mobility Promotion Scheme continue to drive the growth of this sector [93, 96].

The Indian government continues to focus on enhancing EV infrastructure, providing fiscal incentives, and promoting research and development in the electric mobility sector. Key features of India's EV policies are depicted in Fig. 4.7.

In order to reduce the usage of gasoline and diesel automobiles, the National Incentive FAMEII (Faster Adoption and Manufacturing of Hybrid and Electric Vehicles) initiative was launched on 1st April 2015. In June 2021, the Department of Heavy Industry launched the Production-Linked Incentive for Advanced Chemistry Cell Battery Storage (PLI-ACC Scheme) to promote investments in the automotive manufacturing sector and enhance domestic production of advanced automotive technology (AAT) products [97]. Also, the government introduced the Battery Swapping Policy to establish uniform battery standards for EVs across India. As per the Union Budget 2023–2024, customs duties are now waived for the import of capital goods and machinery essential for producing lithium-ion cells, which power these EV batteries. The goods and services tax (GST) on EVs have been reduced from 12% to 5%. Similarly, the GST on chargers and charging stations for electric vehicles has been reduced from 18% to 5% [98]. The Ministry of Heavy Industries, Government of India, reported that 1,433,545 electric vehicles (all types) out of a total of 22,717,562 were registered during the year 2023 [99]. The ratio of EVs compared to the total number of vehicles stands at approximately 0.0631. Furthermore, the Ministry of Heavy Industries has recently announced the Electric Mobility Promotion Scheme 2024, set to run from 1 April to 31 July 2024, with a total budget

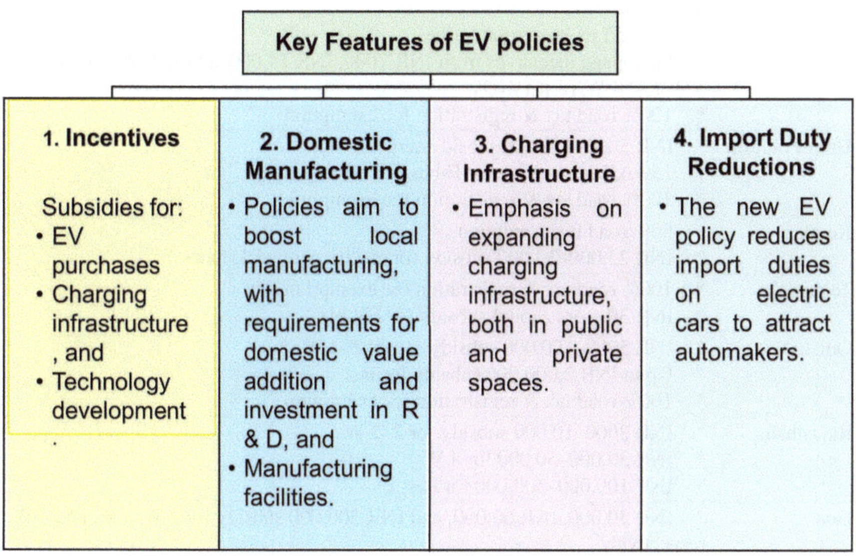

Fig. 4.7 Key features of India's EV policies

of INR 500 crores, equivalent to USD 60.3 million. The scheme provides demand incentives and subsidies to encourage faster adoption of electric two-wheelers (e-2 W) and three-wheelers (e-3 W) [100]. Some analysts expect a substantial uplift in India's EV industry due to the recently implemented e-vehicle policy that permits companies to set up manufacturing facilities in the country. Additionally, as reported by Reuters, the Government of India has revealed plans to lower import duties on specific EVs produced by carmakers willing to invest a minimum of USD 500 million and begin domestic manufacturing within three years [101]. The government offers various types of financial incentives to make EVs more affordable [102].

Moreover, the state-level EV targets and incentive policies for consumers offered by some major state governments in India to boost EV adoption are illustrated in Table 4.11 [103]. As of 2023, 26 states and 4 union territories in India have launched

Table 4.11 States level consumer subsidies under their EV policies as of mid-2025

State	Key incentives
Delhi	* INR 5000/kWh subsidy for E-2 W (maximum INR 30,000) * INR 30,000/kWh subsidy for E-3 W * INR 10,000/kWh subsidy (up to INR 1.5 lakhs for 4 W) * Scrapping incentives of INR 5000 and INR 7500 for 2 W and 3 W, respectively * 100% Road tax & registration fee waiver
Gujarat	* INR 10,000/kWh subsidy across E-2 W–E-4 W * Up to INR 10 lakh for charging infrastructure * Full registration waiver
Tamil Nadu	* INR 10,000/kWh for 2 W, 3 W, and 4 W. INR 20,000/kWh for E-bus * 25% subsidy for charging infrastructure * 100% Road tax & registration fee exemption
Maharashtra	* INR 5000/kWh subsidy for 2 W. INR 30,000 for e-3 W and INR 150,000 for e-4 W * 15% off ex-showroom price * Scrappage incentives from INR 7000, INR 15,000, and INR 25,000 for 2 W, 3 W, and 4 W, respectively * 100% road tax & registration fee exemption
Uttar Pradesh	* INR 5000–1.0 lakh subsidy across 2–4 W * 15%/kWh incentive for E-bus * 100% road tax & registration fee exemption
Kerala	* 50% road tax exemption * INR 25,000–30,000 subsidy for electric auto-rickshaws
Telangana	* 100% road tax & registration fee exemption * INR 30 crore capital subsidy for EV biz
Odisha	* INR 5000–100,000 subsidy across 2–4 W * Up to INR 2,000,000 subsidy for bus * 100% road tax & registration fee exemption
Rajasthan	* INR 2000–10,000 subsidy for 2–3 W * INR 30,000–50,000 for 4 W * INR 100,000–500,000 for bus
Goa	* INR 30,000, INR 60,000, and INR 300,000 subsidy for 2 W, 3 W, and 4 W, respectively * Scrapping incentive of INR 10,000 across 2–4 W

their own EV policies, complementing the national initiatives and promoting local manufacturing and adoption of electric vehicles. As of mid-2025, 21 Indian states have notified EV policies, and 15 offer direct incentives to buyers.

The EV sales have experienced significant growth in recent years because of increased consumer awareness and government policies and incentives at the federal and state levels. The EV sales growth since FY 2013–14 is displayed in Fig. 4.8. In FY 2013–14, India's EV sales were very low, with near–negligible impact on the overall automotive market. The Indian government's focus in the early years of EV promotion was primarily on electric 2 and 3 W. Passenger EV car sales were extremely limited, with only a few hundred electric cars sold. In FY 2015–16, India's EV market saw a notable uptick in sales, owing to growing awareness and the early impact of green mobility policy [104]. In 2017–18, electric car sale dropped due to lack of infrastructure [105]. The projected India's new EV sale in FY 2025–26 is expected to expand by 9% (excluding 3 W and commercial vehicles), compared to FY 2024–25 [106]. Charging infrastructure is expanding rapidly, with over 25,500 public stations and 60,000 connectors nationwide. The Prime Minister e-Bus Sewa scheme aims to deploy 10,000 e-buses across 169 cities in FY 2025–26. Overall, India's EV market is experiencing strong growth, driven by a combined effect of multiple factors and strongly supported by government initiatives.

In summary, this timeline highlights the determinations of the Indian government to foster a sustainable electric vehicle ecosystem, reflecting its commitment to environmental sustainability and energy independence.

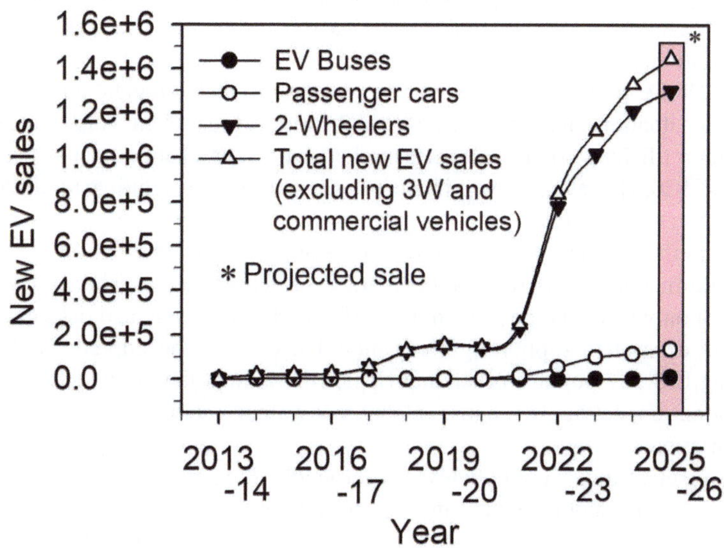

Fig. 4.8 Status of EV sales (excluding 3 W and commercial vehicles) in India since FY 2013–14. The financial year starts from 1st April to 31st March

4.9 Conclusion

Roads played a vital role for transporting goods and connecting local and rural communities and will continue to play their role for the foreseeable future. The transition to EVs is critical for mitigating GHG emissions and averting climate disaster. The transition to zero-emission transportation represented a significant move towards decarbonizing road transport. Numerous studies indicated that increasing the adoption rates of EVs is a key strategy for reducing GHG emissions from the transportation sector.

Several variables contribute to the decarbonization of road transport, including government energy policies, the development of regulatory standards, essential infrastructure investments and developments, research and innovation, and advancements in EV and battery technologies. Various policies/strategies aimed at decarbonizing road transportation are categorized into different groups, each addressing different aspects of the transition.

In EU member states, a variety of incentives at the local, regional, and national levels, such as lower taxes or free public parking for EVs, are utilized to promote electric mobility. Key elements like financial incentives, the development of charging infrastructure, the availability of a wide range of EV models, etc. are critical to overcome major challenges related to affordability, convenience, and awareness of EVs. All stakeholders are making efforts to establish and implement comprehensive regulatory frameworks aimed at accelerating the shift to zero-emission vehicles, integrating renewable energy sources, and promoting sustainable transportation, thereby achieving carbon neutrality across different sectors, including transportation.

The US EV policies have evolved in response to technological advancements, economic situations, and shifting political priorities. Electric car sales increased to 1.6 million till June 2025, with their market share skyrocketing over 400%, compared to 2020. However, growth in electric car sales saw a significant slowdown in 2024, increasing by only 10% compared to 40% in 2023. In 2025, the future of EV policies in the U.S. has entered a turbulent phase, with major shifts on the horizon that could fundamentally reshape the market for years to come. This policy uncertainty not only hampers the growth of the EV market share but also risks disrupting an entire ecosystem supporting clean mobility. Clearly, the unpredictability surrounding U.S. EV policies is sending ripples throughout the global clean mobility ecosystem, impacting the entire global ecosystem established for sustainable transportation.

Conversely, China's EV policies are profoundly impacting the global automotive industry, shaping supply chains, and transforming international trade dynamics and market competition. This influence is expected to persist in the near future. Chinese manufacturers have made a significant impact on both the domestic and global EV markets, establishing China as a global leader in production and sales. The country's EV program plays a major role in shaping the automotive landscape, impacting trade, technology, and energy sectors. While this presents challenges for some regions, it also offers opportunities for developing countries, encouraging innovation and competition across the world.

The Indian government continues to focus on enhancing EV infrastructure, providing fiscal incentives, and promoting research and development in the electric mobility sector. Favorable initiatives in this sector highlight the determinations of the Indian government to foster a sustainable EV ecosystem, reflecting its commitment to environmental sustainability and energy independence.

In summary, uncertainty surrounding EV policies not only slows down the growth of the EV market but also threatens to destabilize the entire clean mobility ecosystem. A reversal of EV initiatives could impede the global transition towards sustainable transportation. Consistency and reliability in EV policies are paramount for the ongoing viability of this sector. A sustained financial commitment to support incentives and infrastructure programs for this transition could help to achieve carbon neutrality goals.

References

1. CO_2 emissions in 2023. International Energy Agency (IEA) (Feb 2024). https://iea.blob.core.windows.net/assets/33e2badc-b839-4c18-84ce-f6387b3c008f/CO2Emissionsin2023.pdf
2. IEA (2022) World energy outlook 2022. IEA, Paris. https://www.iea.org/reports/world-energy-outlook-2022
3. Patil G, Pode G, Diouf B, Pode R (2024) Sustainable decarbonization of road transport: policies, current status, and challenges of electric vehicles. Sustainability 16(18):8058. https://doi.org/10.3390/su16188058
4. Rho motion view: over 20M EVs to be sold globally in 2025. Rho Motion (28 Jan 2025). https://rhomotion.com/news/rho-motion-view-over-20m-evs-to-be-sold-globally-in-2025/
5. Noussan M, Hafner M, Tagliapietra S (2020) Policies to decarbonize the transport sector. In: The future of transport between digitalization and decarbonization, Springer briefs in energy. Springer, Cham. https://doi.org/10.1007/978-3-030-37966-7_4
6. Reducing Greenhouse Gas Emissions from the Transportation Sector through Climate Planning (EPA-420-F-24-042) (December 2024) Climate Pollution Reduction Grants (CPRG) program. Environment Protection Agency. https://www.epa.gov/system/files/documents/2024-12/420f24042.pdf
7. Barth M, Wu G, Boriboonsomsin K (2015) Intelligent transportation systems for improving traffic energy efficiency and reducing GHG emissions from roadways. White paper, November 2015. National Center for Sustainable Transportation University of California, Riverside. https://escholarship.org/content/qt1hw130h2/qt1hw130h2_noSplash_b69be7af1a424ca3c21b5fd06726d489.pdf?t=pnc8kp
8. Zhao P, Li Z, He Z, Chen Y, Xiao Z (2025) Reducing the road freight emissions through integrated strategy in the port cities. Nat Commun 16:2563. https://doi.org/10.1038/s41467-025-57861-z
9. Reducing greenhouse gas emissions from heavy-duty vehicles in Europe. European Environment Agency (2022). https://doi.org/10.2800/066953
10. Nasri S, Mansouri N, Mnassri A, Lashab A, Vasquez J, Rezk H (2025) Global analysis of electric vehicle charging infrastructure and sustainable energy sources solutions. World Electric Vehicle J 16:194. https://doi.org/10.3390/wevj16040194

11. Zhao X, Li X, Jiao D, Mao Y, Sun J, Liu G (2024) Policy incentives and electric vehicle adoption in China: from a perspective of policy mixes. Transp Res A Policy Pract 190:104235. https://doi.org/10.1016/j.tra.2024.104235
12. IEA (2021) Global EV outlook 2021. IEA, Paris. https://www.iea.org/reports/global-ev-outlook-2021
13. Bibri SE, Krogstie J, Kärrholm M (2020) Compact city planning and development: emerging practices and strategies for achieving the goals of sustainability. Dev Built Environ 4:100021. https://doi.org/10.1016/j.dibe.2020.100021
14. Mohanty M (2020) Sustainable urban planning and making sustainable cities. In: Leal Filho W, Azul A, Brandli L, Özuyar P, Wall T (eds) Sustainable cities and communities, Encyclopedia of the UN sustainable development goals. Springer, Cham. https://doi.org/10.1007/978-3-319-71061-7_62-1
15. Jin L, Slowik P (2017) Literature review of electric vehicle consumer awareness and outreach activities. Working paper 2017-03. International Council on Clean Transportation. https://theicct.org/sites/default/files/publications/Consumer-EV-Awareness_ICCT_Working-Paper_23032017_vF.pdf
16. Patil P (2019) Innovations in electric vehicle technology: a review of emerging trends and their potential impacts on transportation and society. Rev Contemp Bus Anal 2(1):20–32
17. Shahed MT, Harun-ur Rashid ABM (2024) Battery charging technologies and standards for electric vehicles: a state-of-the-art review, challenges, and future research prospects. Energy Rep 11:5978–5998. https://doi.org/10.1016/j.egyr.2024.05.062
18. Moran G (2024) Navigating EV business models: what will drive EV penetration in 2024? Forbes Business Council, Council Post. Forbes Media LLC. 17 May. https://www.forbes.com/councils/forbesbusinesscouncil/2024/05/17/navigating-ev-business-models-what-will-drive-ev-penetration-in-2024/
19. Huang Y, Qian L, Soopramanien D, Tyfield D (2021) Buy, lease, or share? Consumer preferences for innovative business models in the market for electric vehicles. Technol Forecast Soc Chang 166:120639. https://doi.org/10.1016/j.techfore.2021.120639
20. H.R. 6049—110th Congress (2007–2008) US Passed the Bill in 2008. https://www.congress.gov/bill/110th-congress/house-bill/6049/text#toc-id28E82AC4F03540D4B7192D47AF6C089B. Accessed 8 Apr 2024
21. Zhang X, Liang Y, Yu E, Rao R, Xie J (2017) Review of electric vehicle policies in China: content summary and effect analysis. Renew Sust Energ Rev 70:698–714. https://doi.org/10.1016/j.rser.2016.11.250
22. Liu X, Sun X, Zheng H, Huang D (2021) Do policy incentives drive electric vehicle adoption? Evidence from China. Transp Res Part A Policy Pract 150:49–62. https://doi.org/10.1016/j.tra.2021.05.013
23. Electrification Trends Worldwide. European Automobile Manufacturers' Association (ACEA) (25 Nov 2022). https://www.acea.auto/news/electrification-trends-worldwide/. Accessed 21 Mar 2024
24. Carey N (2024) Global EV Sales up 69% y/y in Jan, down 26% vs Dec—Rho Motion. Reuters. 14 Feb. https://www.reuters.com/business/autos-transportation/global-ev-sales-up-69-yy-jan-down-26-vs-dec-rho-motion-2024-02-14/. Accessed 10 Apr 2024
25. Q1 2024: over 3 million electric vehicles sold globally. Rho Motion (Apr 2024). https://rhomotion.com/news/q1-2024-over-3-million-electric-vehicles-sold-globally/. Accessed 23 Apr 2024
26. Electric Road Vehicles in the European Union: Trends, impacts and policies (3 Apr 2019). https://www.europarl.europa.eu/RegData/etudes/BRIE/2019/637895/EPRS_BRI(2019)637895_EN.pdf. Accessed 10 Apr 2024
27. Wappelhorst S, Hall D, Nicholas M, Lutsey N (2020) Analyzing policies to grow the electric vehicle market in European cities. International Council on Clean Transportation . https://theicct.org/sites/default/files/publications/EV_city_policies_white_paper_fv_20200224.pdf. Accessed 12 Mar 2024

28. Van-Wieringen K (2024) The future of European electric vehicles (PE 762.873). European Parliament. Nov. https://doi.org/10.2861/1724777
29. Electric vehicles. Mobility and Transport, European Commission, Brussels. https://transport.ec.europa.eu/transport-themes/clean-transport/clean-and-energy-efficient-vehicles/green-propulsion-transport/electric-vehicles_en
30. EAFO policy paper: Europe on the electrification path towards clean transportation. European Alternative Fuels Observatory, European Commission (9 Feb 2022). https://alternative-fuels-observatory.ec.europa.eu/general-information/news/eafo-policy-paper-europe-electrification-path-towards-clean-transportation
31. Electric road vehicles in the European Union Trends, impacts and policies. European Parliamentary Research Service, European Commission (Apr 2019). https://www.europarl.europa.eu/RegData/etudes/BRIE/2019/637895/EPRS_BRI%282019%29637895_EN.pdf
32. European E-mobility at a Crossroads. E-Mobility Europe (Jan 2025). https://www.emobility-europe.org/wp-content/uploads/2025/02/NA-E-ME-Report-EU-2025-forecast-FINAL-1.pdf
33. Schlacke S, Wentzien H, Thierjung E-M, Köster M (2022) Implementing the EU climate law via the 'Fit for 55' package. Oxford Open Energy 1:oiab002. https://doi.org/10.1093/ooenergy/oiab002
34. Lacoume S, von Berg C, Sielewicz G. Electric vehicles: competition between China and Europe in an age of mobility transition. Coface (11 Mar 2025). https://www.coface.com/news-economy-and-insights/electric-vehicles-competition-between-china-and-europe-in-an-age-of-mobility-transition
35. CCCEU Statement on EU's Provisional Countervailing Duties. China Chamber of Commerce to the EU, Brussels (4 Jul 2024). http://en.ccceu.eu/2024-07/04/c_4375.htm#:~:text=On%20July%204%2C%202024%2C%20the%20European%20Commission%20announced,BYD%2C%2019.9%25%20for%20Geely%2C%20and%2037.6%25%20for%20SAIC
36. Zero emission vehicles: first 'Fit for 55' deal will end the sale of new CO_2 emitting cars in Europe by 2035. European Commission—Press release. Brussels (28 Oct 2022). https://ec.europa.eu/commission/presscorner/api/files/document/print/en/ip_22_6462/IP_22_6462_EN.pdf
37. Katanich D, Sanchez-Alonso A (2025) Transition to zero-emission mobility: the EU's plan to save Europe's car industry. Euronews, Lyon. 5 Mar. https://www.euronews.com/business/2025/03/05/transition-to-zero-emission-mobility-here-is-the-eus-plan-to-save-europes-car-industry
38. CO_2 emissions: EP adopts flexibility measures for carmakers. News European Parliament (8 May 2025). https://www.europarl.europa.eu/news/en/press-room/20250502IPR28225/co2-emissions-ep-adopts-flexibility-measures-for-carmakers
39. Here's how EU legislation accelerates the EV revolution. Virta, Helsinki (28 Jun 2024). https://www.virta.global/blog/this-is-how-eu-regulation-accelerates-the-electric-vehicle-revolution
40. Incentives and Legislation, Austria. European Alternative Fuels Observatory, Directorate-General for Mobility and Transport, European Commission, Brussels (18 Apr 2025). https://alternative-fuels-observatory.ec.europa.eu/transport-mode/road/austria/incentives-legislations
41. Electric vehicles. European Environment Agency, European Union, Copenhagen (5 Dec 2024). https://www.eea.europa.eu/en/topics/in-depth/electric-vehicles?activeTab=e3e6b879-fef4-4a88-9436-5f0064698270&activeAccordion=4268d9b2-6e3b-409b-8b2a-b624c120090d
42. The European Green Deal. European Commission. https://commission.europa.eu/strategy-and-policy/priorities-2019-2024/european-green-deal_en
43. Towards a green and smart mobility: the EU's transport strategy. EU Tourism Platform, European Union (14 Aug 2024). https://transition-pathways.europa.eu/system/files/2024-09/2021-mobility-strategy-and-action-plan_1.pdf
44. The future of European sustainable mobility. AECC, Brussels (2024). https://www.aecc.eu/the-future-of-european-sustainable-mobility/

45. Energy Performance of Buildings Directive. Energy, European Commission. https://energy.ec.europa.eu/topics/energy-efficiency/energy-performance-buildings/energy-performance-buildings-directive_en
46. Bryant J (2025) The regulatory roadmap to decarbonized mobility: policy frameworks for Europe. 16 Jan. https://www.europeanfutureenergyforum.com/the-regulatory-roadmap-to-decarbonized-mobility-policy-frameworks-for-europe/
47. 5 key European policy trends that will shape sustainable mobility in the coming years. EU Business News (20 Aug 2024). https://www.eubusinessnews.com/5-key-european-policy-trends-that-will-shape-sustainable-mobility-in-the-coming-years/
48. Editorial Team. Greening Europe's urban mobility: strategies for decarbonising public transport. EU Tourism Platform, European Union (5 Jan 2025). https://transition-pathways.europa.eu/articles/greening-europes-urban-mobility-strategies-decarbonising-public-transport
49. Overview—electric vehicles: tax benefits & purchase incentives in the European Union (21 Sept 2022). https://www.acea.auto/fact/overview-electric-vehicles-tax-benefits-purchase-incentives-in-the-european-union-2022/. Accessed 5 May 2024
50. First 'fit for 55' proposal agreed: the EU strengthens targets for CO_2 emissions for new cars and vans. Council of the EU (27 Oct 2022). https://www.consilium.europa.eu/en/press/press-releases/2022/10/27/first-fit-for-55-proposal-agreed-the-eu-strengthens-targets-for-co2-emissions-for-new-cars-and-vans/. Accessed 7 May 2024
51. E.U. plans for only electric new vehicles by 2035 'without precedent'. Washington Post (28 Oct 2022). https://www.washingtonpost.com/climate-solutions/2022/10/28/eu-electric-cars-2035/. Accessed 7 May 2024
52. Shaping the future of fast-charging EV infrastructure. McKinsey and Company (25 Oct 2021). https://www.mckinsey.com/capabilities/operations/our-insights/shaping-the-future-of-fast-charging-ev-infrastructure#/. Accessed 10 May 2024
53. European Electric Vehicle Charging Infrastructure Masterplan. European Automobile Manufacturers' Association (28 Mar 2022). https://www.acea.auto/files/Research-Whitepaper-A-European-EV-Charging-Infrastructure-Masterplan.pdf. Accessed 10 May 2024
54. New Car Registrations, European Union. The European Automobile Manufacturers' Association (ACEA) (20 Feb 2024). https://www.acea.auto/files/Press_release_car_registrations-January_2024.pdf. Accessed 12 Mar 2024
55. EU formally bans new ICE vehicles, aims to slash emissions from trucks and buses. https://electrek.co/2023/02/15/eu-bans-new-ice-vehicles-aims-to-slash-emissions-from-trucks-and-buses/. Accessed 10 May 2024
56. Everything you need to know about the UK's switch to electric. Electric Car Guide. https://electriccar.guide/transition-to-electric. Accessed 4 Mar 2024
57. European new-car market could need decade to recover, but electric cars will dominate. https://autovista24.autovistagroup.com/news/european-new-car-market-recover-but-electric-cars-will-dominate/. Accessed 15 Mar 2024
58. Hayashida S, Croix SL, Coffman M (2021) Understanding changes in electric vehicle policies in the U.S. states, 2010–2018. Transp Policy 103:211–223. https://doi.org/10.1016/j.tranpol.2021.01.001
59. Electric vehicle policies of US. EV a2z, Sydney (8 May 2024). https://ev-a2z.com/policy-standards-regulations/electric-vehicle-policies-of-us/
60. Federal EV policy. Electrification Coalition (2025). https://electrificationcoalition.org/work/federal-ev-policy/
61. Wen L (2025) Act now before the$7,500EV tax credit disappears: what drivers need to know before September 30, 2025. Evdance, Shenzhen, Guangdong Province. 9 Jul. https://evdances.com/blogs/news/act-now-before-the-7-500-ev-tax-credit-disappears-what-drivers-need-to-know-before-september-30-2025

62. Mathrani A (2024) The rise of electric vehicles in the US and the road ahead. RaboResearch, RaboBank, Amsterdam, CB Utrecht. 5 June. https://www.rabobank.com/knowledge/d011429876-the-rise-of-electric-vehicles-in-the-us-and-the-road-ahead
63. Electric Vehicle (EV) and Fuel Cell Electric Vehicle (FCEV) tax credit. https://afdc.energy.gov/laws/409. Accessed 15 Mar 2024
64. Bui V (2023) The road to an electric vehicle future. Department of Energy. 23 June. https://www.energy.gov/articles/road-electric-vehicle-future. Accessed 21 Mar 2024
65. EY research: nearly half of US car buyers intend to purchase an EV. Ernst & Young Global Limited, Americas (27 Jun 2023). https://www.ey.com/en_us/news/2023/06/ey-research-nearly-half-of-us-car-buyers-intend-to-purchase-an-ev. Accessed 21 Mar 2024
66. Goodbye to electric cars in the US in 2024. Automakers are fed up and want to go back to combustion. La Grada (Spanish Agency) (13 Feb 2024). https://lagradaonline.com/en/electric-cars-usa-automakers-combustion/. Accessed 23 Apr 2024
67. The EV future—prepare to meet growing demand while you still can. https://www.urban-science.com/evfuture/#:~:text=Both%20auto%20buyers%20and%20dealers%20recognize. Accessed 21 Mar 2024
68. US automakers race to build more hybrids as EV sales slow. Reuters (15 Mar 2024). https://www.reuters.com/business/autos-transportation/us-automakers-race-build-more-hybrids-ev-sales-slow-2024-03-15/. Accessed 20 Mar 2024
69. IEA (2025) Global EV outlook 2025. IEA, Paris. https://www.iea.org/reports/global-ev-outlook-2025
70. H.R. 1: One Big Beautiful Bill Act. July 9, 2025 (Passed Congress). H.R. 1 (ENR)—An Act To provide for reconciliation pursuant to title II of H. Con. Res. 14. https://www.govinfo.gov/content/pkg/BILLS-119hr1enr/pdf/BILLS-119hr1enr.pdf
71. Walton R (2025) US electric vehicle sales are slowing amid policy shifts: BNEF. Utility Dive, Informa TechTarget, Newton, MA. 18 Jun. https://www.utilitydive.com/news/us-electric-vehicle-sales-are-slowing-amid-policy-shifts-bnef/751079/. Accessed 11 Jul 2025
72. Jenkins J, Farbes J, Jones R, Cheng A, Luo Q, Haley B (2025) A fork in the road: impacts of federal policy repeal on the U.S. energy transition. REPEAT Project. https://doi.org/10.5281/zenodo.15497751
73. With 'Big Beautiful Bill,' U.S. to Reverse Course on Clean Energy. Yale Environment 360, Yale School of Environment, New Haven, CT (7 Jul 2025). https://e360.yale.edu/digest/big-beautiful-bill-renewable-energy-china
74. Millard F (2025) Reversing into tomorrow—Trump vs. EVs. Electric & Hybrid Vehicle Technology International, Dartford, Kent. 8 Jul. https://www.electrichybridvehicletechnology.com/features/feature-reversing-into-tomorrow-trump-vs-evs.html
75. Singh C (2025) How Trump's 'big, beautiful bill' could help China win EV race. USA Today. 9 Jul. https://www.usatoday.com/story/money/cars/2025/07/09/trums-big-beautiful-bill-china-ev-market/84505321007/
76. Jin S-H (2025) U.S. ends electric vehicle tax credit, hurting domestic battery manufacturers amid Chinese competition. ChosunBiz, Seoul. 8 Jul. https://biz.chosun.com/en/en-industry/2025/07/08/FO6I2CASEFE5JLO2IE7BK3WH4I/
77. Zhou G, Ou X, Zhang X (2013) Development of electric vehicles use in China: a study from the perspective of life-cycle energy consumption and greenhouse gas emissions. Energy Policy 59:875–884. https://doi.org/10.1016/j.enpol.2013.04.057
78. Sinocharged: The Bright Future of China's Electric Vehicle Market (A special Report on Electrification). 2021 KPMG Huazhen LLP, a People's Republic of China. https://assets.kpmg.com/content/dam/kpmg/cn/pdf/en/2021/01/2020-china-leading-autotech-50.pdf. Accessed 21 Mar 2024
79. China extends NEV tax reduction and exemption policy to 2027. China Briefing, Dezan Shira & Associates (28 Jun 2023). https://www.china-briefing.com/news/china-extends-nev-tax-reduction-and-exemption-policy-to-2027
80. Sandalov D, Meidan M, Andrews-Speed P, Hove A, Qiu SY, Downie E (2022) Guide to Chinese climate policy 2022. The Oxford Institute for Energy Studies, Oxford. https://chine-

seclimatepolicy.oxfordenergy.org/wp-content/uploads/2022/11/Guide-to-Chinese-Climate-Policy-2022.pdf
81. IEA (2023), Global EV outlook 2023, IEA, Paris. https://www.iea.org/reports/global-ev-outlook-2023. Accessed 21 Mar 2024
82. China Unveils $72 Billion Tax Break for EVs, Other Green Cars to Spur Demand. Reuters (21 Jun 2023). https://www.reuters.com/business/autos-transportation/china-announces-extension-purchase-tax-break-nevs-until-2027-2023-06-21/. Accessed 18 Mar 2024
83. China considers extending its EV subsidies to 2023 (Updated). China Briefing (27 Jun 2023). https://www.china-briefing.com/news/china-considers-extending-its-ev-subsidies-to-2023/. Accessed 18 Mar 2024
84. EV incentives in 2024: Europe, the US and China. Carvibz (9 Feb 2024). https://www.car-vibz.com/news/ev-incentives-in-2024-europe-the-us-and-china/#google_vignette. Accessed 22 Jul 2025
85. Liu Y, Zhao X, Lu D, Li X (2023) Impact of policy incentives on the adoption of electric vehicle in China. Transp Res Part A Policy Pract 176:103801. https://doi.org/10.1016/j.tra.2023.103801
86. Zhao Y, Jian Z, Du Y (2024) How can China's subsidy promote the transition to electric vehicles? Renew Sustain Energy Rev 189(Part B):114010. https://doi.org/10.1016/j.rser.2023.114010
87. China vows to help its EV champions succeed overseas despite threats of sanctions. CNN Business (7 Feb 2024). https://edition.cnn.com/2024/02/07/cars/china-ev-global-push-intl-hnk/index.html
88. Shen L, Chen Z, Dou X, Xu X, Cao Z, Liao S (2024) Restricting factors for promoting electric vehicles: evidence from China. Transp Policy 148:234–245. https://doi.org/10.1016/j.tranpol.2024.01.017
89. Lin C (2024) 3 drivers of China's booming electric vehicle market. Harvard Business Review. 3 Jan. https://hbr.org/2024/01/3-drivers-of-chinas-booming-electric-vehicle-market
90. The automotive market is changing. Can China show us the future? World Economic Forum (2 Jul 2021). https://www.weforum.org/agenda/2021/07/future-automative-market-china/
91. The history of EVs in India: 9 key milestones shaping the future of electric vehicles. ChargeMOD, Trivandrum (23 Sept 2024). https://chargemod.com/blog/The-History-of-EVs-in-India1727096717iz7#:~:text=The%20history%20of%20EVs%20in%20India%20began%20in,and%20the%20growing%20demand%20for%20cleaner%20mobility%20solutions
92. EV central policy: tracing key milestones. JMK Research and Analysis, Gurugram (12 Jun 2020). https://jmkresearch.com/ev-central-policy-tracing-key-milestones/
93. Electrifying India's roads: the rise of EVs. PIB Research Unit, Ministry of Heavy Industries, Government of India (16 Jul 2024). https://static.pib.gov.in/WriteReadData/specificdocs/documents/2024/jul/doc2024716350301.pdf
94. Government approves E-vehicle policy to promote India as a manufacturing destination for E-vehicles. Ministry of Heavy Industries. Press Release, Press Information Bureau, Delhi (15 Mar 2024). https://www.pib.gov.in/PressReleasePage.aspx?PRID=2014874
95. Pandey PC (2025) Indian Government sets target to achieve 30% electrification of the country's vehicle fleet by 2030. Climate Scorecard. 13 Mar. https://www.climatescorecard.org/2025/03/indian-government-sets-target-to-achieve-30-electrification-of-the-countrys-vehicle-fleet-by-2030/
96. India's EV revolution: policies and incentives driving growth. Kingsway (Owned by KK Motors Inc), Ghaziabad (25 Apr 2025). https://kingswayaccessories.com/blog/indias-ev-revolution-policies-and-incentives-driving-growth/
97. Das PK, Bhat MY (2022) Global electric vehicle adoption: Implementation and policy implications for India. Environ Sci Pollut Res 29:40612–40622. https://doi.org/10.1007/s11356-021-18211-w

98. Government's EV policies driving India's Green Revolution. In Invest India (9 Sept 2023). https://www.investindia.gov.in/team-india-blogs/governments-ev-policies-driving-indias-green-revolution
99. Total 22,717,562 vehicles registered during 2023 including electric, petrol and diesel vehicles. Ministry of Heavy Industries, Government of India (18 Dec 2023). https://pib.gov.in/PressReleaseIframePage.aspx?PRID=1987802
100. Ministry of Heavy Industries, India (13 Mar 2024). https://www.pib.gov.in/PressReleaseIframePage.aspx?PRID=2014366
101. Electric vehicle plans of Indian Automakers. Reuters (16 Mar 2024). https://www.reuters.com/business/autos-transportation/electric-vehicle-plans-indian-automakers-2024-03-15/
102. Electric vehicle incentives. NITI AAYOG, Government of India. https://e-amrit.niti.gov.in/electric-vehicleincentives
103. Nair M, Minocha A (2023) Greening India's automotive sector: EV policies, categories and subnational trends. Council on Energy, Environment and Water. Vasant Kunj Institutional Area New Delhi, New Delhi. https://www.ceew.in/gfc/solutions-factory/publications/CEEW-CEF-Greening-Indias-Automotive-Sector.pdf
104. Soman A, Kaur H, Jain H, Ganesan K (2020) India's electric vehicle transition: can electric mobility support India's sustainable economic recovery post COVID-19? Council on Energy, Environment and Water (CEEW) Report, New Delhi. Nov. https://www.ceew.in/sites/default/files/CEEW-Indias-EV-Transition-Post-COVID-19-22Dec20_0.pdf
105. Parashar S (2018) Electric car sales in India fall in FY18, electric two-wheelers record healthy growth. Car & Bike, Mahindra First Choice Wheels Limited, Mumbai. 19 Oct. https://www.carandbike.com/news/electric-car-sales-in-india-fall-in-fy18-electric-two-wheelers-record-healthy-growth-1934323
106. Anand S (2025) EV sales in India projected to grow 40% to 1.38 lakh units in 2025: report. ETEnergyWorld, India. 28 Jun. https://energy.economictimes.indiatimes.com/news/power/indias-ev-market-set-for-40-surge-by-2025-key-insights-and-trends/122122994

Chapter 5
Transition to Zero-Emission Vehicles: Case Studies of Norway and China

Contents

5.1 Introduction

In 1989, two members of the popular Norwegian synth-pop band A-Ha, Morten Harket and Magne Furuholmen, were in Switzerland to attend the electric vehicle (EV) conference. They were inspired by the demonstration of a Fiat Panda that was converted to run on electricity at an exhibition. Subsequently, they imported an electric Fiat Panda and embarked on a series of trips across Norway, refusing to pay road tolls and parking illegally to incur fines. Their aim was to gain publicity for EVs and push the government to provide incentives for their adoption [1]. The media coverage of their efforts and public attention are said to have sparked a cultural shift towards EVs. As a result, the Norwegian government has been taking

R. Pode, B. Diouf, *Decarbonization of Road Transportation*, Green Energy and Technology, https://doi.org/10.1007/978-3-032-11388-7_5

actions to accelerate the EV transition since the 1990s. Now, it has become the world's electric car capital, with the highest electric vehicle adoption rate in the world.

At the beginning of the century, China's leadership laid a vision to dominate the technologies of the future. When examining the origins of China's EV supremacy, many analysts often credit Wan Gang, a German-trained engineer who held the position of Minister of Science and Technology from 2007 to 2018 [2]. Having transitioned from a nation known for its bicycles, China has now become the global leader in the EV market.

Subsequently, China rapidly expanded its EV market, becoming the largest EV producer and consumer [3]. Key factors include: (i) *government subsidies* for EV manufacturers and buyers, (ii) *investment in battery technology*, improving efficiency and affordability, and (iii) *electrification of public transport*, with cities like Shenzhen converting their entire bus fleet to electric. A record number of 24% of total new car sales were reported as EVs in China in 2023.

Here, we presented the case studies of Norway and China for the successful adoption of EVs. Each country has a particular set of policies, each with its own advantages and disadvantages regarding the implementation of specific measures. There is no universally ideal policy combination to ensure the effective decarbonization of the transport sector. The effectiveness of the management of policies, financial incentives, and infrastructure development seems to be driving the growth of EVs.

5.2 Case Study of Norway

In Norway, EVs made up over 20% of all passenger vehicles and accounted for more than 80% of new vehicle sales in 2022 [4]. Additionally, in 2023, more than 82% of new car sales were EVs, increasing to 90% when plug-in vehicles are included. Norway stands at the forefront of global EV adoption, with over 90% of new car sales being electric. This success is driven by strong government incentives (including tax exemptions and toll reductions), extensive charging infrastructure (making EVs convenient for daily use), and public awareness campaigns (encouraging consumers to switch to electric) [5]. Figure 5.1 shows the share of new passenger car sales along with the share of the total passenger car fleet [6]. In 2025, zero-emission cars reached over 90% of new car sales in Norway.

The Norwegian Public Roads Administration (known as Opplysningsrådet for veitrafikken, or OFV) recently reported that there were 104,590 new registrations for electric cars and 10,169 plug-in hybrids from January to December 2023 [7]. Furthermore, in January 2024, Norway saw the registration of 4717 new electric cars, according to the Norwegian Public Roads Administration.

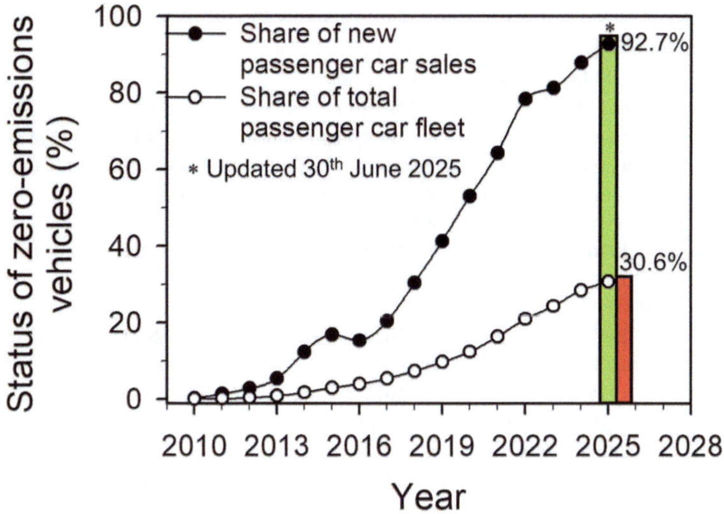

Fig. 5.1 Share of zero-emission vehicle sales in Norway from 2010 to June 2025

5.2.1 Incentives and Subsidies

Norway has implemented a comprehensive set of policies and incentives to promote the adoption of EVs, aiming for all new cars (passenger cars, light vans, and city buses) sold by 2025 to be zero emission [8, 9]. The government started promoting the purchase of EVs back in the 1990s, but it wasn't until Tesla's Model S (introduced in 2012–2013) and other EV models became available around 2012 that sales truly took off [10]. Øyvind Solberg Thorsen, CEO of the Norwegian Road Federation (OFV), persuaded the Norwegian government to provide subsidies and exclude the electric car from all automotive taxes, including value-added and import taxes. The Think City electric car, produced by Norwegian carmaker Think Global and first put on the road in 2011, was small, reasonably affordable, and had a relatively short range of 160 km and a top speed of around 100 km/h, whereas the Tesla S offered approximately 426 km on a single charge with a top speed of 209 km/h. Monetary and non-monetary incentives have been instrumental in the very high adoption rate and popularization of EVs in Norway. Table 5.1 outlines the timeline of key policies and incentives offered by the government since 1990. These initiatives encompass tax exemptions, reduced fees on tolls and ferries, access to bus lanes, and free or reduced parking [11]. Although these incentives are being gradually reduced, Norway is committed to ensuring that choosing an EV remains an appealing and attractive choice [12–14]. Additionally, Norway plans to transition to electric power for its ferries, buses, semi-trucks, and even construction gear in the future. The country aims for 75% of new long-distance buses and 50% of new trucks to be zero-emission, facilitating emission-free goods transport in major cities by 2030 [5].

Table 5.1 Timeline of key policies and incentives offered by Norwegian Government in the past and present (passenger vehicles)

Phase	1. Early Incentives (1990s–2000s)	2. Expansion (2008–2017)	3. Refinement (2018–Present)
Policies and incentives	* **1990**: Exemption from purchase and import taxes on EVs	* **2008**: Oslo launched the world's first municipal charging system	* **2018**: Toll road and ferry discounts limited to 50% of the total cost
	* **1996**: Exemption from annual road tax for EVs	* **2009**: Reduced ferry fares for EVs	* **2018**: Company car tax reduction reduced to 40%
	* **1997**: Exemption from toll road charges	* **2012**: Sets carbon neutrality goal by 2050	* **2021**: Reduced annual road tax for EVs (NOK 455)
	* **1999**: Free or reduced municipal parking fee for EVs	* **2015**: Continued VAT exemption (25%) on EV leasing	* **2022**: Full annual road tax reinstated
	* **2000**: Reduced company car tax for EVs	* **2016**: Local authorities allowed to limit bus lane access to EVs with passengers	* **2022**: Public procurement requires zero-emission vehicles
	* **2000**: Reduced company car tax for EVs * **2001**: VAT exemption on EV purchases	* **2017**: Norwegian Parliament set the 2025 target for all new cars to be zero-emission	* **2023**: Some purchase tax based on car weight introduced on new EVs
	* **2005**: Access to bus lane for EVs	* **2017**: Charging right established for apartment dwellers	* **2023**: VAT of 25% introduced on EV purchases exceeding NOK 500,000
			* **2025**: Target for all new cars sold to be zero emission
			* Ongoing: Support for charging infrastructure development and public procurement of emission vehicles

The aim of achieving zero carbon emissions from all public transportation by the end of 2024 has been set. In 2024, EVs reached 88.9% of new car sales, and over 9000 fast chargers were available nationwide. It's an electric utopia for the future. In Norway, the charging of EVs may be a third of the price of gasoline with hydropower-generated electricity. In February 2024, almost 100% (more than 98%) of electricity production in Norway came from renewable energy sources (hydropower 87.8%, wind power 10.9%, and thermal power merely 1.3%) [15]. The advantage of hydropower electric charging is that it is inexpensive, renewable, and available. Additionally, the Norwegian government set the goal for 2025 that all new passenger cars should have zero emissions [16]. To realize the stated goal, the government has put in some incentives like a zero-registration tax, free tolls, free parking, a zero value-added tax, access to bus lanes, etc. [17]. This has a huge impact, and eight out of ten new cars are zero-emission vehicles. The remaining 20% are plug-in hybrids and some gasoline cars as well. Polluting cars have heavy taxes in Norway, whereas EVs are tax-free. Furthermore, the process of buying EVs is also simplified by reducing the paperwork. As a result, EVs in Norway present the advantage of lower cost of ownership compared to their ICE vehicle counterparts.

Norway has the most comprehensive set of policies in order to promote electric vehicles. Norway leads the world in electric vehicle adoption rate [18]. In 2022, around 80% of the new cars sold in the Norwegian automobile market were EVs, and EVs have already supplanted more than 20% of the passenger cars, or vehicles with up to seven seats [19]. The government of Norway has a plan to restrict the sale of electric cars only by 2025. Norway's share of the EV market grew dramatically to 79.3% in 2022, and previously it was 64.5% in 2021. More than 80% of the passenger cars are charged at home, which shows their successful home charging schemes. The Norwegian government provides a set of incentives and subsidies to promote EVs in comparison with conventional vehicles, such as exemption from registration tax, VAT, traffic congestion fees, annual vehicle tax, and road tolls. In the case of Norway, both financial and preferential policies have contributed to the success of EV adoption.

5.2.2 Advancement in Charging Infrastructure

Besides rolling out EV incentives, Norway has also invested in first-class charging infrastructure to support the transition to more sustainable vehicles. The electric car is affordable to use since the electric bill is lower than the gasoline bill. EV charging is subsidized as well. Earlier, it was heavily subsidized; now some rural areas get some additional benefits. The country has both privately owned and publicly owned EV charging stations. Nonetheless, most of those who purchase private EVs can charge them at home, whilst commercial vehicles and city buses charge at a depot. In Norway, almost every EV has a dedicated home charger, and many Norwegians charge their EVs at home [20, 21].

Norway's journey with EV charging infrastructure is a masterclass in forward-thinking policy and rapid adoption. The timeline of public EV charging infrastructure and how it unfolded from 2010 to 2025 is depicted in Table 5.2 [22–31]. The largest operators include Recharge, Mer, Eviny, Ionity, Ishavsveien, Tesla, and Circle K. From well-established networks like Fortum Charge & Drive and Tesla Supercharger Network to emerging players like Mer, the country offers a robust infrastructure to support the growing number of EV users. Selecting the appropriate network requires evaluating factors such as coverage, speed, cost, and user experience, as well as leveraging government incentives designed to encourage sustainable transportation. Norway's extensive and efficient EV charging networks make it a global leader in electric vehicle adoption. The Norwegian government has mostly encouraged a market-driven development of the charging infrastructure and

Table 5.2 Timeline of public EV charging infrastructure growth

Year	Key developments	References
2008	* First municipal charging system in Oslo	[22]
2010	* Introduction of popular EVs like Mitsubishi i-MiEV and Nissan Leaf sparked demand	[23]
2011	* Launch of NOBIL, Norway's national EV charging station database	[24]
	* Norway had approximately 3000 public charging stations, including around 150 fast chargers	[25]
2015	Over 10,000 charging stations installed nationwide	[26, 27]
2017	* The number of fast chargers along the road increased to 1500 Charging infrastructure development expanded to remote Northern region in 2019	[23, 25, 28]
2020	* The number of fast chargers had grown to nearly 1600	[25]
	* ABB Ltd. installed its 1000th DC fast charging station in Bergen (Norway)	
2022	* The total number of public charging stations had reached just under 23,800	[29]
	* Installed 5041 public fast charging stations, with approximately 700 funded through public subsidies by 2022	
2023	* More than 25,000 charge points available across the country	[26, 27]
	* Fast charging stations number reached 9000	
2024 (2nd quarter)	* Public charging points totaled 27,500	[30]
2025	* Mer to install 215+ new chargers at Coop stores, boosting retail infrastructure	[31]
Estimates for 2030	* Expansion of fast charging stations:	[6]
	10,000–14,000 (for passenger cars and vans)	
	1500–2500 (for heavy vehicles)	

intervened where there appeared to be a market failure through its policy supports to cover the development costs; there were no subsidies on operations of charging stations.

Initially, Oslo had been at the heart of testing new infrastructure, and in 2008 it launched the first municipal EV charging system [32]. In 2015, the Norwegian government (Ministry of Climate and the Environment) initiated a nationwide campaign to ensure that all major public roads had fast chargers spaced no more than 50 km apart [33]. In recent years, the expansion of public electric car charging stations has been considerable, totaling around 23,800, with 15,400 of those being Level 2 charging stations. This infrastructure supports around 817,500 registered electric and plug-in hybrid cars in Norway as of 2022 [94]. There were 503 Tesla Superchargers available that year. Additionally, there were 5041 publicly accessible rapid chargers (over 50 kW) for light vehicles (passenger cars and vans) in Norway as of the end of September 2022 [34, 35].

Earlier in 2017, Erik Lorentzen et al. reported on the status of public and private charging infrastructure investments and the user experiences from the 2017

Norwegian EV owners survey report, with a focus on the fast charging infrastructure [23]. The presence of impressive charging infrastructure (447 chargers per 100,000 people) in Norway is instrumental in alleviating range anxiety and supporting widespread EV adoption [20].

From being non-existent 10 years ago, there were 5041 publicly accessible rapid chargers (over 50 kW) for light vehicles (passenger cars and vans) in Norway as of the end of September 2022 [34]. It has over 27,500 public EV charging points as of mid-2024. The majority of charging stations consist of standard AC chargers, accounting for approximately 67.8% of the total. The country has set a goal to establish fast charging stations every 50 km on major highways [30]. As of the end of 2024, there were more than 9000 car chargers that can fast-charge at the same time. Out of 14 selected European countries, Norway has the fastest average electric vehicle charging speed, at 81 kW. This extensive network stretches over 1700 km from the southern cities to the northern Arctic Circle, ensuring comprehensive coverage and alleviating range anxiety for EV drivers. The salient features of charging infrastructure are highlighted in Fig. 5.2. Norway's charging infrastructure features a wide network of charging points, access to home charging points for almost all EVs, strong collaboration between public and private sectors, and availability of fast-charging stations at 50 km intervals [36–38].

The vehicle-to-public-charge-point ratio is about 24.7, which is below global averages due to Norway's strong private and workplace charging network. The ratio of fast charge points is higher than in many other countries, with 40% of chargers being DC. Almost 30% of chargers are high-power points (HPCs), with public funding devoted to installing HPCs every 50 km on main roads [39]. Norway is leading the EV charge segment of the clean e-mobility ecosystem. Generous incentives, a strong grid ensuring reliable and clean electricity (90% + hydropower), and

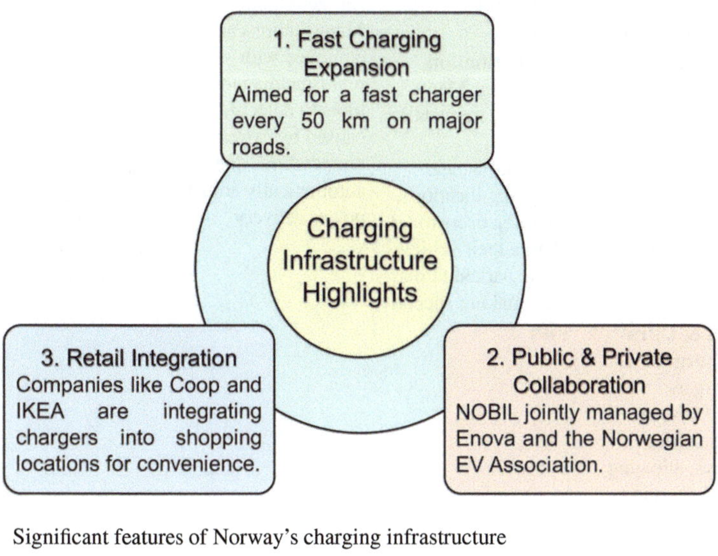

Fig. 5.2 Significant features of Norway's charging infrastructure

public acceptance of EVs (nearly 90% of total cars sold in 2024 were electric) are contributing to its standing as an EV charge leader. Norway's EV charging infrastructure is a blend of cutting-edge tech and practical innovation, tailored to its unique geography and climate. The breakdown of the key technologies powering the EV charge role is illustrated in Table 5.3.

Zaptec and Amina use dynamic load balancing to optimize electricity use, adjusting charging speed based on grid capacity and time-of-day pricing [40]. Electreon, an Israel-based company, launched the world's first inductive charging road for electric buses in Trondheim, Norway [41]. Further, Norway's EV infrastructure is built to withstand some of the harshest winter conditions in the world. EV chargers undergo rigorous testing to ensure performance in sub-zero temperatures.

Oslo Airport V2G Pilot: EV fleet providing grid support and peak-load management. Demonstrates financial and operational benefits [42]. In recent years, Oslo Airport tested the V2G (vehicle-to-grid) pilot program [42]. Here, EV fleets provide grid support and peak-load management, which is demonstrating financial and operational benefits. V2G helps decrease price volatility, reduce peak and valley

Table 5.3 Key technologies driving EV charge eco-system

Core charging technologies	State-of-the-art additions	Designed & built for harsh environments	Grid & energy integration
* **Smart Charging Systems**: Use dynamic load balancing to optimize electricity consumption by adjusting charging speeds according to grid capacity and time-of-day pricing	* **Wireless Charging Roads**: Introduced the world's first inductive charging roadway designed for electric buses. This reduces need for stationary charging	* **Cold-Weather Flexibility**: Chargers are designed to function reliably in snowy, icy conditions, with rigorously testing to guarantee performance in sub-zero temperatures Many stations are equipped with real-time diagnostics tools to detect faults caused by extreme temperatures and automatically adjust power delivery	* **Hydropower-Driven Grid**: More than 90% of Norway's electricity is generated from hydropower, ensuring that EV charging is incredibly clean and sustainable
* **RFID & App Integration**: Enable seamless access across networks by supporting RFID cards, mobile applications, QR codes, and SMS-based start/stop functions	* **Destination Charging**: Many chargers are installed at homes, workplaces, and shopping locations, enabling users to charge their vehicles while parked rather than making special trips		* **Vehicle-to-Grid (V2G) Potential**: As it continues to develop, Norway is exploring bidirectional charging technology that allows EVs to feed energy back to the grid during times of peak demand
* **Fast & Ultra-Fast Chargers**: DC fast chargers (50–350 kW) are prevalent along highways, allowing for rapid charging during long-distance travel			

electricity prices, reduce renewable curtailment, and reduce the cost of operation in the energy system [43]. Simultaneous EV charging, especially during peak hours, can overload transformers and cables. Furthermore, V2G demonstrates its effectiveness as a tool for increasing the integration of wind and solar PV generation into the grid, substantially reducing the need for curtailment. V2G also contributes to reducing system emissions. However, frequent charging/discharging may affect battery life, though smart algorithms can mitigate this.

With the expansion of technology and charging networks even further, the future looks bright for EV owners in Norway. Norway's approach isn't just about plugging in; it's about creating a seamless, sustainable, and scalable system. Its comprehensive strategy in EV charging infrastructure underscores its role as a global leader in sustainable mobility. Through strategic planning, robust infrastructure development, and supportive policies, Norway continues to pave the way for widespread adoption of electric vehicles while setting an example for other countries to follow in the quest for cleaner and more sustainable transportation solutions.

5.2.3 Domestic Impact of Policies on EV Adoption

Back in the early 1990s, the Norwegian parliament realized that electrifying road traffic alone would mean a huge cut in Norwegian climate emissions and a way to reduce climate emissions [44]. Norway's EV policies have had a transformational impact on adoption, turning the country into a global trailblazer for electric mobility [45, 46].

Norway's success in rapidly increasing the adoption of fully EVs is fundamentally tied to its progressive policies and incentives [47]. The country has implemented a robust approach to deter the use of petrol and diesel vehicles by imposing high taxes, while also actively promoting EV adoption. This combination of favorable incentives and policies has a diverse and significant impact, as displayed in Fig. 5.3.

Undeniably, large subsidies, financial incentives, charging infrastructure advancement, and vision played an important role in attracting more and more customers to buy EVs, which in turn aimed at reducing pollution [48]. In 2024, EVs in Norway accounted for 88.9% of new cars sold, up from 82% in 2023, according to the Norwegian Road Federation. In January 2025, the car registration figure hit 98% of new cars that were electric, with fossil fuel car sales almost evaporating. Norway is within a touching distance of its 2025 goal (set in 2017) to have 100% of new cars powered by electricity, the first country in the world to do so. However, Costanza Cincotta et al. warned in their recent study that undoing the incentive with the largest effect (the EV exemption from purchase taxes) would reduce the EV market share to 25% from the 66% observed in 2021, increase CO_2 emissions of new cars sold by 167%, reduce their total weight by 22%, and reduce the number of new cars sold by 10% [49]. These findings are based on the data in 2021. Anni Yang et al. showed that the BEV adoption in Norway has a significant clustering pattern.

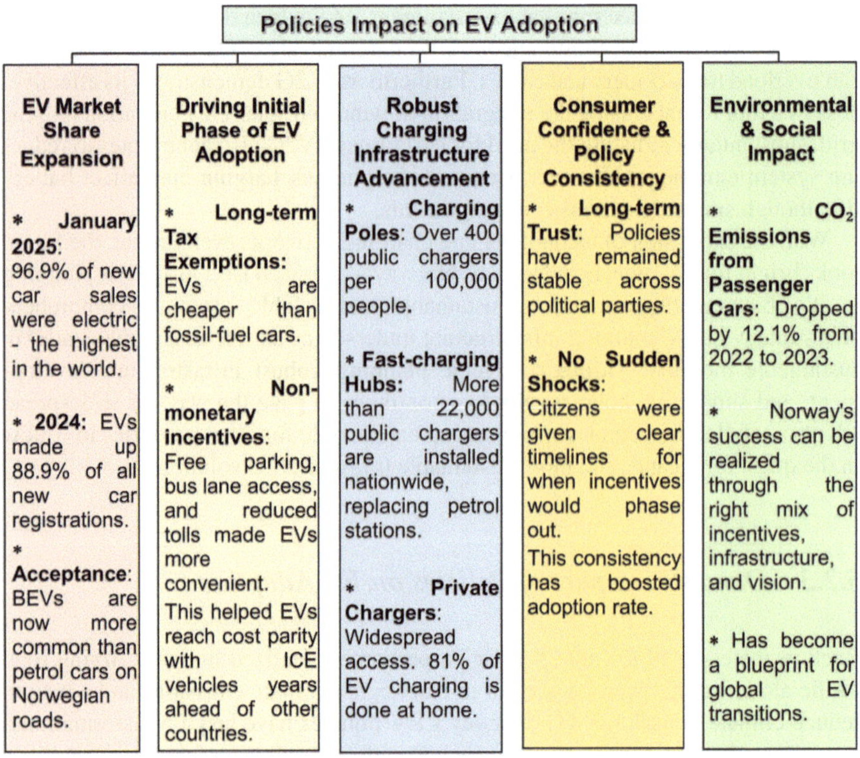

Fig. 5.3 Domestic impact of Norway's policies on EV adoption

Norwegians living in urban areas that offer more charging stations, as well as those with higher income levels and greater travel demand, are more inclined to adopt battery EVs. Conversely, municipalities characterized by lower minimum temperatures and a higher percentage of elderly residents tend to experience a lower rate of battery EV adoption [50].

The transition to battery EVs in Norway has had a considerable positive influence on public health and the environment, but there are still issues that must be addressed. On the bright side, BEVs lower harmful pollutants and greenhouse gas emissions, resulting in better air quality and health outcomes, particularly in vulnerable communities. This move also helps to support global efforts to mitigate climate change and its associated health hazards [51].

A long-term financial commitment to fund incentives and infrastructure programs that supported the transition. Norway has demonstrated that the transition to EVs is not only possible but also follows familiar patterns of technological diffusion. In summary, Norway didn't just adopt EVs; it redefined what's possible. Norway could become a role model for decarbonizing road transport for other countries.

5.2.4 A Global Inspiration

As Norway aims to achieve its 2025 goal of 100% zero-emission new car sales, the country serves as a blueprint for other nations seeking to transition to fully electric vehicles. Norway's achievements are a testament to the effectiveness of coordinated policies, financial incentives, and infrastructure development in driving the adoption of fully electric vehicles. The European Union's decision to ban the sale of carbon-emitting vehicles by 2035 echoes Norway's pioneering efforts, with many countries looking to replicate its success. As the country approaches closer to its 2025 goal of achieving zero emissions, it serves as an inspiring model for others globally. By focusing on environmental sustainability and increasing the accessibility of fully electric vehicles, Norway is not only shaping the future of transportation but also accelerating its arrival.

5.2.5 Next Phase of Norway's EV Transition

According to the Institute of Transport Economics, users of EVs have to deal with up to 20 different operators, 20–30 apps, and 13 payment systems to access all charging infrastructure in Norway [52]. To make payments, users can either create an account or download an app for accessing chargers. While most public charging stations accept credit card payments, some only take RFID (Radio Frequency Identification) cards. Customers often find it inconvenient to manage multiple mobile apps and payment systems. In 2022, the Ministry of Culture and Equality provided funds to Standards Norway to prepare a report on universal design for charging infrastructure. The purpose of this study aims to ensure that all drivers, including those with special needs, can easily use EVs. Charging a vehicle should be as simple as refueling a conventional vehicle [45]. Furthermore, it is important to open more supercharger networks that can accommodate all types of EVs in Norway.

Although Norway has achieved remarkable success and positively influenced consumer behavior regarding EVs, the penetration of EVs in other regions remains considerably lower. The high market shares of EVs in Norway are in stark contrast to the situation in other countries. In 2024, EV market share was just 13.6% in Europe and 6% in the US [53].

Consistency and certainty in EV policies could significantly influence the EV market and its impact on the environment. According to a recent report by Costanza Cincotta et al., reversing the most effective incentive, the EV exemption from purchase taxes, would reduce the market share of EVs from 66% in 2021 to just 25%. This change would also increase CO_2 emissions by 167% from new cars sold, a 22% reduction in their overall weight, and a 10% decrease in the number of new cars sold in Norway [49].

5.3 Case Study of China

China's EV industry has become a global powerhouse, driven by innovation, government incentives, and growing demand for sustainable transportation [54]. Electric car sales in China increased by almost 40% year-on-year in 2024, further driving up China's share of global electric car sales [55]. In recent years, sales of plug-in hybrid electric cars have been growing faster than sales of battery electric cars in China.

China has set ambitious long-term policies to support its EV transition beyond 2025. Policies for EV transition, such as (i) carbon neutrality goals [56], (ii) NEV tax exemptions [57], (iii) investments in renewable energies [58], (iv) charging infrastructure expansion [58], and (v) coal consumption reduction strategy [58], are depicted in Fig. 5.4.

To achieve carbon neutrality by 2060, EVs play a crucial role in reducing transportation emissions. Providing financial incentives for EV adoption helps to ensure clean energy vehicles on the road. Further, a sustainable clean power supply for EV charging reduces the carbon footprint. Grid modernization and expansion of EV charging networks to support the growing number of EVs ensure facile charging and reduce the consumers' anxiety. Although coal is still a major component of China's energy mix, the government aims to gradually eliminate it in favor of renewable energy sources, ensuring a cleaner energy supply for EVs. All these policies reflect China's commitment to maintaining its leadership in the global EV industry while ensuring a sustainable and low-carbon future.

Key policies influencing EVs transition				
(1) Carbon Neutrality Goals	(2) Extended NEV Tax Exemptions	(3) Renewable Energy Investments	(4) Grid & Charging Infrastructure Expansion	(5) Coal Reduction Strategy
Aims to reach peak carbon emissions before 2030 and attain carbon neutrality by 2060. EVs are vital in lowering emissions from the transportation sector	NEV purchase tax exemption extended until 2027, providing financial incentives for EV adoption	Continues to invest heavily in clean energy, ensuring a sustainable power supply for EVs.	Significant funding allocated to grid modernization and EV charging networks to support the growing number of EVs	Government plans to gradually phase out coal from its energy mix in favor of renewables, ensuring a cleaner energy supply for EVs.
Leading to sustainable and low-carbon future				

Fig. 5.4 Key policies fueling the EV transition in China

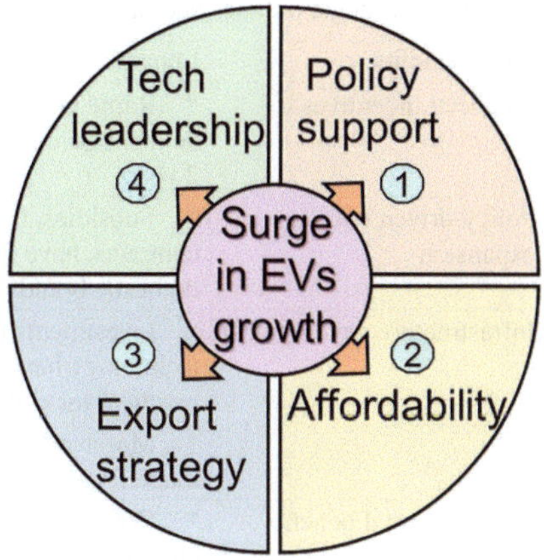

Fig. 5.5 Key factors driving surge in EV growth in China

5.3.1 Incentives and Subsidies Impact on EV Growth

In 2024, a research study by Farheen Ehsan et al. extensively demonstrated the significant and positive impact of two distinguished types of EV policies, including financial policies and preferential policies, on consumers' intention to adopt EVs by implementing an extended and improved version of the theory of planned behavior (TPB) [59]. They showed the significance of policies and their role in the popularization of EVs in China, mainly in Shanghai.

China's EV market has experienced *explosive growth* from 2015 to 2025. The driving surge in the EV ecosystem is mainly due to the policy support, including technology advancement and affordability, as illustrated in Fig. 5.5 [55–60]. Long-term subsidies, emissions targets, local-level mandates, innovation in battery technology, ultra-fast charging infrastructure, affordability compared to their gasoline counterparts, global export strategy, etc. are some of the attributes responsible for driving the surge in EV growth.

China's EV policies have had a transformative impact on adoption, making the country the largest EV market globally. These NEV policies impact on the growth of EVs is illustrated in Table 5.4 [61–63]. The country sold a record-breaking one million EVs (fully battery-electric and plug-in hybrids) in May 2025 alone [64].

5.3.2 Advancement in Battery Range

The impact of policies and subsidies on the growth of the domestic EV market is clearly illustrated in Fig. 5.6. Batteries make up approximately 40% of an EV's cost, highlighting their fundamental importance to the overall notion of EVs [65]. China has made significant advancements in EV battery technology over the past decade [63, 66–68]. The battery range on a single charge is vital to avoid the consumers' anxiety. Advancement in battery range in China (on a single charge) is shown in

Table 5.4 Strategic policies and incentives impacting the growth of EV market share in China

Tools/benefits	Impact	References
Financial incentives	* Rapid Growth in EV Sales: EV sales, with annual figures rising from 1.3 to 6.8 million in just 2 years	[61]
Policy-driven market expansion	* Subsidies, tax exemptions, and procurement contracts, have fueled demand and encouraged domestic brands to innovate.	[61]
Infrastructure development	* Investments in charging stations and battery technology have made EVs more accessible and practical for consumers	[62]
	* Massive investments in battery recycling have made EVs more practical for everyday use	
Environmental benefits	* Contributed to reducing air pollution and aligning with China's net-zero emissions goals	[62]
Global influence; trade & export strategies	* China's dominance in EV production has positioned it as a leader in climate policy, influencing other nations to adopt similar strategies	[62]
	* Leveraged its EV leadership to strengthen international trade agreements	
Battery technology	* Provided substantial subsidies to both manufacturers and consumers, making EVs more affordable and boosting production	[63]
Government procurement & fleet electrification	* Ensure stable demand and market growth	[64]

Fig. 5.6a. China's EV battery industry continues to be dominated by CATL and BYD, with CATL holding a 45.9% market share.

5.3.2.1 Advancement in Battery Charging Time

In 2015, electric vehicle (EV) charging in China was still in its early stages, and charging times varied widely depending on the type of charger and vehicle. Advancement in the battery charging time in China is illustrated in Fig. 5.6b. The battery charging time reduced from 1–2 h to just 30–40 min. Recently, researchers in China reported integrated assessment models and real-world EV charging data to study the operating patterns of existing fast and potential future ultrafast charging stations [69]. For more typical stations (e.g., 8–9 chargers and 120 kW/charger), upgrading chargers to 350–550 kW while allowing managed dynamic waiting strategies (of ~1 min) can reduce overall charging times to ~9 min. A 5-min charge for an EV, virtually as quick as a gasoline fill-up, has long seemed a quixotic dream. Companies like CATL and BYD have unveiled batteries capable of adding 250–320 miles of range in just 5 min, using advanced lithium-iron-phosphate (LFP) chemistry [70]. In 2025, BYD unveiled a new EV platform and charger combo capable of delivering up to 292 miles (470 km) of range in just 5 min [71].

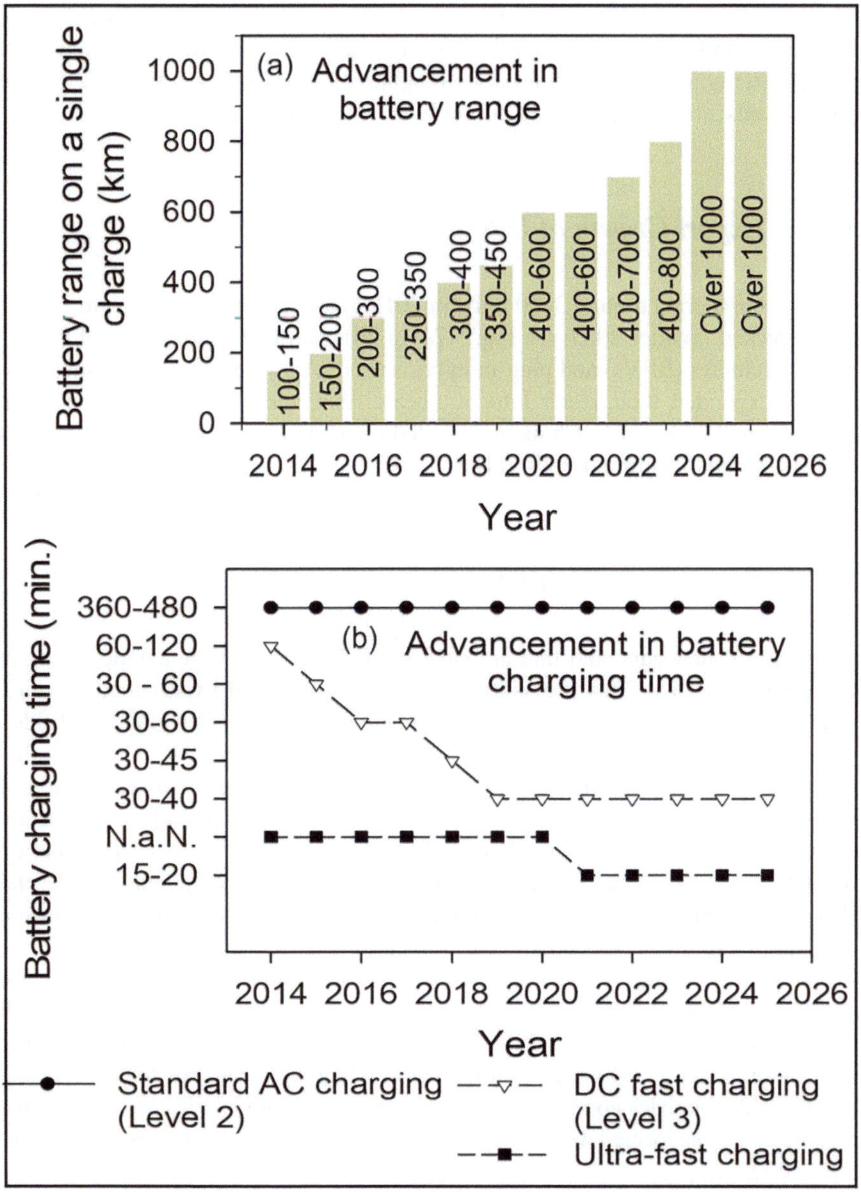

Fig. 5.6 Policies and subsidies impact on the growth of the domestic EV market. (**a**) Battery range advancement (on a single charge) in China from 2014. (**b**) Advancement in the battery charging time in China from 2014

In cities such as Shanghai, drivers can swap EV batteries in less than 3 min at automated Nio stations. Coupled with their low maintenance and operating expenses, EVs have become the norm for urban professionals and drivers. Further,

China is integrating EVs into its power grid. Through pilot programs in cities like Shenzhen, BYD is enabling vehicle-to-grid (V2G) tech—meaning EVs don't just pull electricity, they can push it back. Imagine millions of EVs acting like mobile power plants, stabilizing the grid during high demand [60].

5.3.3 Export Growth of EVs

China just exported around 3000 units in 2015, but later the export growth significantly picked up after 2020; see Fig. 5.7a. In 2019, around 25,000 units of battery electric vehicles (BEVs) and plug-in hybrids (PHEVs) were exported. Chinese EV makers expanded their reach into Southeast Asia, the Middle East, and Latin America in 2024 [72, 73]. China is projected to export around 2.3–2.4 million EVs in 2025, based on current trends [74].

5.3.4 Domestic Sales of EVs

The share of new cars sold that are electric in China (or EV penetration rates in China) from 2013 to 2024 is displayed in Fig. 5.7b [75]. 2014 was called the "first year" of the development of Chinese EVs (fully battery-electric and plug-in hybrids), with annual sales reaching 74,763, 3.2 times higher than in 2013. Domestically, China sold around 75,000 EVs and hybrids in 2014, making it the second-largest EV market globally at the time [60]. As of May 2025, China has sold approximately 7.2 million electric vehicles, marking a 28% increase compared to the same period in 2024 [76]. As of early 2025, China's NEV segment, comprising BEVs and PHEVs, accounts for nearly half of all passenger vehicle sales. Prominent domestic companies like BYD, NIO, XPeng, and Geely are leading the charge in the industry, inventing new battery technologies, smart driving solutions, and strategies for global expansion. It is anticipated that sales of new energy vehicles in China will exceed 16 million units by 2025, as the market enters into a stable growth phase with more moderate annual growth rates [77]. Ongoing incentives for the replacement of older vehicles combined with declining prices for electric cars suggest that electric cars are expected to reach approximately 60% of total car sales in the country by 2025 [55].

5.3.5 Affordability of EVs

The EV prices are lower in China than in other countries, making them more accessible to the average consumer. The average cost of EVs in China from 2014 to 2025 reduced by 50%, as shown in Fig. 5.7c. At the same time, EVs performance and technical advancement over the years improved significantly. Affordability is

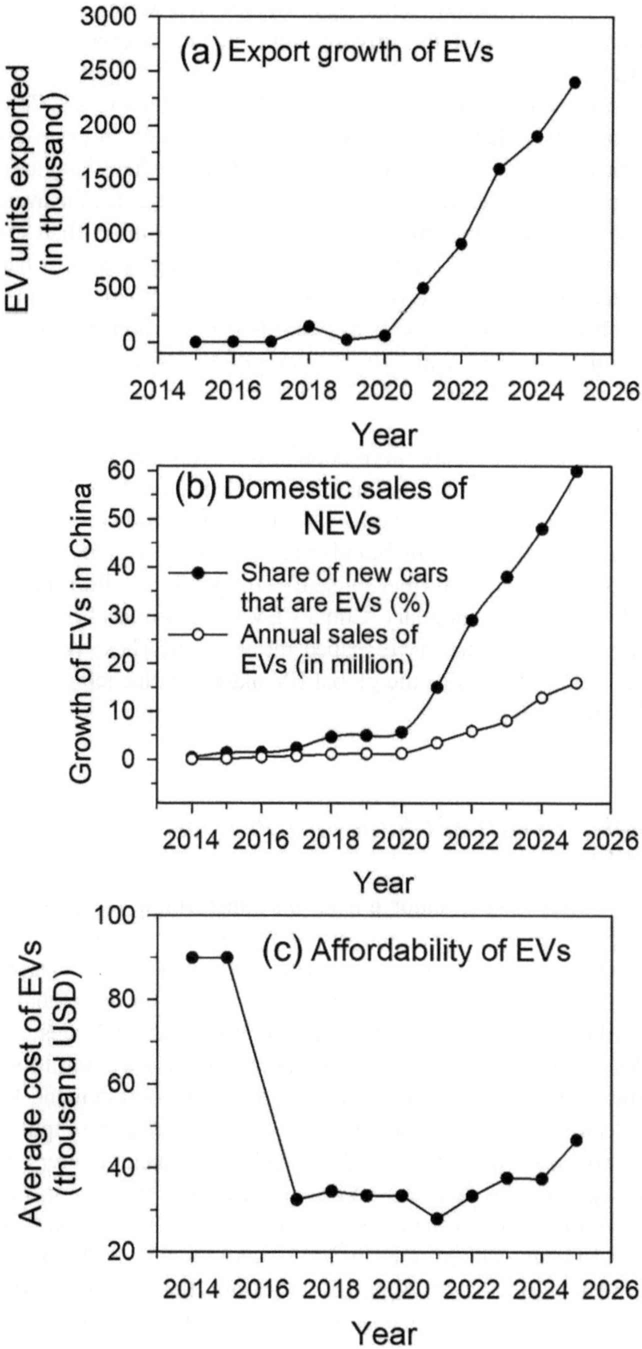

Fig. 5.7 Impact of policies and subsidies on the growth of the domestic EV market. (**a**) Export growth of Chinese EVs since 2014. (**b**) Annual sales of EVs (fully battery-electric and plug-in hybrids) in the Chinese market since 2014. (**c**) Affordability and average cost of EVs in China since 2014

accompanied by the advanced technology. Models like XPeng's Mona Max offer voice commands, autonomous driving, and in-car entertainment for under $20,000 [78].

As per Jato Dynamics, the average price of EVs in China was €67,000 in 2015, which has currently reduced to €32,000, a 50% reduction in price over the last 7 years. On the contrary, EV prices in the USA and Europe have increased by 12.5% and 17%, respectively, over the same time period [79]. In 2016, the average cost of an EV in China was estimated to be around $15,000 more than conventional gasoline-powered vehicles [80]. This price gap was largely due to battery costs, limited production scale, and early-stage technology. In 2020, the upfront cost of electric vehicles (EVs) in China was $5000–$17,000 higher than comparable gasoline-powered cars [81]. Similarly, the price of an EV in China is 33% cheaper than a petrol vehicle, while it is 27% and 43% higher in the USA and Europe, respectively, in 2023.

Currently, passenger cars are more popular than commercial vehicles, accounting for 95.1% of EV sales. The commercial vehicle sector is on the verge of widespread EV adoption; more than 30 cities are planning to fully electrify their public transit systems by 2028. Domestic brands like BYD lead, with an 81% share, while startups like Xpeng and Nio further intensify competition, solidifying China's EV production dominance. In summary, China's EV transition has been driven by several high-impact policies that have shaped the industry. These policies have propelled China to the forefront of the global EV market, influencing other nations to adopt similar strategies.

5.3.6 Global Impact of China's EV Program

China's EV policies have a major impact on transforming the global automotive industry, shaping supply chains, and reshaping international trade dynamics and market competition and will continue to do so in the near future also [82]. Indeed, Chinese manufacturers have made a significant impact on the global EV market and positioned China as a global leader in both production and sales [83–85]. This surge in EVs is impacting the global oil market and also leading to scrutiny and trade tensions with other countries. The IEA anticipates oil demand for cars will continue to decline, with the transition to EVs altering refinery outputs, electricity generation, and grid requirements [72, 86]. Chinese manufacturers are opening new overseas production facilities and continuing their expansion into overseas markets. Brazil may witness the commencement of EV production facilities from both BYD and Great Wall Motors in 2025. Additionally, BYD plans to initiate production in Hungary in 2025. Figure 5.8 illustrates the key impacts on the global market, influenced by government policies, technology advancements, and domestic demand.

China is well placed as a world leader in EV production and the world's largest exporter of EVs. The influx of affordable Chinese EVs into the international market is fueling competition with the local manufacturers, leading to trade tension. The

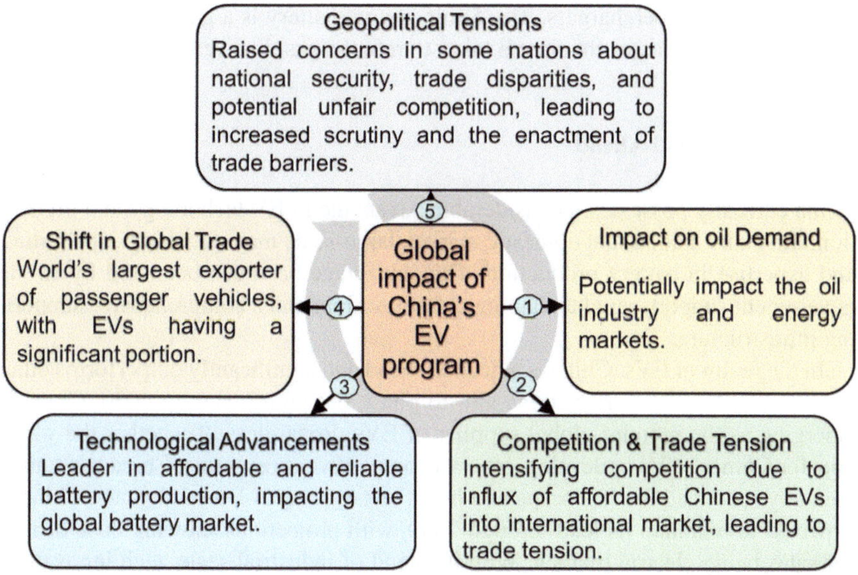

Fig. 5.8 Key impacts on the global auto ecosystem because of Chinese government EV policies and technology advancements

U.S., EU, and Canada have imposed high tariffs on Chinese EVs, citing concerns over subsidies and market overcapacity. China has responded with countermeasures, escalating trade disputes [87]. Subsequently, China imposed export restrictions on rare earth elements, which are critical for EV batteries and motors, causing concerns among automakers worldwide. Several manufacturers like Ford and GM have feared that these restrictions could slow production and increase costs [4]. China focused on LFP battery technology and is the leader in affordable and reliable battery production chains. Its dominance in battery technology and manufacturing efficiency continues to shape global EV trends [65]. Additionally, China's aggressive push for EV adoption is forcing traditional automakers to accelerate their transition to electric models. As a result, companies worldwide are investing in alternative battery technologies to reduce reliance on Chinese supply chains [88].

In conclusion, China's EV program is a major force shaping the global automotive industry, impacting trade, technology, and energy markets. While it presents challenges for some, it also offers opportunities for developing countries and encourages innovation and competition worldwide.

5.3.7 Technology Advancement and Future

In early 2025, BYD revealed a groundbreaking battery capable of charging 400 km (249 miles) of range in just 5 min. Featuring its innovative "Super e-Platform," this technology enables a charging capacity of 1000 kW, which is double the power of

Tesla's top-tier Superchargers. The 5-min charge battery is a game-changer, offering speed that surpasses the time it takes to refuel a gasoline vehicle [60].

5.3.7.1 The Road Ahead

China currently possesses a considerable advantage in EV technology and production, driven by a dominant domestic market, large-scale manufacturing capabilities, and expertise in battery production. This advantage has been bolstered by strong government support, rapid technological innovation, and a comprehensive supporting infrastructure.

In the realm of EVs, China is a decade ahead and significantly outperforms other countries. While Western nations aim to phase out petrol cars by 2030, China is emerging as the premier global supplier of EVs. Regardless of whether the world opts for Chinese EVs or develops its own competitive ecosystems, China has fundamentally redefined the landscape and holds an advantage of several years. China is expected to maintain its lead through 2030, with projections showing 80% of new car sales being electric by then. With its blend of industrial scale, tech innovation, and policy alignment, China isn't just shaping its own EV future, it is reshaping the global auto industry. The Chinese companies envision a future where they dominate car manufacturing for the global market. However, the government subsidies fueling China's EV growth are viewed as unfair by countries seeking to safeguard their automotive industries. In response, the US, Canada, and the European Union have imposed considerable import tariffs on Chinese EVs.

5.4 Conclusion

Norway's effective alignment of policies, financial incentives, and infrastructure expansion is propelling the adoption rate of fully EVs. This approach serves as a powerful model for other countries worldwide. Norway aims to achieve its 2025 goal of 100% zero-emission new car sales and is poised to serve as a blueprint for other nations seeking to transition to fully EVs. In this regard, Norway serves as a prime example, demonstrating that the transition to EVs is not only achievable but also aligns with established patterns of technological diffusion. Norway hasn't simply adopted EVs; it has redefined the possibilities surrounding them. Its methodology could position Norway as a role model for other countries aiming to decarbonize road transport.

China is not only defining its own EV landscape, but is also transforming the global auto industry. The Chinese envision a future where they produce all the cars for the world market. The government subsidies fueling China's EV growth are perceived as unjust by nations striving to protect their automotive industries. In response, the US, Canada, and the European Union have all enacted significant import tariffs on Chinese EVs.

China, now the dominant force in EV manufacturing, exported a record $15 billion in vehicles in 2023. The US is risking ceding global leadership in electrified transport to China. The rollback of U.S. electric vehicle incentives under the *One Big Beautiful Bill Act* is creating a vacuum in global clean mobility leadership, and China is swiftly stepping in to fill it.

China has emerged as the dominant force in EV manufacturing, exporting a record $15 billion in vehicles in 2023. On the global stage, China is at the forefront of the drive for fleet electrification. This dominance poses a risk for the US, which may find itself ceding its leadership in the electrified transport sector to China. China is not only poised to become a global leader in clean mobility; it is actively shaping that future. However, as China surges ahead in the clean mobility competition, it may encounter challenges.

In conclusion, there is no universally ideal policy combination to ensure the effective decarbonization of the transport sector. The effective coordination between policies, financial incentives, and infrastructure development are driving the adoption rate of electric vehicles.

References

1. Axon G (2021) How a pop group brought EVs to Norway—Axon's Automotive Anorak. Goodwood Road & Racing. 29 Jan. https://www.goodwood.com/grr/road/news/how-a-pop-group-brought-evs-to-norway%2D%2Daxons-automotive-anorak/
2. Liang A, Marsh N (2025) Tracing the origins of China's dominance in the electric vehicle market. Energy in Demand. 27 Jun. https://energyindemand.com/2025/06/27/tracing-the-origins-of-chinas-dominance-in-the-electric-vehicle-market/
3. CICC Research, CICC Global Institute (2024) China's auto industry to grow from large to strong. In: The reshaping of China's industry chains. Springer, Singapore, pp 273–290. https://doi.org/10.1007/978-981-97-1647-0_14
4. Luman R, Manthey E (2025) China's export restrictions on rare earths causes alarm for automotive industry. ING, Bank N.V. 10 June 2025. https://think.ing.com/articles/chinas-crackdown-on-rare-earth-causes-alarm-automotive-industry/
5. Norway is electric. Ministry of Transport Government of Norway. 15 Jan. https://www.regjeringen.no/en/topics/transport-and-communications/veg/faktaartikler-vei-og-ts/norway-is-electric/id2677481/
6. Status for zero-emissions vehicles. The Norwegian Public Roads Administration (Statens Vegvesen). https://www.vegvesen.no/en/fag/fokusomrader/baerekraftig-mobilitet/zero-emission-targets/
7. Randall C (2024) Norway Saw 4,717 New Electric Cars in January 2024. 2 Feb. https://www.electrive.com/2024/02/02/norway-saw-4717-new-electric-cars-in-january-2024/
8. Norway's evolving incentives for zero emission vehicles (2022) International programme for action on climate-policies in practice. OECD. 7 Nov. https://www.oecd.org/en/publications/ipac-policies-in-practice_22632907-en/norway-s-evolving-incentives-for-zero-emission-vehicles_22d2485b-en.html
9. Zeniewski P (2017) Electric vehicle policy in Norway. ClimateXChange, Scotland's Centre of Expertise on Climate Change, Scottish Government. Feb. https://www.climatexchange.org.uk/wp-content/uploads/2023/09/eu_case_studies_norway_transport.pdf

10. Bello C (2023) Europe's electric future: how Norway came to lead the charge in the EV revolution. Euro News. 31May. https://www.euronews.com/next/2023/05/31/europes-electric-future-how-norway-came-to-lead-the-charge-in-the-ev-revolution
11. Norwegian EV policy. The Norwegian EV Association. https://elbil.no/english/norwegian-ev-policy/
12. Kuruthukulangara J (2025) Seeking alpha, EV success: Norway on track to be first country to go fully electric. 28 Jan. https://seekingalpha.com/news/4399127-ev-success-norway-on-track-to-be-first-country-to-go-fully-electric#comments
13. Murray A (2025) Reporting for BBC from Oslo. Norway on track to be first to go all-electric. 13 Jan. https://www.bbc.com/news/articles/cg52543v6rmo
14. Norway to be first to go all-electric for cars. Natural World Fund, England. 14 Feb. https://naturalworldfund.com/norway-to-be-first-to-go-all-electric-for-cars/
15. Electricity, Statistic Norway. 28 Feb 2024. https://www.ssb.no/en/energi-og-industri/energi/statistikk/elektrisitet
16. Bu C (2022) What Norway can teach the world about switching to electric vehicles. TIME. 7 Jan. https://time.com/6133180/norway-electric-vehicles/
17. OECD (2022) Norway's Evolving Incentives for Zero-Emission Vehicles. International Programme for Action on Climate. Policies in Practice. 7 Nov. https://www.oecd.org/en/publications/2021/09/ipac-policies-in-practice_1a65968e/norway-s-evolving-incentives-for-zero-emission-vehicles_c74c1e9d.html#:~:text=In%202021,%20about%20two-thirds%20of%20new
18. Asif M, Jajja MSS, Searcy (2021) A review of literature on the antecedents of electric vehicles promotion: lessons for value chains in developing countries. IEEE Trans Eng Manag 70 3697-3710. DOI: https://doi.org/10.1109/tem.2021.3099070
19. Vega-Perkins J, Newell JP, Keoleian G (2023) Mapping electric vehicle impacts: greenhouse gas emissions, fuel costs, and energy justice in the United States. Environ Res Lett 18(1):014027. https://doi.org/10.1088/1748-9326/aca4e6
20. Schulz F, Rode J (2022) Public charging infrastructure and electric vehicles in Norway. Energy Policy 160:112660. https://doi.org/10.1016/j.enpol.2021.112660
21. Dnistran I (2024) Here's why Norway hasn't had trouble with winter EV charging. Inside EVs. 19 Jan. https://insideevs.com/news/705338/norway-winter-ev-charging-no-trouble/
22. Figenbaum E, Kolbenstvedt M (2013) Electromobility in Norway—experiences and opportunities with electric vehicles. Norwegian Centre for Transport Research. TØI Report: 1281/2013. https://www.toi.no/getfile.php?mmfileid=33828
23. Lorentzen E, Haugneland P, Bu C, Hauge E (2017) Charging infrastructure experiences in Norway—the worlds most advanced EV market. EVS30 International Battery, Hybrid and Fuel Cell Electric Vehicle Symposium, Stuttgart, Germany, 9–11 October 2017. https://elbil.no/wp-content/uploads/2016/08/EVS30-Charging-infrastrucure-experiences-in-Norway-paper.pdf
24. Welcome to the charging station database NOBIL (which is developed in a cooperation between Enova and The Norwegian EV Association). https://info.nobil.no/english
25. Yatsura A (2025) Driving change: how Norway's EV experience can guide global transitions. In Mostafa R, Raz G (eds) Lawrence National Centre Policy Paper 25201, Ivey Business School, Canada. https://www.ivey.uwo.ca/media/5lobf4h5/western-drivingchange-14-sm-final.pdf
26. National Charging Strategy. Norwegian Ministry of Transport. Publica number: N-0582E. https://www.regjeringen.no/contentassets/26d4c472862342b69e8d49803b45c36a/en-gb/pdfs/national-charging-strategy.pdf. Accessed 8 July 2025
27. National Charging Strategy. Ministry of Transport, Government of Norway. https://www.regjeringen.no/en/dokumenter/national-charging-strategy/id2950371/?ch=3. Accessed 8 July 2025
28. Norway leads charge in EV adoption. Wallbox, Barcelona. https://wallbox.com/en/blog/how-norway-became-a-global-ev-leader

29. Carlier M (2025) Number of electric car charging stations in Norway by type. Statista. 30 Jun. https://www.statista.com/statistics/696548/number-of-electric-car-charging-stations-in-norway-by-type/. Accessed 8 Jul 2025.
30. How many EV charging stations in Norway. AMPPAL (Zhejiang Anfu New Energy Technology Co., Ltd.) Wenzhou, Zhejiang. 20 Apr 2025. https://anfuenergy.com/how-many-ev-charging-stations-in-norway/
31. Munoz D (2025) Mer to install over 215 EV chargers at Coop Stores across Norway in 2025. EVMagz. 7 Jan. https://evmagz.com/mer-to-install-over-215-ev-chargers-at-coop-stores-across-norway-in-2025/
32. Figenbaum E, Assum T, Kolbenstvedt M (2015) Electromobility in Norway: experiences and opportunities. Res Transp Econ 50:29–38. https://doi.org/10.1016/j.retrec.2015.06.004
33. Enova and the climate and energy fund—policies. International Energy Agency. 6 Nov 2019. https://www.iea.org/policies/2617-enova-and-the-climate-andenergy-fund?s=1
34. National Charging Strategy. Norwegian Ministry of Transport. Publication number: N-0582E. https://www.regjeringen.no/contentassets/26d4c472862342b69e8d49803b45c36a/en-gb/pdfs/national-charging-strategy.pdf
35. National Transport Plan 2018–2029. Meld. St. 33 (2016–2017) Report to the Storting (White Paper), Norwegian Ministry of Transport and Communications. https://www.regjeringen.no/contentassets/7c52fd2938ca42209e4286fe86bb28bd/en-gb/pdfs/stm201620170033000engpdfs.pdf
36. EV charging infrastructure best practice: what we can learn from Norway. Mer, London. 15 Aug 2024. https://uk.mer.eco/news/ev-charging-infrastructure-best-practice-learnings-from-norway/
37. Yousuf AKM, Wang Z, Paranjape R, Tang Y (2024) An in-depth exploration of electric vehicle charging station infrastructure: a comprehensive review of challenges, mitigation approaches, and optimization strategies. IEEE Access 12:51570–51589. https://doi.org/10.1109/ACCESS.2024.3385731
38. EV charging index: expert insight from Norway. Roland Berger GmbH, Munich. 15 Aug 2023. https://www.rolandberger.com/en/Insights/Publications/EV-Charging-Index-Expert-insight-from-Norway.html. Accessed 9 Jul 2025
39. Guttman B (2024) EV charging index: expert insight from Norway. Roland Berger GmbH, Munich. 28 Aug. https://www.rolandberger.com/en/Insights/Publications/EV-Charging-Index-Expert-insight-from-Norway-2024.html
40. Zaptec Pro is a smart, flexible charging station for larger parking spaces. Business Norway. 21 Oct 2022. https://businessnorway.com/solutions/zaptec-charging-system-for-multiple-electric-cars
41. Khan F (2024) Norway got its first-ever wireless charging road. MyElectricSparks [MES], Islamabad. 24 Jul. https://myelectricsparks.com/norway-got-its-first-ever-wireless-charging-road/
42. V2G Hub. V2G around the World. https://www.v2g-hub.com/insights/. Accessed 9 Jul 2025
43. Nagel NO, Jåstad EO, Martinsen T (2024) The grid benefits of vehicle-to-grid in Norway and Denmark: an analysis of home- and public parking potentials. Energy 293:130729. https://doi.org/10.1016/j.energy.2024.130729
44. Haug T, Rotevatn MN (2024) How policies changed the game for electric vehicles in Norway. Clean Mobility Shift, India. 25 Nov. https://cleanmobilityshift.com/ecosystem/how-policies-changed-the-game-for-electric-vehicles-in-norway/
45. Prendergast R (2025) Why is Norway's EV adoption ahead of the rest of Europe? Power Technology, New York. 17 Apr. https://www.power-technology.com/features/why-is-norways-ev-adoption-ahead-of-the-rest-of-europe/
46. Nolan S (2025) Norway's EV Dominance: a roadmap for global success. EV Magazine. 20 Mar. https://evmagazine.com/news/norways-ev-dominance-a-roadmap-for-global-success

47. Nearly 90% of Norway's new cars are fully electric vehicles. Innovation New Net Work, Pan European Networks Ltd UK. 15 Jan 2025. https://www.innovationnewsnetwork.com/nearly-90-of-norways-new-cars-are-fully-electric-vehicles/
48. Asgarian F, Hejazi SR, Khosroshahi H (2023) Investigating the impact of government policies to develop sustainable transportation and promote electric cars, considering fossil fuel subsidies elimination: a case of Norway. Appl Energy 347:121434. https://doi.org/10.1016/j.apenergy.2023.121434
49. Cincotta C, Thomassen Ø (2025) Evaluating Norway's electric vehicle incentives. Energy Econ 146:108490. https://doi.org/10.1016/j.eneco.2025.108490
50. Yang A, Liu C, Yang D, Lu C (2023) Electric vehicle adoption in a mature market: a case study of Norway. J Transp Geogr 106:103489. https://doi.org/10.1016/j.jtrangeo.2022.103489
51. Okesanya J, Saclolo JMB, Mia KBP, Ntacyabukura B, Corman V, Ahmadi A, Nugraha RR, He J, Telin JM, Tshering UU, Abdullahi Y, Ogaya JB, Delos Santos FE, Pedrajas-Mendoza SA, Magramo MM, Lucero-Prisno DE 3rd, Kouwenhoven MBN (2024) Norway's battery electric vehicles and public health- findings from the literature. Environ Health Insights 18:11786302241238171. https://doi.org/10.1177/11786302241238171
52. Wangsness PB, Figenbaum E (2022) Norwegian Center for Transport Research, Institute of Transport Economics, Oslo 2022. TØIReport 1867/2022. https://www.toi.no/getfile.php/1373853-1664456499/Publikasjoner/T%C3%98I%20rapporter/2022/1867-2022/1867-Summary.pdf. Accessed 23 Apr 2024
53. Revving up for a green future: mapping India's EV Landscape & Progress. Inc42, 22 Apr 2023. https://inc42.com/features/revving-up-for-a-green-future-mapping-india-ev-landscape-progress/. Accessed 24 Mar 2024
54. Zhao X, Li X, Jiao D, Mao Y, Sun J, Liu G (2024) Policy incentives and electric vehicle adoption in China: from a perspective of policy mixes. Transp Res A Policy Pract 190:104235. https://doi.org/10.1016/j.tra.2024.104235
55. IEA (2025) Global EV outlook 2025. IEA, Paris. https://www.iea.org/reports/global-ev-outlook-2025
56. Jessen J (2025) The role of EVs in China's 2035 decarbonisation plans. Sustainability, EV Magazine. 24 Apr. https://evmagazine.com/news/energy-evs-how-will-china-hit-its-2035-carbon-targets
57. China extends NEV tax reduction and exemption policy to 2027. China Briefing, Dezan Shira & Associates. 28 Jun. https://www.china-briefing.com/news/china-extends-nev-tax-reduction-and-exemption-policy-to-2027
58. IEA (2025) World energy investment 2025. IEA, Paris. https://www.iea.org/reports/world-energy-investment-2025
59. Ehsan F, Habib S, Gulzar MM, Guo J, Muyeen SM (2024) Innocent Kamwa. Assessing policy influence on electric vehicle adoption in China: an in-depth study. Energy Strategy Rev 54:101471. https://doi.org/10.1016/j.esr.2024.101471
60. Pocuca R (2025) How China's EV boom is charging past tesla and redefining the Global Auto Market. MOTORWATT (⚡ Green Energy Community). 7 Apr. https://motorwatt.com/ev-news/china-ev-boom
61. Yang Z (2023) How did China come to dominate the world of electric cars? MIT Technology Review. 21 Feb. https://www.technologyreview.com/2023/02/21/1068880/how-did-china-dominate-electric-cars-policy/
62. Abudu H, Sai R, Hossin MA, Botah EB (2024) China's electric vehicles adoption: implications for sustainable electricity, transportation, and net zero emissions. Front Sustain Energy Policy 3:1457743. https://doi.org/10.3389/fsuep.2024.1457743
63. IEA (2023) Global EV outlook 2023. IEA, Paris. https://www.iea.org/reports/global-ev-outlook-2023
64. Lewis M (2025) China breaks records as global EV sales hit 7.2 million in 2025. Electrek. 12 Jun. https://electrek.co/2025/06/12/china-breaks-records-as-global-ev-sales-hit-7-2-million-in-2025/

65. Yol VK (2024) The rise of China's EV sector and its impacts. Belt and Road Initiative Association for Economy and Culture, İstanbul-Türkiye. 24 Dec. https://www.kusakveyol.org/en/the-rise-of-chinas-ev-sector-and-its-impacts/#
66. Jin L, Wang S, He H (2023) Real-world performance of battery electric passenger cars in China. International Council on Clean Transportation (ICCT) Report, Washington. Apr. https://theicct.org/wp-content/uploads/2023/04/Passenger-cars-BEVs-real-world_final.pdf
67. China's electric vehicle batteries: an analysis of 2023 from January to June. CJO Global, Beijing. https://www.cjoglobal.com/2023/07/31/electric-vehiclechinas-electric-vehicle-batteries-an-analysis-of-2023-from-january-to-june/
68. 7 long-range electric cars from China; 2024 update. Foshan Guangcai Auto Trading Company Ltd. Guang Dong Sheng, Guangdong Province. 8 Aug 2024. https://guangcaiauto.com/chinese-electric-cars-with-longest-range/
69. Zhao Y, Chen X, Liu P, Nielsen CP, McElroy MB (2025) Future ultrafast charging stations for electric vehicles in China: charging patterns, grid impacts and solutions, and upgrade costs. Engineering 48:309–322. https://doi.org/10.1016/j.eng.2025.01.015
70. Pierce B (2025) China's 5-minute EV charging is a game-changer. EVinfo.net. https://evinfo.net/2025/04/chinas-5-minute-ev-charging-is-a-game-changer/
71. Khollam A (2025) Tesla on notice: BYD's new EV battery delivers 292 miles range in 5-minute charge. Interesting Engineering. 17 Mar. https://interestingengineering.com/energy/china-byd-unveils-5-minute-fast-charging-ev-battery-tech
72. Huang B, Xia L (2024) China: EV sector: forging ahead amid intensifying headwinds. BBVA Research, Hong Kong. 27 Jun.
73. Weihuan Z, Gao H (2024) Major economies are taking aim at China's EV industry. Here's what to know. World Economic Forum. 16 Sep. https://www.weforum.org/stories/2024/09/major-economies-are-taking-aim-at-china-s-ev-industry-here-s-what-to-know/
74. Gershom SN (2025) China's EV exports grow by 19% in 2025—what it means for the UAE & GCC. DUBI CARS. 10 Jun. https://www.dubicars.com/news/china-ev-exports-2025-growth-uae-gcc-impact.html
75. International Energy Agency (2025) Global EV outlook 2025. Processed by Our World in Data. Share of new cars that are electric. Global EV outlook 2025. IEA, Paris. https://www.iea.org/reports/global-ev-outlook-2025
76. China Electric Vehicle Industry Report, 2014–2015. Research in China, Beijing. Apr 2015. http://www.researchinchina.com/htmls/report/2015/10092.html
77. Si M (2025) NEV sales in China seen topping 16m this year. China Daily, Hong Kong. 1 Apr. https://www.chinadailyhk.com/hk/article/608387
78. How China Made Electric Vehicles Mainstream—and took over the Global EV Race. AllAroundWorlds. 23 Jun 2025. https://www.allaroundworlds.com/china-electric-vehicles-leader-2025/
79. EV Price Gap: a divide in the Global automotive industry. JATO Dynamics, Uxbridge. 2023. https://www.jato.com/hubfs/Q4%202023%20%20EV%20Price%20Gap%20Report/JATO%20mini%20Paper%2c%20EV%20price%20gap%2c%20A%20divide%20in%20the%20global%20automotive%20industry.pdf
80. Lu J (2018) Comparing U.S. and Chinese Electric Vehicle Policies. Environmental and Energy Study Institute, Washington, DC. 28 Feb. https://www.eesi.org/articles/view/comparing-u.s.-and-chinese-electric-vehicle-policies
81. Lutsey N, Cui H, Yu R (2021) Evaluating electric vehicle costs and benefits in China 2020–2035 time frame. White Paper. International Council on Clean Transportation, Washington, DC. Apr. https://theicct.org/sites/default/files/publications/EV-costs-benefits-china-EN-apr2021.pdf
82. China's thriving EV industry takes green innovation global. Xinhua News, Chengdu. 10 Sept 2024. https://english.www.gov.cn/news/202409/10/content_WS66e00fa4c6d0868f4e8eacb5.html

83. Campbell C (2025) The shift east: how China's EV boom powers its tech rise. Time, USA LLC. 29 May. https://time.com/7288660/shift-east-china-electric-vehicles-economy-technology-trump-tariffs-ai/
84. He H, Sun F, Wang Z, Lin C, Zhang C, Xiong R, Deng J, Zhu X, Xie P, Zhang S, Wei Z, Cao W, Zhai L (2022) China's battery electric vehicles lead the world: achievements in technology system architecture and technological breakthroughs. Green Energy Intell Transport 1(1):100020. https://doi.org/10.1016/j.geits.2022.100020
85. Liang A, Marsh N (2025) How China made electric vehicles mainstream. BBC, UK. 23 June. https://www.bbc.com/news/articles/c2d5ld8y8pwo
86. Williment C (2025) IEA: how EV adoption is changing global oil demand. Energy Digital. 29 May. https://energydigital.com/articles/iea-how-ev-adoption-is-changing-global-oil-demand
87. Saptakee S, How EV (2025) Adoption is reshaping global oil demand: IEA's 2025 outlook and 2030 forecast. Carbon Credit Inc., Houston. https://carboncredits.com/how-ev-adoption-is-reshaping-global-oil-demand-ieas-2025-outlook-and-2030-forecast/
88. Song W, Viisainen V, Patel A (2024) Q&A: the global 'trade war' over China's booming EV industry. Carbon Brief Ltd. 28 Aug. https://www.carbonbrief.org/qa-the-global-trade-war-over-chinas-booming-ev-industry/

Chapter 6
Technology and Subsidies in the Depreciation of Electric Vehicles

Contents

6.1 Introduction

EVs have progressed into a dependable technology, leading certain countries to aim for 100% EV registrations for new sales by 2035 [1, 2]. EV technology is now advanced and progressively competitive. The performance of EVs matches that of ICE vehicles, although EVs still fall short in certain aspects, such as range and quicker depreciation specifically, that affects resale value. Nevertheless, according to the literature review [3–5], it appears that faster depreciation of EVs, compared

R. Pode, B. Diouf, *Decarbonization of Road Transportation*, Green Energy and Technology, https://doi.org/10.1007/978-3-032-11388-7_6

to ICE vehicles, has limited impact on the vehicle purchasing decision-making process. However, upon closer examination, purchasers indirectly factor in the depreciation of EVs among their primary motivations or apprehensions about buying an EV. In fact, they take into account aspects like range, charging network accessibility, and clearly acquisition cost before all other considerations. The wider EV market continues to be influenced by changing consumer preferences. These fluctuations create times of high demand and other times when sales are significantly lower. Despite a strong brand reputation and advanced technology, interest in EVs can be influenced by external factors. These factors include economic conditions, fluctuations in energy prices, ongoing issues like accidental fires, or the introduction of new models. For example, when gas prices rise sharply or environmental regulations tighten significantly, there can be an important surge in the demand for EVs, resulting in higher prices and slower depreciation. Conversely, if incentives decrease or consumer focus turns to the newest technology cars, the interest of potential buyers in EVs might wane, resulting in a rapid decline in value.

Another important element leading to the decline in the value of EVs is the concern of buyers about the time after the warranty expires. This is primarily due to the fact that, in contrast to ICE vehicles, whose maintenance and repair costs are generally more predictable and understood. EVs carry a significant amount of uncertainty regarding repair costs, especially concerning the remaining useful life of the battery or its potential replacement when needed. These elements are the primary variables influencing the pace of depreciation of EVs. This can be comprehended by examining how the loss of value of EVs affects drivers' choices to purchase an electric car. Therefore, it is essential to investigate the reality of EVs and the depreciation of popular models. Manufacturers and governments promote EVs via decarbonization strategies for road transport because of reduced ownership expenses, ignoring the variable depreciation that occurs.

Governmental structures provided subsidies for EVs to encourage their uptake [6–8] and lessen reliance on oil simultaneously. EVs are eco-friendlier compared to ICE cars [9–11], and they generate zero emissions. Conversely, subsidies may serve as a means to enhance the local EV manufacturing sector.

Subsidies for acquisitions have led to a distortion in the market [12–14]. The anticipated removal of subsidies is likely to inject new energy into the secondhand EV market because of a greater price disparity between new and used EVs. However, if subsidies focus solely on operation and not on acquisition, the second-hand market will probably see only minimal impact. This will stay valid unless there is significant advancement in battery technology that makes range less dependent on battery age or energy density so crucial that capacity reduction will be so slight it will not considerably influence the anticipated range. Battery presents the double role of driving the EV market and being a main player in EV depreciation. In fact, recognized for their impressive energy density and compactness, Li-ion batteries have outperformed earlier battery technologies regarding power generation, charge sustainability, and diminished environmental harm [15–17]. In the last almost 30 years, Li-ion batteries have seen an increase in efficiency of 5–10% and a decrease in cost of 15–18% annually [18, 19]. This rapid expansion has resulted in an extraordinary rise in the worldwide total of EVs, which expanded more than 1000-fold from 2010 to 2024 [18]. Forecasts from the International Energy Agency (IEA) indicate a possible increase of 12–21 times in EV stocks from 2021 to 2030 [18]. Even with this hopeful outlook, numerous obstacles still need to be tackled, such as safety concerns; problems regarding mineral availability and their social and policy consequences; environmental challenges tied to mineral extraction; infrastructure challenges; and disposal challenges. All these may contribute directly or indirectly to the depreciation of EVs.

At the moment, it is unclear what the long-term effects of the subsidy reduction will be. However, it appears clear that certain EV companies could be significantly impacted. As subsidies come to a close, a few of them might struggle to remain competitive. Producers with sufficient influence might take in the subsidy reduction, whereas lesser competitors could encounter significant challenges. Consequently, just a handful of big corporations might control the market down the line. Nonetheless, various factors may still propel the expansion of the EV market, such as increasing fuel costs, governmental regulations, and advancements in technology. Should these factors develop positively, the EV market could keep expanding swiftly in the years ahead without financial aid.

The EV market after subsidies could introduce new energy in the worldwide second-hand EV market.

The decline in value of EVs affects both owners and manufacturers. Currently, ICE vehicles maintain the same appreciation more effectively compared to EVs [3, 20–22]; nevertheless, trends and advancements in technology will likely result in similar depreciation rates in the near future as EV markets reach maturity. Tesla cars are a good illustration, as they depreciate less rapidly primarily because of the brand's strong reputation and established cutting-edge technology. As technology advances and consumer demand rises, it is anticipated that EVs will hold their value more effectively, resulting in a reduced pace in loss of value and a bigger used car market.

The decline in value of EVs is a genuine issue when evaluating the average period of 4–5 years when initial owners sell their cars. The swift loss of value might postpone the time when EVs reach the used car market.

The research on depreciation concentrated primarily on calculation techniques related to different types of products. The existing literature on depreciation methods has identified a variety of approaches that allow public and private organizations to compute asset depreciation more precisely and in accordance with their unique requirements. These techniques have developed throughout the years, adjusting to various settings. The straight-line method is among the easiest and most widely utilized techniques. Alternative methods encompass component depreciation, which analyzes distinct parts of the asset individually, recoverable amount, revalued cost, and variable depreciation [23, 24]. The declining balance method speeds up depreciation during the initial years of the asset's lifespan. The double declining balance method is notably aggressive, whereas the declining balance paired with straight-line depreciation permits a shift to linear depreciation in the subsequent years. Additional variations consist of variable declining balance, stepped declining balance, adjusted declining balance, and component declining balance [25, 26]. The units of production method for depreciation relies on the asset's actual usage, like operating hours or the number of units created. This method has variations such as adjusted production units depreciation and component production units depreciation [27, 28]. The sum-of-years digits method is an alternative accelerated depreciation approach that allocates a greater expense during the initial years. Variants consist of the altered sum-of-years digits, reverse sum-of-years digits, component-based sum-of-years digits, and productivity-modified sum-of-years digits [29, 30]. The general accelerated depreciation method encompasses methods like the modified accelerated cost recovery system. Other options include alternative accelerated depreciation and customized accelerated depreciation, designed to meet particular requirements [31]. There exist various less frequent yet equally beneficial methods and approaches in particular situations. Included are the replacement value method, which determines depreciation based on the expense of replacing the asset instead of its initial value; the parabolic-Kuentzle method, which modifies depreciation following a parabolic curve, suitable for assets that experience non-linear wear; the Ross method, applied in cases where the asset's depreciation occurs irregularly throughout its useful life; and accelerated depreciation for intangible assets, which emphasizes the swift reduction in value of assets like patents or copyrights [32–34].

This manuscript explores the interplay between the different factors affecting the value of EVs and the secondhand market. The main factors that affect EV depreciation are subsidies, progress in technology, including batteries, and battery deterioration. The battery technology affects EVs in different aspects: performance that makes EVs a practical reality and makes older models obsolete, cost that affects the price of new EVs, and deterioration that impacts depreciation. Clearly, the speed of depreciation of EVs does not benefit a significant second-hand market, especially in the initial phases of an innovative industry like EVs. It is fair to anticipate that once the elements causing the depreciation of EVs are managed, especially subsidies and improved battery technology, and price parity is reached between EVs and ICE vehicles, the used electric car market will expand significantly.

The approach to sustainability in this manuscript is the ability to maintain or support over time subsidies of all sorts before reaching maturity of markets and the price parity between new EVs and new ICE vehicles. Particularly in the context of dependence on natural resources like fossil fuels and environmental issues as global warming. It emphasizes the need to use financial resources to foster technological progress in a way that meets current needs to decarbonize road transport without compromising the ability to develop a necessary future large second-hand market.

The study of depreciation of EVs in this manuscript attempts at understanding how their value declines over time, particularly in the first 3–5 years, compared to their initial purchase price. The goal of this study includes:

- Analysis of resale value trends across different EV makes and models in different countries.
- Comparison of depreciation rates of EVs versus ICE vehicles.

- Assessment of the impact of governments' policies or market conditions on resale values.
- Identification of other major factors affecting EV depreciation, such as progress in battery technology and battery degradation, swift progress in technology such as AI and brand perception.
- The evolution of the second-hand EV market

Regarding the monetary analysis, the depreciation in this manuscript is generally presented within a country based on a given currency. Even though the depreciation rates can be compared between two different countries or economic zones.

6.2 Non-subsidy EV Market Drivers from Consumers' Perspective and Depreciation in the Cost of Ownership

The total cost of ownership of an automobile comprises the buying price, any grants, and operational costs like fuel, insurance, maintenance, and depreciation. Nevertheless, depreciation is often omitted or entirely disregarded in comparisons between EVs and ICE cars.

The reduced ownership costs of EVs compared to ICE vehicles are frequently highlighted as a key factor in countries with significant EV adoption [1]. Reduced operational costs are accompanied by subsidies and rapid technological advancements to enhance the shift to EVs. The latter two elements contribute to the widespread acceptance of EVs and the quick depreciation of EVs because of swift obsolescence. Premium EVs, such as Tesla, featuring advanced technology and artificial intelligence (AI) integration, experience a lower depreciation rate due to higher market demand compared to more affordable EVs, exemplified here by the Nissan Leaf. These simpler cars experience a steeper rate of depreciation. There is continuous progress in battery technology, featuring greater energy density that enables extended range at reduced costs as battery prices consistently decline [35, 36]. The growing technological incorporation in the automotive sector, especially in EVs, is AI; AI aids in machine learning, safety, efficiency, battery management, driver interaction, and intelligent driving. AI is increasingly becoming a key component of vehicle technology, and its incorporation in EVs is growing at a fast pace.

The expense of ownership is a key factor in the EV sector, especially in nations where EV adoption is at its peak, as shown in Table 6.1. The cost of ownership primarily relates to the price of electricity in comparison to gasoline. Nations that have the cheapest electricity and the most expensive gasoline are likely to see the greatest EV adoption rate, especially in urban regions. In Table 6.1 it is indicated that the average EV driver is a person with a high income in a nation where electricity costs significantly less than gasoline, resides in an urban area within a low-density community, and where EVs have gained a degree of popularity.

Table 6.1 Factors influencing the EV market excluding subsidies in the top 20 markets

Country	% of e-cars [37]	1 L Gasoline ($ US) [38]	1 kWh ($US) [39]	1 L/1 kWh	GDP/ Capita [40]	Density pop/km^2 [41]	Urban Pop (%) [42]
Norway	86	2.1	0.13	16.2	89,154	43	84
Iceland	72	2.2	0.15	15.4	68,728	4	94
Sweden	43	1.9	0.36	5.1	61,029	60	88
Denmark	35	2.1	0.58	3.6	68,008	352	88
Finland	31	2.2	0.45	4.9	53,655	43	86
Netherlands	30	2.0	0.49	4.0	57,768	520	93
Germany	26	1.9	0.55	3.5	51,204	605	78
Switzerland	22	2.0	0.23	8.6	91,992	545	74
Portugal	20	1.8	0.29	6.1	24,567	289	67
United Kingdom	19	1.8	0.41	4.4	46,510	715	84
France	19	2.0	0.22	9.1	43,659	303	82
Belgium	18	1.8	0.52	3.4	51,247	985	98
China	16	1.2	0.08	15.0	12,556	385	64
Italy	9	2.0	0.58	3.4	35,657	509	72
Spain	8	1.7	0.37	4.6	30,103	243	81
Greece	7	2.0	0.27	7.3	20,193	205	80
Canada	7	1.3	0.12	10.9	51,988	11	82
South Korea	6	1.2	0.10	12.2	34,998	518	81
United States	5	1.0	0.18	5.6	70,249	91	83
New Zealand	4	1.6	0.19	8.4	48,781	49	87

Norway, Iceland, and Sweden therefore exhibit the most important levels of EV adoption (Table 6.1).

In 2022, EV sales accounted for 14% of worldwide sales [43], and various studies estimate that by 2040, EV sales will comprise 54% of new vehicles available, with forecasts indicating approximately 70 million EVs sold by that year [44, 45].

The main consideration for car buyers when opting for an EV is affordability, followed by the vehicle's range and the availability of public charging stations [1]. From the perspective of the owners, numerous other criteria can be articulated in terms of cost-effectiveness. This suggests a potentially thriving second-hand EV market if the cars are priced competitively against new models and the battery preserves a considerable part of its initial capacity to satisfy range requirements. Subsidies have a major influence on buying costs, thereby impacting the gap between new and pre-owned EVs.

Norway and Iceland, securing the first and second spots in EV sales percentage among the top 20 nations in 2022, are ranked second and fourth for GDP per capita in the list presented in Table 6.1. These countries also display other remarkable traits: the highest gasoline rates and one of the most significant ratios of gasoline price to electricity. Norway has the highest ratio among nations, followed closely by Iceland.

This indicates that in the absence of subsidies, the operational expenses of an EV would be significantly more advantageous in these countries than in any of the other nations on the list.

Ultimately, the markets that exhibit the greatest EV adoption are those where owning ICE vehicles is costlier and in nations with more substantial subsidies.

Many of the other factors are closely linked to batteries, including vehicle cost, EV performance, safety, and longevity. Despite the competitive maturity of EV technology, further research is still required due to market demands. As time goes on, more research is needed to develop batteries with enhanced performance that can compete better regarding cost, energy density, longevity, and simpler material recycling.

The non-subsidy parameters listed in Table 6.1 above clearly apply to both new EVs and used EVs. The primary distinction between owners of new EVs and those of used EVs is probably related to budget.

The hidden factor of the reduced ownership expense of EVs is their faster depreciation rate, which is often overlooked when comparing EVs and ICE vehicles to potential buyers. When considering depreciation, the ownership expense of an EV can be considerably higher than that of an ICE vehicle of the same brand and model.

For instance, in 2023, the cost of an electric Kia Ray is USD 21,000, whereas a gasoline Kia Ray costs USD 11,000 [46]. Table 6.2 below compares the actual ownership expenses, including depreciation, of an EV Kia Ray and a gasoline Kia Ray following 5 years of usage, assuming maintenance is covered by the manufacturer's warranty.

Table 6.2 juxtaposes EV and ICE Kia Rays, considering the vehicles' depreciation. An in-depth evaluation of the swift decrease in EV values should lead to an extended ownership period for EV buyers to maintain reasonable ownership expenses.

Evidently, the depreciation rate of electric vehicles will differ by country depending on various influencing factors. For instance, in countries with lower subsidies, the appetite for pre-owned EVs will rise, and the rate of depreciation will fall.

Attaining equal cost of ownership between electric and internal combustion engine vehicles will offer a notable financial advantage to support the shift towards full EV adoption alongside a more dynamic used EV marketplace.

Table 6.2 Comparison of cost of ownership between an EV and gasoline Kia Rays

	EV Kia Ray	Gasoline Kia Ray
Purchase cost	$21,000	$11,000
Fuel	$ 0.131/kWh	$1/l
Average energy consumption	10.5kWh/100 km	8.5 l/100 km
Total cost of energy 5 years	$ 687.75	$ 4250
Total distance traveled in 5 years	50,000 km	50, 000 km
5 years value loss	50%	35%
Cost of ownership after 5 years	$ 11,187.75	$ 8100

Subsidies can significantly reduce the duration required to reach total cost of ownership parity between EVs and ICE vehicles.

6.3 Depreciation of Electric Cars

6.3.1 Value Loss of Well-Known EV Models

Electric cars lose value faster than all other types of vehicles. In 2023, the average electric car had lost 49.1% of its value following 5 years of use [47]. Several prominent brands that saw considerable value reductions after their first 5 years in business include the Chevrolet Bolt EV, which decreased by −51.1%, and the Nissan Leaf, which declined by −50.8% [48].

Fast technological progress and effective incentives that have lowered the cost of new EVs are two elements influencing the quick depreciation of EVs relative to comparable ICE vehicles (Tables 6.3 and 6.4). This program could continue until the EV market matures and no longer requires substantial incentives that correspond with consumers' confidence in their long-term ownership costs. As the market and technology advance, phasing out subsidies will increase the demand for pre-owned EVs, thus enhancing their worth.

A research project focused on improving the comprehension of the differences in depreciation between electric vehicles and internal combustion engine cars. The study centered on vehicles that have completed their lease contracts [47]. The study investigated the off-lease vehicle models that lost the most value by the end of a 3-year contract. The study discovered that the typical model decreased in value by 39.1% of its original worth when it became 3 years old. Among the off-lease car models analyzed in the study, the Audi A6 emerged as the most notable loser. The luxury car depreciated by 55.8% of its original selling price [48].

Table 6.3 Electric cars value loss from leased vehicles study over the first 3 years [48]

Model (brand)	Value loss (%)
A6 (Audi)	55.8
Fusion Hybrid (Ford)	54.9
3 Series (BMW)	53.4
S60 (Volvo)	53.2
E-Class (Mercedes-Benz)	52.7

Table 6.4 Five EV models with the highest depreciation rates after 3 years in 2023 [48]

Vehicles (brand)	Depreciation %
i3 (BMW)	60.4
LEAF (Nissan)	60.2
Soul EV (Kia)	58.7
Ioniq Electric (Hyundai)	47.7
Bolt (Chevrolet)	47.5

Out of those five vehicles, the Ford Fusion Hybrid and the Volvo S60 are electric vehicles. The study also investigated the depreciation rates of off-lease electric vehicles. It found that 3-year-old EVs usually lost 52% of their initial worth when compared to new ones [48].

The used market with high resale prices benefits leasing companies. Their advantage stems from a market marked by low depreciation, allowing them to sell their cars at a favorable price after a number of years. These companies significantly impact secondhand markets due to their large number of vehicles available for a short period. They generally retain their vehicles for less than 3 years, unlike 3–5 years for a personal household. They can also greatly affect the market for new automobiles; for example, in 2022, leasing companies accounted for over 20% of new vehicle sales in Europe. Here are the five electric vehicle models that experience the highest depreciation rates after 3 years of use.

The i3 (BMW), LEAF (Nissan), and Soul EV (Kia) face greater depreciation rates than the Audi A6.

6.3.2 EVs Vs. ICE Cars Depreciation

The preceding Fig. 6.1 shows that between 2020 and 2023, the average depreciation rate for EVs was 51%, whereas it was 31% for ICE vehicles.

A study indicates that, on average, 3-year-old electric vehicles usually lose 52% of their original value, which surpasses the 39.1% decline seen in traditional vehicles. Table 6.5 illustrates the depreciation rates of well-known EV brands and models in the United Kingdom.

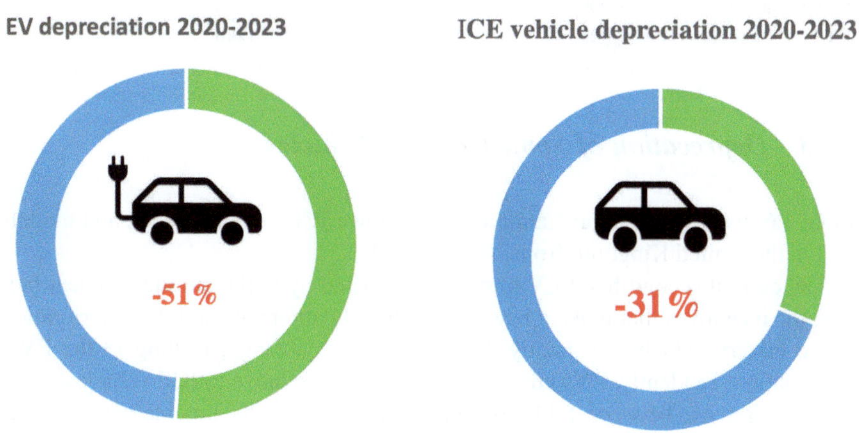

Fig. 6.1 Loss of value of EVs (**a**) vs. ICE cars (**b**) [3, 49, 50]

Table 6.5 Depreciation, over 3 years, of famous EV brands and models, without including inflation [51]

Model (brand)	Original range (km)	Price new in 2020 (Pound UK)	Value in 2023 (Pound UK)	Depreciation (Pound UK)	Depreciation (%)
Ionic (Hyundai)	295	32,950£	15,500£	17,450£	−53
ID3 (VW)	405	39,425£	23,000£	17,425£	−42
Leaf (Nissan)	340	26,995£	14,000£	12,995£	−48
Mini Electric	310	27,500£	17,500£	10,000£	−36
E (Honda)	335	28,000£	22,000£	6000£	−21
Zoe (Renault)	315	29,000£	14,000£	15,000£	−52
E-nero (Kia)	385	36,145£	24,000£	15,145£	−34
Polestar 2	470	49,900£	38,000£	11,900£	−24
Tesla 3	420	40,500£	25,000£	15,500£	−38
Tesla S	575	75,000£	50,000£	25,000£	−33

Table 6.6 Depreciation rates, over 3 years, of well-known ICE vehicle brands and models [51]

Model (brand)	Price new in 2020 (Pound UK)	Value in 2023	Depreciation (UK pounds)	Value loss (%)
i30 (Hyundai)	22,000£	15,000£	7000£	−32
Golf TSI (VW)	26,000£	16,500£	9500£	−37
Micra (Nissan)	17,200£	11,50£	5700£	−33
3 Series (BMW)	32,000£	23,000£	16,670£	−28
Mini Clubman	21,000£	16,500£	4500£	−21.5
Jazz (Honda)	20,000£	11,000£	9000£	−45
Clio (Renault)	17,595£	11,500£	6095£	−35
Ceed (Kia)	25,450£	14,500£	11,450£	−43
F Type (Jaguar)	54,000£	42,000£	12,000£	−22
E Class (Mercedes)	39,000£	32,000£	7000£	−18
Range Rover	85,000£	65,000£	20,000£	−24

6.3.3 Deprecation of Some Gasoline Vehicles

Table 6.6 displays the depreciation of well-known ICE vehicle brands and models [51] in the United Kingdom from 2020 to 2023.

The current second-hand EV market is impacted by two main independent factors: the incentives that make a new EV financially attainable and the deterioration of the battery, which can greatly diminish the initial range, leading to the EV's depreciation. Evidently, EVs are still in their infancy relative to ICE vehicles.

The difference between Tables 6.5 and 6.6 underscores the notable gap in depreciation between EVs and ICE vehicles.

6.4 Origins of EV Depreciation

EVs are transforming the automobile industry, providing an environmentally friendly option compared to conventional ICE cars. Nonetheless, EVs lose value as time passes. The rate and causes of depreciation for EVs can vary greatly from those of ICE vehicles, influencing consumer preferences and long-term financial strategies. Grasping how the depreciation of EVs operates is essential for purchasers, vendors, and industry participants [52]. It influences not only individual purchasing decisions but also molds broader economic and environmental regulations. Electric vehicles are relatively new in the market, which complicates the evaluation of depreciation compared to internal combustion engine vehicles. For example, in the USA, electric vehicles depreciate by an amount equal to the tax incentive provided for that vehicle. The matter of depreciation becomes even more complex for used cars.

Experts indicate that the reasons for EV depreciation vary from a surge of new models to a drop in battery efficiency. Facing "the drop in EV value in the initial years of ownership is the most expensive part of owning a new vehicle." One possible result is that as more potential EV buyers become aware of this reality, a smaller number will express interest in buying an EV.

In a 3-year ownership cost evaluation examining similar Ford and Hyundai EV/ICE models currently available, Car and Driver analyzed maintenance, refueling, and depreciation expenses with an annual driving distance of 24,000 km. The research revealed that the Hyundai Kona and Ford F-150 EVs excelled compared to their internal combustion engine counterparts in terms of fuel and maintenance costs, yet depreciation stood out as the EVs' weak point. It is evident that EVs lose value faster than gasoline ones; the research indicates that the Kona ICE saw a decline in value of $9795, whereas the Kona Electric faced a decrease of $15,305. The Ford F-150 experienced a decline in value of $13,981, while the F-150 Lightning EV lost $15,738 in value [53].

6.4.1 Different Factors Affecting EV Depreciation

6.4.1.1 Government Incentives

Subsidies for EVs from governments are implemented globally to promote the uptake of EVs and foster industry growth. These incentives, appearing in various forms, primarily focus on acquisition subsidies realized through purchase rebates, tax exemptions, and credits, as well as extra benefits like the possibility to use bus tracks and preferential conditions for some car-related services. Subsidies that focus solely on new EVs tend to have a greater negative impact on depreciation rates. For instance, in China, where used EVs depreciate faster than elsewhere, the incentives for EVs were tied to their range, which is primarily influenced by battery capacity.

As a result, vehicles with greater range qualified for larger subsidies. A previous thorough analysis performed by the authors regarding EV incentives is available in reference [2].

Incentives for electric cars manifest in different forms, including tax credits, rebates, and discounts offered by manufacturers. These incentives reduce the initial expense, making EVs more accessible to a broader range of purchasers. Nonetheless, their impact on the secondary market, particularly concerning resale values, is a complex and evolving story. Subsidies and tax credits may decrease the cost of an EV, possibly leading to a lower resale value because these incentives factor into the depreciation calculation.

Conventional ICE vehicles have established depreciation rates, whereas EVs have yet to establish theirs. Incentives that reduce acquisition costs result in quicker depreciation, especially during the initial years, though this may stabilize as the market matures.

These incentives facilitate the process for new EV owners, but they will undoubtedly impact the used EV market. As long as depreciation rates are not properly adjusted, more affordable new EVs render used ones less attractive.

When new EVs receive significant subsidies, the perceived value of pre-owned models decreases for buyers. Consumers will favor new incentivized EVs over used cars. This choice by consumers is expected to lead to a supply of used EVs that exceeds demand.

Although the short-term effects of incentives may appear detrimental to resale prices, the long-term implications are more ambiguous. As an increasing number of drivers choose EVs, the market for used EVs expands. This enhances, as time progresses, the acceptance and uptake of EVs, potentially resulting in greater demand and improved resale values.

In the future, the increasing demand for EVs is expected to enhance their residual values in the years ahead compared to the present. In 2023, EVs retained a resale value of only 41%, while ICE vehicles maintained 66% of their worth. Such a scenario is anticipated to get better for EVs.

Professionals predict that with the growth of the EV market, resale prices will level off and become more predictable. Research suggests that electric vehicles coming back to the market after a standard 36-month lease are expected to retain roughly 53% of their value by 2027. This represents a significant improvement over the 41% retention recorded in 2023.

Improved incentives for electric vehicles undoubtedly influence their resale values, creating both challenges and chances. While these incentives might at the beginning reduce resale prices, the growing acceptance and adoption of EVs are anticipated to stabilize the market over time. Understanding these dynamics is crucial for both consumers and industry players to make informed decisions in the evolving EV conditions [54].

6.4.1.2 Technology

AI is accompanying EV technology in its evolution, resulting in a more intelligent transportation environment. AI enhances charging plans, customizes the driving experience, and also aids in battery management. Such partnership enhances efficiency, increases driving range, sets the stage for self-driving vehicles, and prepares to delve into the future of eco-friendly transportation. The speed of technological progress is among the key elements affecting EV depreciation. Advancements in battery technology result in extended ranges and reduced charging times, which can create a significant difference between two iterations of the same EV model. The incorporation of AI in advanced driver-assistance systems, along with other technological advancements, may enhance appreciation of EVs.

With the swift progress of EV technology, older vehicles may swiftly become obsolete, resulting in quicker depreciation. The demand for EVs is also significant in influencing their resale value.

Figure 6.2 above illustrates the decline in Li-ion battery prices from 2010 and the forecast for 2030. Improved battery longevity and reduced costs for replacement or upgrade will enhance the value of used EVs. AI supports machine learning, safety, efficiency, battery control, driver engagement, intelligent driving … AI is increasingly essential to vehicle technology, and additionally, its incorporation in EVs is accelerating quickly, as illustrated in Fig. 6.3.

The swift advancement of AI, ML (Machine Learning), and DNN (Deep Neural Networks) in EVs leads to a higher depreciation rate of these vehicles, caused by the

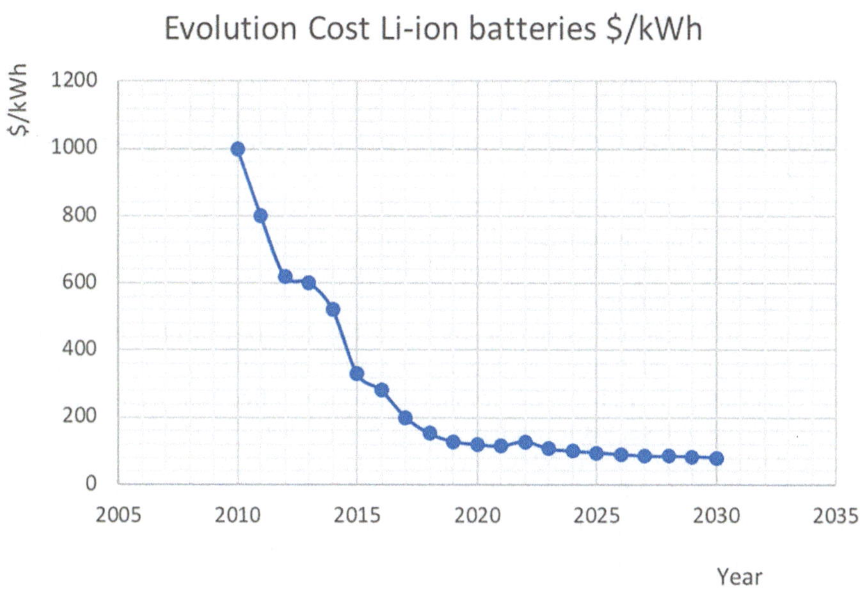

Fig. 6.2 History and forecast of lithium-ion battery prices [55]

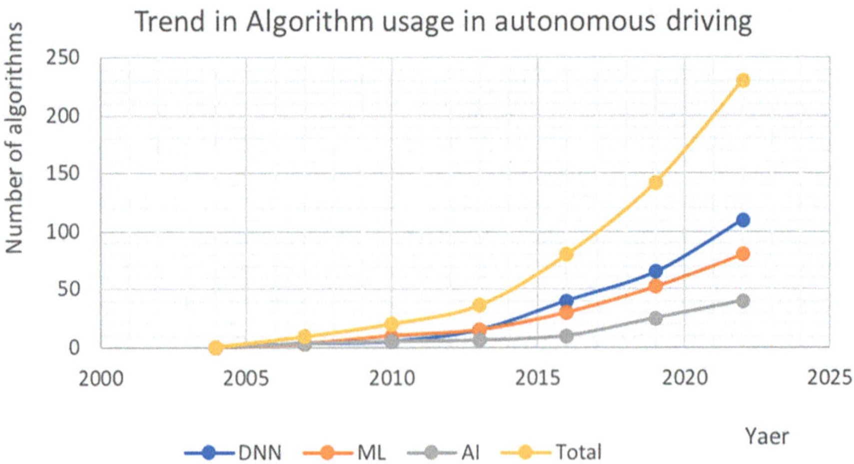

Fig. 6.3 Usage of algorithms related to evolution technology for autonomous vehicles [56]

variations between identical models from different years. The overall rise in the quantity of algorithms is approximately 820% from 2010 to 2022 (Fig. 6.3). Currently, there is no change in the progression of technology within the EV sector. A stagnation in the adoption of technology in electric vehicles could likely lead to a resurgence in the used EV market.

6.4.1.3 The battery

The environmental effects and sustainability of the EV transition will heavily rely on the advancements in battery technology. Li-ion batteries seem to have a bright future even with advancements in alternative technologies like sodium-ion (Na-ion) [57–59].

The state of health (SOH) of batteries remains acceptable between 100% and 80–70% [60]. When the capacity drops below 80–70%, it raises some issues, and the battery must be replaced for reliable EV functioning. The cost of battery replacement remains exorbitant.

Certain technical and economic issues in EV batteries involve the tradeoff between cost and the limited availability of certain elements like cobalt. Lithium Iron Phosphate (LFP) [1, 60–62] technology, which is free from cobalt, is commonly utilized in EVs, yet it still has a lower energy density compared to Lithium Nickel Manganese Cobalt (LNMC) technology [1, 43, 59].

At present, typical EV batteries lose 2% to 3% of their capacity annually, which means that in 7–10 years, their capacity declines to 70–80%, necessitating a battery replacement [60, 63]. A 10-year-old electric vehicle would not appeal much to buyers unless its battery is replaced.

In 2015, batteries accounted for 57% of the cost of EVs; by 2020, this dropped to 33%, and it is projected to reach 20% by 2025 [64]. Additionally, enhanced battery longevity will alleviate consumer concerns regarding replacement expenses and after-warranty support. Customer concerns include not just battery expenses but also quicker charging, extended range, and increased energy density for lighter automobiles.

Another function of the decrease in battery prices is the future price parity between EVs and ICE vehicles. It will be an ideal scenario to reduce subsidies [64].

Advancements in battery technology have been crucial for the uptake of EVs. Significant advancements were made in safety, performance, and range, utilizing eco-friendlier materials, increased longevity, and batteries with greater energy density. The peak in technical advancement has not yet been achieved, as significant progress remains attainable in energy density for batteries that are lighter and offer longer range, especially aiming towards generalized solid-state batteries [65]. In spite of all the significant accomplishments that have established the EV as the leading option for decarbonizing road transportation, the battery appears to be the primary instrument that enables EVs to become a feasible reality, supported by the network of charging stations. Indeed, without appropriate ranges and a well-distributed charging infrastructure, the EV would not have served as a viable mode of transportation, making its adoption less appealing to drivers. However, the primary factor that distinguishes the EV driving experience from that of ICE vehicles extends beyond just the powertrain: it is probably the technology that introduces the most significant innovation within the EV experience, and this reality becomes increasingly evident over time with the growing integration of AI, ML (machine learning), and DNN (deep neural networks). These algorithms progressively establish autonomous driving as a norm in the EV experience, along with enhanced battery management, increased safety, improved driver-vehicle interaction, and soon greater automobile internet, similar to the evolution of mobile phone technology from 2G to 3G and 4G to 5G. Moving from a basic phone to a smartphone involves more than merely making phone calls. The swift advancement in technology appears to be the key influence in the EV industry, but it also significantly contributes to the quick decline in EVs' value. The influence of the battery on EV depreciation seems to stem from its declining cost, which makes EVs more affordable, along with a movement towards price equality with ICE vehicles. The price of batteries decreased by roughly a factor of 10 from 2010, which signifies the start of the mainstream EV industry, to 2023. The majority of manufacturers provide warranties that cover either 100,000 miles or 70% battery health, whichever occurs first, significantly lowering risks for owners. Governments implemented various types of subsidies to initiate the industry, benefiting both consumers and manufacturers, since EVs remain pricier than their ICE vehicle equivalents. Increased investment will further enhance battery technology to provide future EVs with improved range, reduced weight, and longer lifespan to surpass the vehicles. Certain Tesla models already assert a lifespan of 500,000 miles before requiring a battery pack replacement, highlighting the battery's potential to outlive the vehicle [66]. China is the first nation to perceive the reduction of subsidies as an indicator of market maturity; such a choice may not be essential and will occur in all markets as battery costs keep decreasing, leading to an inevitable parity with ICE vehicles.

Moreover, although inflation has risen, EV prices didn't rise as anticipated; in fact, they decreased in certain instances. As an illustration, in the USA, from 2010 to now, there has been a cumulative price rise of 44.03%, which aligns with an annual dollar inflation rate of 2.64%. The purchasing power of $1 in 2010 is now roughly equivalent to $1.44, showing an increase of $0.44 over a span of 14 years [67].

Since 2010, the typical expense of Li-ion batteries in EVs has decreased by nearly 100%, accounting for inflation [68]. For example, in the US, Nissan Leaf prices rose from $28,120 in 2011 with a 24 kWh battery to $38,839 in 2022 with a 62 kWh battery, marking a 258.3% increase in battery capacity [69].

The majority of EV manufacturers provide warranties for batteries that last between 8 and 10 years or cover up to 100,000 miles [70]. In the United States of America (USA), federal regulations require car manufacturers to guarantee EV and HEV batteries for at least 8 years or 100,000 miles. California enhances consumer protection by mandating a warranty of 10 years or 150,000 miles [71].

Nonetheless, EV batteries frequently have a lifespan that surpasses this duration and are anticipated to endure for as long as 20 years. This fact is expected to improve as a result of advancements in EV battery technology. The cost of EV batteries has decreased significantly, being reduced by a factor of 10 since 2010 [72]. Moreover, the technology is continually advancing, providing extended lifespan and improved performance, especially in areas directly associated with energy density. However, substituting an EV battery can be costly. Certain manufacturers provide battery replacement kits that come with installation. Nissan Leaf, a leader in the EV market, provides a battery replacement program for a fixed cost of £5000 in the UK, and vehicle owners get an additional £1000 for their used battery [73]. However, the Nissan Leaf is among the EV models experiencing some of the quickest depreciation rates. Clearly, the expense of replacing EV batteries can be significantly higher, yet it is typically anticipated to drop as battery technology advances and the need for cost-effective battery replacement services rises.

A warranty appears to be typical for EVs, and it doesn't significantly affect maintaining vehicle value, especially in the initial 3 years when the coverage is complete and there is still considerable operational time remaining.

Technology influences consumers' preference for EVs instead of ICE vehicles. The ongoing decline in battery prices is expected to create parity in pricing between EVs and ICE vehicles by 2030 [72, 74].

There will be two primary diverging paths associated with EV batteries. Should the battery life and performance facilitate a vibrant second-hand EV market, significant export opportunities will arise, and developing nations may require investments in the recycling sector, with extracted materials returned to the manufacturing nations. However, many developing nations have not yet implemented a genuine electric mobility policy. Conversely, if used EVs are not widely exported to developing nations, the recycling sector will grow in many countries with a significant penetration of EVs.

The state of the battery greatly influences the worth of EVs. The more aged the battery and the greater the mileage, the lesser the worth of the EV, just like with ICE vehicles. The greater the distance a car is driven, the lower its value becomes. This applies to EVs, where greater mileage and older age lead to quicker depreciation.

The residual worth of an EV is influenced by multiple factors, with the battery being arguably one of the key elements. Concerning the state of health (SoH) of a battery, it is what determines the remaining value of the EV. Currently, there is no established method for calculating the SoH of EV batteries in diagnostics, with the range being the primary point of reference.

Battery status and lifespan Lithium-ion batteries are reported to last between 8–15 years or 150,000 to 300,000 km, influenced by usage as well as charging habits and techniques. Different manufacturers provide guarantees for their batteries, typically ensuring an SoH of no less than 70% for 8 years or 160,000 km. These warranties can enhance the residual value of an electric vehicle, as they offer assurance and security. Similarly, well-regarded EVs maintain a greater residual value. The subsequent factors are significant regarding service duration.

The battery of an EV gradually and naturally deteriorates over time, even when it is not in use. The remaining capacity of the battery is calculated as a percentage of its initial value. Tesla has released information indicating that its batteries retain approximately 80% of their initial capacity after 240,000 km. Likewise, various sources suggest that one can cover several hundreds of thousands of kilometers before the battery energy diminishes considerably. Clearly, from the viewpoint of the manufacturers, it is crucial for this number to stay as elevated as possible for the maximum duration, preferably for the entire lifespan of the vehicle.

The battery cycle is a crucial factor in understanding the state of the battery. A battery's longevity is defined by its possible number of cycles, which is clearly constrained. Under specified usage conditions, the number of cycles a battery has undergone reflects its reduction in capacity.

Vehicles charged mainly at rapid-charging stations might undergo faster battery deterioration, impacting their resale worth. Quick charging might seem appealing, but it continues to be a key factor that adversely impacts a battery's lifespan. It places a significantly greater demand on the battery compared to slow charging. Batteries that often experience rapid charging tend to decrease in capacity more quickly.

To gain a better understanding of an EV's worth, batteries require assessment and analysis. Typically, owners change their cars every 5 years or after 90,000 km, before they enter the used car market. Currently, there is no universally recognized approach to assess the residual value of a second-hand EV [75].

6.4.1.4 Brand Reputation and Loyalty

Cars from well-known brands generally hold their value more effectively. For example, Tesla vehicles exhibit reduced depreciation rates because of the brand's solid market presence and ongoing software updates. Customer loyalty can also influence resale prices. As buyers grow more acquainted with and devoted to certain EV brands, the resale market for those brands could improve. This commitment can reduce some of the depreciation impacts observed with heightened incentives.

The Tesla Model S is an adequate illustration [76]. The Tesla Model S is a luxury sedan that is entirely electric and full-sized. It was crucial in building Tesla's reputation as a manufacturer of high-performance and long-range EVs. Since its launch in 2012, the Model S has received considerable attention and has played a key role in driving the wider automotive sector towards electrification. The Tesla Model S was revealed in 2009 and started deliveries in 2012. It came after the Tesla Roadster, which was the company's initial production car. The Model S demonstrated that EVs can offer extended range, superior performance, and numerous technological advancements while still maintaining luxury. The Model S presented the prominent central touchscreen interface, which manages the majority of the car's features, setting a trend that numerous automakers have since embraced. Over-the-air updates enable Tesla to upgrade the vehicle's software remotely, adding new features, enhancements, and repairs. The Model S has undergone several updates since it was first released. By 2023, Tesla had delivered a cumulative total of 630,063 Model S/X automobiles globally.

According to research cited in reference [43], a Tesla Model S will lose 35% of its initial value over 5 years, which is impressive for an alternative fuel vehicle, as the majority of EVs and plug-in hybrids struggle to retain their worth. Tesla's depreciation trajectory indicates that it declines in value substantially during the first 3 years but levels off considerably in subsequent years. This may stem from buyers of new Tesla vehicles traditionally receiving substantial government incentives for purchasing an EV, occasionally up to $7500 at the time of purchase. Taking everything into account, the Model S stands alone in the high-end EV segment, particularly in terms of performance, making it hard to surpass.

In the technological advancement, Table 6.7 above illustrates that for the Tesla Model S, the range improved from 426 km in the 2012 model to 603 km in the 2022 model with an autopilot feature, while the catalog price decreased from $115,050 to $99,990.

Table 6.7 Technical evolution of Tesla model S [76]

Model S	2012 85 RWD	2022 long range dual motor
Price (long range)	$115,050	$99,990
Price (inflation adj.)	$144,069	$99,990
Range (miles)	265	375
0–60 MPH	4.3 s	3.1 s
Autopilot/FSD	None—no cameras (Autopilot was introduced in 2014)	Autopilot standard, FSD Upgrade $12,000
Charging rate (kW)	150 (12.55 km/min max)	250 (17.7 km/min max)
Gaming	Chess, Backgammon, Some 80s arcade games	Equivalent to modern game console. Steam client support planned

6.4.1.5 Other Factors

- Low gas prices
  The price of gasoline and diesel can influence the depreciation rate of EVs. Although fuel prices will never match EV charging costs, they can drop sufficiently to encourage people to remain with ICE vehicles. The consequent decline in interest will diminish the worth of EVs, particularly the older versions.
  Certainly, the reverse is also feasible. Indeed, gas prices are significantly higher than the cost of EV charging at the midpoint of 2022
- Decreasing sticker prices on brand-new EVs
  Each year, automobile makers launch new EV models. Many have been creating EVs for budget-conscious drivers to broaden their market reach.
  The reduced sticker prices on affordable choices not only motivate drivers to purchase an EV but also raise the depreciation rate.
- Range anxiety
  A lot of individuals worry that their EVs do not compare to the range of ICE vehicles. Depending on the kind of EV, their worries could be justified.
  Hybrid and plug-in HEVs have gasoline or diesel engines that can energize their electric motors when their batteries are depleted. Nonetheless, all-electric vehicles (AEV) depend solely on their batteries. There are fuel cell vehicles that utilize hydrogen fuel.
  AEVs can solely recharge their batteries at a charging station. Even though the US has over 46,000 charging stations, gas stations still exceed their numbers nationwide. Regarding hydrogen stations, there are under 60, with the majority located in California [77].

6.4.2 Evolution of EV Depreciation

As the EV market develops and expands, various factors are expected to affect their depreciation rate. Here are several factors that could influence the future of EV depreciation:

- Market share
  The demand in the electric vehicle market is expected to rise as more individuals become EV owners and these cars gain wider acceptance. This increased demand will positively influence the rate of EV depreciation.
- Accessibility of charging infrastructure
  The presence of charging infrastructure will mitigate concerns regarding EV range and thereby enhance the perception and value of electric vehicles. This supports the market for pre-owned EVs.
- Appreciation from consumers
  Greater public awareness, acceptance of EVs, and acknowledged advantages of EVs will assist in maintaining the value of used EVs.

- The impact of the used EV market on depreciation
  As the EV market expands, the portion of the used EV market will also become more prominent, and these pre-owned vehicles will experience improved depreciation rates.
  Here are several points concerning the influence of the secondhand EV market on the value decline of pre-owned EVs.
- Availability of used electric vehicles
  After over a decade of operation, the initial wave of EVs will enter the used car market. This market will continuously expand. This growing availability of used EVs could result in a type of saturation, where supply exceeds demand, causing a quicker depreciation rate.
- Economic aspects
  The provision of government incentives for new EVs makes them more cost-effective and renders used EVs less appealing to buyers unless their price is considerably reduced.
- Consumer Awareness and Trust in EV Technology
  As EV technology matures, the chances of consumers purchasing used EVs will become more realistic, and their worth will increase.

6.4.3 The Way Forward: Solid State Li-Ion Batteries in EVs

Lithium batteries using solid electrolytes without a separator, or solid-state Li-ion batteries, are considered the batteries of the future, offering improved energy density and greater safety [78, 79]. Solid-state lithium batteries employ a solid material rather than a potentially combustible gel. These materials are neither volatile nor flammable.

Lithium will very likely continue to hold an important role in the electric car industry. Innovative battery technology is being developed rapidly by car manufacturers, which includes solid-state batteries. The higher energy density offered may improve the range of EVs by about 30% to one-half. A prototype was created by Toyota for a vehicle that utilizes a solid-state battery, with manufacturing expected to begin in 2024 [80].

Energy storage in electric transportation demands an extended lifespan, lower degradation rates, enhanced energy density, decreased sensitivity to elevated temperatures and their variations, and improved resistance to moisture. A significant level of investigation is underway to improve battery quality and efficiency via experiments, theoretical methods, or simulations [37, 39, 40, 81–83]. The main methods to reach this goal include design and thermal control, material composition, and structure [37, 39, 40, 81–83].

6.5 Depreciation Calculation Methods for EVs

Loss of value arises from comparable factors in almost every market; nonetheless, the impact of these factors differs across markets. It remains broadly recognized that the depreciation rate of EVs is not linear. EVs experience significant depreciation in value during their early years of ownership. Usually, new models can lose about 40–50% of their value within the first 3 years. This first decline in value is usually more pronounced for luxury vehicles, as the market is restricted, and buyers often look for the latest and most sophisticated features.

On the other hand, long-term depreciation usually stabilizes after the first few years, leading to a linear section of the curve. An EV that is well-maintained, remains under warranty, and is roughly 5–7 years old can retain about 40–50% of its original value, depending on its condition, mileage, and aftermarket changes. Some EVs tend to hold their value better over time compared to other models, influenced by brand prestige and features. Older EV models may struggle to compete with the latest technology, leading to higher depreciation rates. In addition, the earlier noted out-of-warranty problems can greatly affect the long-term value retention of EVs.

The role of technology and brand reputation in the depreciation of EVs is evident when comparing two models from different categories released in the past decade: the Nissan Leaf and Tesla Model S in the US market. The Nissan Leaf, being an entry-level EV, experiences a greater depreciation rate, while the Tesla Model S, a premium EV, retains its value more effectively [69, 84, 85].

These distinct types of EVs also highlight how crucial technology integration is in the depreciation of EVs, especially regarding batteries that improve yearly with enhanced energy density and decreasing costs, alongside AI, which is increasingly becoming a standard feature in EVs.

The changes observed today in the automotive industry are akin to those that impacted the smartphone market in the early 2000s. An EV is a device that can be programmed and is regularly and consistently updated; various improvements are continually made to components like sensors, safety systems, battery management systems, electric powertrain systems, and navigation systems, among others.

One of the primary challenges that the EV industry is beginning to encounter is obsolescence, especially for entry-level vehicles. Vehicles are becoming obsolete because rapid technological advancements create a disconnect between new software and old hardware.

From an empirical perspective and according to professionals cited in references [3, 47, 48, 50, 69, 76, 86], as well as analysts, depreciation indeed adheres to a nonlinear trajectory. It is sharper in the initial years before tapering off with a nearly linear trend in the subsequent years.

Various methods can be utilized to evaluate the depreciation of EVs. These models do not include inflation and are only valid in a timeframe where inflation does not affect the value of money over time.

Example of Depreciation: Nissan Leaf [69]

In the EV sector and among the most budget-friendly electric cars available globally. The Nissan Leaf, a trailblazer, is approaching the end of its life cycle. A report from Automotive News states that Nissan is not planning to launch a new generation of the Leaf and might even stop using the name entirely. The Nissan Leaf is undoubtedly a remarkable example of an EV because of its long-standing presence in the market; it was initially launched in 2010, featuring a claimed range of 159 km with a 24 kWh battery. Nissan's 2017 version offered a range of 172 km with a 40 kWh battery; the 2018 version increased to 243 km; the 2019 version reached 346 km with an enhanced 62 kWh battery, while the latest 2021 version is speculated to provide approximately 362 km. Additionally, significant enhancements are occurring in the navigation system, powertrain, and safety features, resulting in a car that is both efficient and advanced.

Since its introduction in December 2010, Nissan has sold more than 650,000 units of the LEAF model globally. The model, presently available in about 50 markets focusing on Japan, the U.S., and Europe, remains highly regarded by consumers worldwide and has received several prestigious awards across various markets [69].

The previously mentioned Table 6.8 displays the evolution of the battery for the Nissan Leaf between 2011 and 2022. The evaluation depends on the official ratings from the United States Environmental Protection Agency, which vary from the numbers given by Nissan.

The 2017 to 2022 second generation includes a 40 kWh battery and has an EPA-estimated range of 243 km. ProPilot Assist, a lane-centering system, is available on the two highest trim levels for an added cost and includes automatic parking in certain markets. Since 2019, a Leaf e + model (referred to as Leaf Plus in North America) has been offered. It includes a larger 62 kWh battery (59 kWh usable), providing an EPA range of 364 km, paired with a new 160 kW motor. In June 2022, the Leaf was reworked for the 2023 version. In the US, the EPA range was slightly reduced to 240 km for the 40 kWh version and 341 km for the 62 kWh version [69].

Table 6.9 shows the depreciation of the Nissan Leaf over a period of 10 years. Caredge, a well-known and seasoned car dealer, conducted research on the depreciation of EVs [86]. The price has remained quite stable over the years, especially when accounting for inflation, while the range has increased largely due to the reduction in Li-ion battery costs and advancements in technology.

The above Fig. 6.4 demonstrates the value decline of the Nissan Leaf over a decade in the US market, according to a study by CarEdge [69]. The curve shown in Fig. 6.4 is nonlinear and aligns with one of the depreciation models discussed in

Table 6.8 Evolution of battery and range (EPA) for Nissan Leaf [69]

Version	Battery size	Range km (EPA)
2011/2012	24 kWh	117
2013	24 kWh	121
2014/2015	30 kWh	121–135
2016	30 kWh	172
2017/2022	40 kWh/62 kWh	243/341

Table 6.9 Leaf (Nissan) depreciation over 10 years [69]

Age (years)	Value loss ($)	Remaining value (%)	~ Resale value ($)	Miles
0	0	100	38,309	0
1	5187	86.5	33,122	12k
2	8930	76.7	29,379	24k
3	18,745	51.1	19,564	36k
4	19,143	50.0	19,166	48k
5	20,855	45.6	17,454	60k
6	22,740	40.6	15,569	72k
7	25,246	4.1	13,063	84k
8	27,207	29.0	11,102	96k
9	27,885	27.2	10,424	108k
10	29,073	24.1	9236	120k

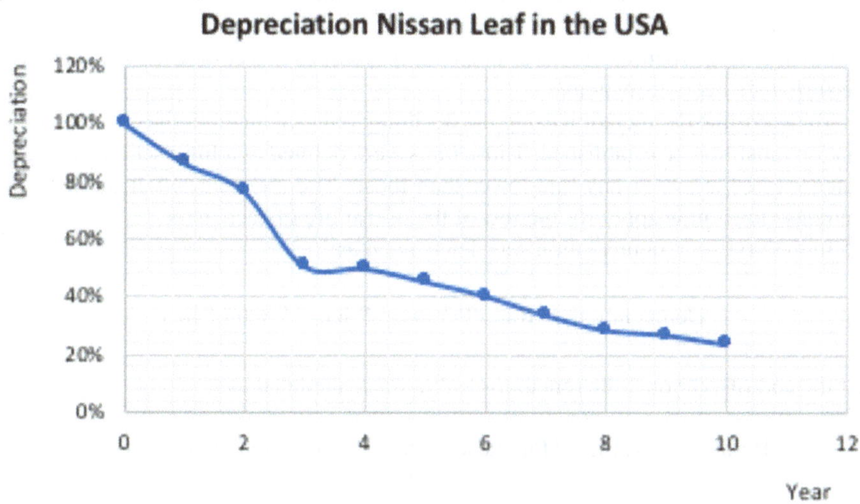

Fig. 6.4 Evolution of Nissan Leaf value left with time of use

Sect. 6.6 of this manuscript, where the initial years are more pronounced than the subsequent ones. The behavior exhibits a linear pattern following 4 years of use.

6.5.1 The Straight-Line Depreciation Method

The straight-line depreciation method is a widely used technique [87]. This simple and commonly utilized approach presupposes a steady depreciation rate annually throughout the vehicle's duration. For example, an EV anticipated to last 10 years and keep 40% of its worth would lose 6% of its initial value each year.

The calculation involves dividing the depreciable balance (purchase price less salvage value) by the number of years of useful life.

$$\text{Depreciation} = (\text{Cost Price} - \text{Residual Value}) / \text{Useful Life in Years}$$

The straight-line approach presumes uniform depreciation annually and suggests that the asset is retained for the full year.

6.5.2 Declining Balance Method

The declining balance approach serves as a different method [88]. The depreciation matters more during the initial years of the vehicle's operation. This may better represent the true wear and tear on the car, along with the swift rate of technological obsolescence in the EV industry. For example, a new EV featuring advanced battery technology might depreciate faster in the initial years as newer models with superior performance become available.

To determine depreciation using the declining balance method, the rate of depreciation must first be established. Those rates were defined as a limit set for different categories of assets or any rate lower than those. The rate selected at the time of purchase must now apply for the rest of the useful life of each asset.

In the first year, depreciation was determined as:

$$\text{Depreciation} = (\text{Rate}(\text{Purchase Price}) / \text{Useful Life})$$

At the end of year one, the remaining value was calculated as:

$$\begin{aligned}&\text{Remaining Value} = (\text{Purchase Price} - \text{Depreciation in year two});\\&\quad \text{depreciation was calculated as : Depreciation}\\&\quad = (\text{Rate}(\text{Remaining Value}) / \text{Useful Life}).\end{aligned}$$

A new residual value was determined by taking the last remaining value and subtracting current depreciation from it. All subsequent years are computed in the same way. Salvage value is not incorporated into the equations at all, yet the remaining value must never drop below the designated salvage value. Additionally, the rate selected and applied in the equations will remain constant over the useful lifespan.

The declining balance calculation needs to be modified for assets owned for less than a complete year. Similar to a straight line, the adjustment procedure is straightforward and entails merely multiplying the total annual depreciation by a fraction that indicates the duration of the year the asset was owned.

This approach might be relevant to the depreciation of EVs if the research was prolonged over a more extended timeframe.

Alternative approaches to assess EV depreciation include the sum-of-years digit technique and the residual value models [89]. These models take into account the anticipated value of the vehicle once its lease term or ownership duration concludes. The vehicle's residual value is determined by considering aspects like expected mileage, predicted condition, and future market trends. For example, an EV anticipated to have a high residual value after 3 years might be more appealing to lease compared to one with a lower residual value, impacting the monthly lease costs.

6.6 The Global Used Car Market and EVs

The typical age of vehicles in the United States, serving as a reference market, has attained a new record lifespan of 12.6 years [90]. The typical age of a pre-owned retail vehicle can be approximated to range from 4–5 years, reflecting the time when initial owners decide to sell their vehicle [91]. Researchers discovered that the typical age of used cars individuals are looking for is 6.47 years [92].

The size of the international used car market was valued at USD 1.67 trillion in 2023 and is anticipated to reach approximately USD 3.05 trillion by 2033 (Fig. 6.5), reflecting a compound annual growth rate of 6.2% throughout the forecast timeframe from 2024 to 2033 [86].

The primary market for pre-owned vehicles is North America, accounting for 30% of the worldwide share, while Europe follows; China, the Middle East, and

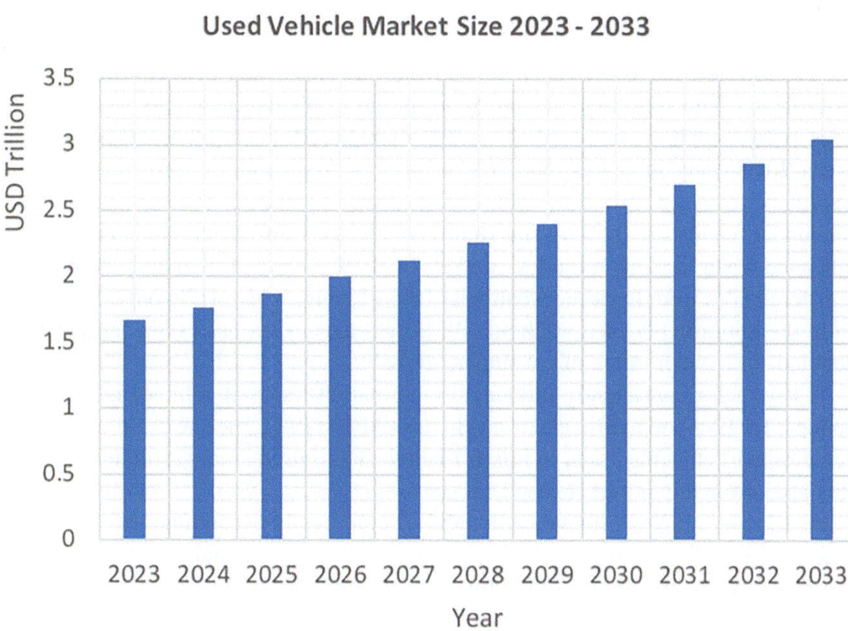

Fig. 6.5 Projection of used car market size between 2023 and 2033 [86]

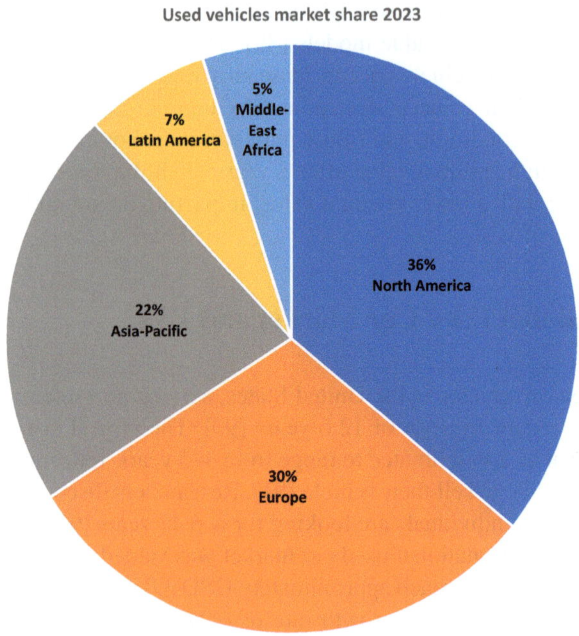

Fig. 6.6 Used car market share by region in 2023 [85]

Africa constitute the least significant markets, holding merely 5% of global sales (Fig. 6.6). At the same time, the United States represented under 10% of all new electric vehicle registrations globally in 2022 [69].

A brand-new car is typically beyond the financial reach of many individuals. Households with high incomes are more prone to owning new cars compared to households with low incomes. In Norway, where the EV penetration rate is the highest, a 2019 study indicates that 31% of new car registrations were made by the wealthiest 10% of families, whereas the lowest income 10% of households accounted for merely 0.6% of registrations [93]. This difference holds greater significance when looking at the EV market. In 2019, Norway saw that 37% of newly registered EVs were from the wealthiest 10% of households, while only 0.7% were from the bottom 10% of households [93]. Less expensive, used EVs can meet the demands of a wider range of consumers globally.

A key factor consistently influencing the market for used vehicles is the rising demand for EVs and hybrid electric vehicles (HEVs) [94]. The spectrum of the used vehicles market is illustrated in Fig. 6.7 below, highlighting that gasoline and diesel cars predominantly dominate used car sales.

As shown in Fig. 6.7 above, although it is continually expanding, the market for used EVs is still quite minimal when compared to conventional ICE vehicles. The growing practicality and cost-effectiveness of used EVs could partially change the market landscape in the years ahead. Additionally, the widespread appeal of premium variants of HEVs and plug-in hybrid electric vehicles (PHEVs) has enhanced

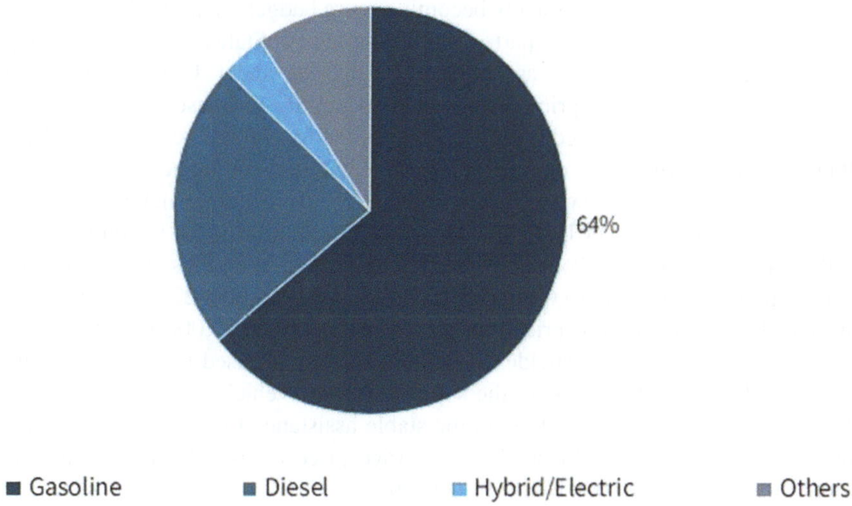

Fig. 6.7 Spectrum of the global used car market by fuel type [95]

the reputation of EV technology. The dynamics of the used EV market might exhibit distinct patterns once the market achieves complete maturity.

Regarding other technologies, the secondhand market for EVs is currently expanding as early adopters transition or enhance their models. Second-hand markets play a vital role in promoting widespread adoption, particularly as new EVs are still costly and used models become more accessible. For ICE vehicles, purchasing used cars is usually the main way to obtain a vehicle in both developing and developed economies, and it is expected to be analogous with EVs. It is estimated that 80% of European Union residents purchase used cars, with this proportion rising to around 90% in low- and middle-income countries [96]. In the U.S., roughly 70% of vehicles sold are used, and just 17% of lower-income families purchase a new car [97].

As key EV markets become established, an increasing number of pre-owned EVs enter the second-hand market. Nonetheless, the volume of the used EV market remains quite limited in comparison to new EV registrations. Nearly 14 million new EVs were registered worldwide in 2023, elevating the total figure on the roads to 40 million. In 2023, nearly 60% of new EV registrations occurred in China, around 25% in Europe, and 10% in the United States, accounting for almost 95% of total global electric vehicle sales. Conversely, estimates indicate that in 2023, the market size for pre-owned EVs hit approximately 800,000 in China, 400,000 in the United States, and over 450,000 when aggregating France, Germany, Italy, Spain, the Netherlands, and the United Kingdom. In the United States, sales of used EVs are projected to rise by 40% in 2024 compared to 2023. Although advancements have been made, the second-hand EV market is still minor compared to the more established second-hand ICE markets that have had decades to develop: 30 million in France, Germany, Italy, Spain, the Netherlands, and the UK combined; nearly 20 million in China; and 36 million in the USA.

Still, used EVs are progressively becoming more budget-friendly, allowing them to rival pre-owned ICE counterparts soon. In the United States, for instance, a significant share of used EVs are currently listed at under USD 30,000 [98]. Additionally, the average price is projected to rapidly decrease to around USD 25,000, the threshold for used EVs to qualify for the federal used car rebate of USD 4000, thereby making them directly competitive with top-selling new and used ICE models. The cost of a pre-owned Tesla in the U.S. fell from more than USD 50,000 in early 2023 to slightly over USD 33,000 in early 2024, allowing it to compete with both used SUVs and numerous new vehicles. In Europe, used EVs are available for prices ranging from EUR 15,000 to EUR 25,000 (USD 16,000–27,000), while pre-owned plug-in hybrids are priced at approximately EUR 30,000 (USD 32,000). Several European nations provide financial incentives for used EVs, including the Netherlands (EUR 2000), where the support for new vehicles has been gradually decreasing since 2020, in contrast to the stable assistance for pre-owned vehicles, and France (EUR 1000). In 2023, the average price of used EVs in China was approximately CNY 75,000 (USD 11,000) [98].

When examining the markets with the greatest EV adoption, such as Norway, the same factors should account for the trends in the second-hand EV market. The primary distinction lies in direct subsidies, which can be crucial in the decision to adopt a new EV. There ought to be a significant distinction between markets supported by subsidies and those after subsidies. Subsidies aren't the sole constraint influencing the used EV market; rapid technological advancements, alongside the performance and depreciation of EV parts like batteries and motors, also affect the success of the worldwide used EV market [63, 99].

The emergence and development of EVs in the worldwide EV market and their role in road decarbonization will probably resemble an S-curve (Fig. 6.8) [1]. It relates to an initial phase where the adoption is sluggish since consumers lack sufficient knowledge and confidence in investing in pre-owned EVs. As the popularity of EVs continues to rise due to subsidies, consumers will gain greater confidence in EV technology and will be more inclined to purchase used EVs as they become familiar with what to expect. This will persist until a balance is reached between new and used EVs in the market each year. The ratio of new to used vehicles in the market varies by country. In the USA, new vehicles accounted for roughly 30% while used vehicles made up around 70% in 2022 [63].

The S-curve displayed above is represented by the following Eq. (6.1):

$$f(t) = \frac{100}{\left[1 + \exp(-\alpha t)\right]^{k}}, \quad \text{where } \alpha > 0 \text{ and } k > 0, \tag{6.1}$$

where f(t) denotes the market penetration rate of used EVs, as t approaches infinity, f(t) approaches saturation.

In the curve above, $\alpha = 0.25$ and $k = 11$, as stated in reference [1], with α and k derived from an empirical approach [1].

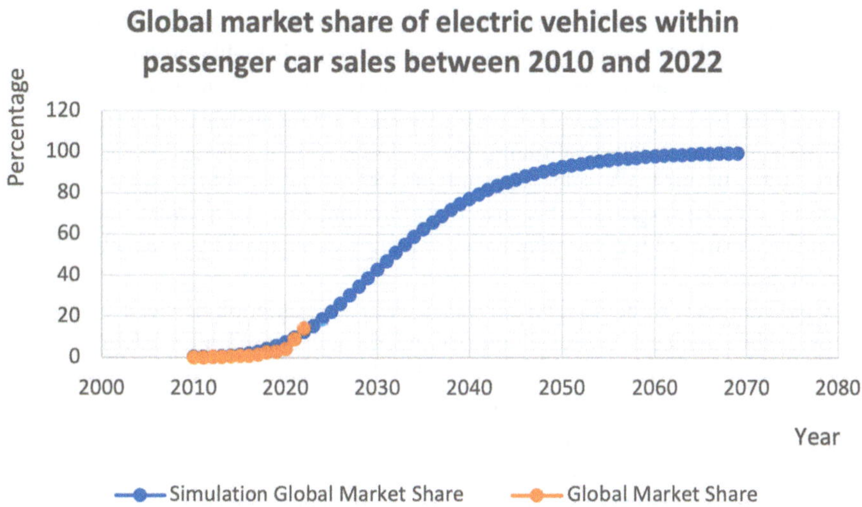

Fig. 6.8 Global EV transition statistics and simulation

A key question that requires a projected answer is the trend that will influence the share of utilized EVs during the transition to electric transportation on a global scale and within individual countries.

The depreciation rate of EVs and advancements in technology will play a significant role. Used EVs are anticipated to significantly enter the market, given that it has been over a decade since EVs were introduced in global road transport. It is anticipated that the proportion of new to used EV registrations will stabilize at x% for new EVs and y% for used EVs. A comparable balance is observed in the majority of non-perishable consumer goods: mobile devices, laptops, bicycles… This balance between innovation and ownership transfer will rely on market conditions. Nations with advanced economies that typically enjoy ample access to financial systems will exhibit a greater share of new vehicles among total new registrations, whereas in the least developed countries, nearly all new registrations consist of used imported vehicles. It is acknowledged that globally, in the initial decade, nearly all EV registrations will consist of new EVs.

In recent years, the resale price of electric vehicles has been rising. In Europe, the resale price of electric vehicles sold after 1 year has consistently risen during the 2017–2022 timeframe, exceeding that of all other powertrains and reaching over 70% by mid-2022. The resale value of EVs sold after 36 months was under 40% in 2017 but has been narrowing the difference with other powertrains, hitting approximately 55% by mid-2022. This outcome stems from several elements, such as rising costs of new EVs and advancing technology that enables vehicles and batteries to hold more value over time; these aspects reduce depreciation and boost interest in used EVs. Comparable patterns have been noted in China.

The issue of used EVs can become more intricate when we factor in the potential future coexistence of EVs and ICE vehicles. Indeed, as shown, energy transition technologies are frequently expected to adhere to S-curves: swiftly bending, then attaining complete market acceptance. Nonetheless, it is highly probable that EVs will persist alongside ICE vehicles, so it is crucial to consider the future equilibrium between the two technologies on a global scale. Certain analysts suggest that electric vehicles may reach a saturation point of 15–30% of sales between 2025 and 2030 [100]. In 2023, EV sales, currently at 15% of global vehicle sales, will continue to advance, especially because of certain policies. Thus, what implications would the restricted EV potential have for energy demand and the capacity of supply and materials?

Cost continues to be a significant obstacle to the adoption of electric vehicles. Thus, in the absence of coercive policies, income will constrain EV adoption, leading to a degree of saturation in the EV market. Information regarding income distributions will establish the threshold for EV adoption. Views on climate are an additional obstacle to the adoption of EVs.

The saturation of the EV market would have significant effects on energy and materials markets because of the considerable demand. The demand for EVs and ICE vehicles will relate to the need for oil, electricity consumption for EVs, lithium requirements for batteries, and the effects on new energy sources and investments.

Tesla cars retain their worth effectively for several reasons. Tesla vehicles receive remote updates throughout their entire service life. This technical choice reduces value depreciation from obsolescence because enhancements to newer models will also advantage older vehicles. Furthermore, an additional factor is the distinctiveness that consumers have begun to associate with Tesla. The company also oversees a network of superchargers. Tesla, like other manufacturers, does not rely as much on buyer incentives.

Tesla holds a strong popularity and reputation among EV brands, recognized as the top-selling brand for new EVs, featuring the most advanced technology and experiencing the least depreciation, which is reflected similarly in the second-hand EV market. Used EVs that experience the largest price reductions are generally older models, which many believe feature outdated technology, especially the Nissan Leaf and Chevrolet Bolt EV. With the Model 3 accounting for 34.9% of the market in 2023, Tesla leads the USA second-hand electric car sales. Models aged 1–5 years face the sharpest decline in value. A report by iSeeCars [101], identified as one of the leading dealers in the USA as a reference market for 2023, details the sales share of models aged 1–5 years:

As indicated by Table 6.10 above, in nearly all markets, the best-selling used EVs will include the Tesla Model 3 and Nissan Leaf for vehicles less than 5 years old.

A detailed model of secondhand EV market is presented in reference [102].

Table 6.10 Proportion of sales for different electric cars in the USA [101]

Brand & model	Market share (%)	Observations
Model 3 (Tesla)	34.9	Reputation & Tech
Model Y (Tesla)	11.9	Reputation & Tech
Bolt EV (Chevrolet)	6.9	Price
Leaf (Nissan)	6.2	Price
Model S (Tesla)	5.6	Reputation & Tech
Model X (Tesla)	5.5	Reputation & Tech
Mustang Mach-E (Ford)	4.6	Symbol
E-tron (Audi)	2.9	Brand name
Taycan (Porsche)	2.4	Symbol
ID.4 (Volkswagen)	2.2	Brand name

6.7 Lessons from China

The EV industry in China, as one of the seven strategically vital emerging sectors, has advanced swiftly in recent years. China formally launched, in 2009, the initiative to promote and showcase 1000 energy-efficient and new energy vehicles annually in ten cities. As of November 2018, China had advanced a total of over 2.8 million EVs and PHEVs, representing over 50% of the global market share of EVs. The adoption of EVs in China did grow without issues. In fact, consumers are especially concerned about the swift decline in value of EVs. As well, buyers risk encountering quality or price fraud and illegal vehicles when purchasing used cars, and both buyers and sellers experience significant information asymmetry, complicating their ability to ensure the fairness of the assessment outcomes. There is no consistent standard for assessing the residual value of EVs. Numerous research efforts have investigated the resale value of EVs [103–105]. In reference [104], a residual value assessment model for new energy vehicles is developed, employing the replacement cost approach to determine the residual value of both battery charging and battery swap EVs. The findings showed that range and lifespan were the two primary factors influencing the residual value of new energy vehicles. It was discovered that the residual rate of PHEVs exceeded that of ICE models during the first 2 years, while in the third year, the residual rate was marginally lower than that of ICE [105].

It was developed a residual value assessment model for EVs, concluding that the replacement cost of the vehicles must be subtracted from government subsidies and tax benefits, while the electric range and vehicle condition were the primary elements influencing the residual rate [106]. Moreover, the usage condition, vehicle state, and market value must be considered as the influencing factors for the residual rate. A research study conducted on the emerging trend of the residual rates for new energy vehicles and their effect on consumers' purchasing behavior [107]. It was discovered that the residual rates for BEV and PHEV were lower than those for ICE without accounting for federal tax. However, when federal tax was deducted, the residual rate for PHEV became comparable to ICE. It was also found that consumers prioritized price, fuel costs, and safety the most.

6.7.1 End of Subsidies: Lessons from China

The electric car industry in China, like in other countries with an important EV penetration, gained from governmental backing during its initial stage as a burgeoning green industry. Since 2009, the Chinese government has provided subsidies to EV manufacturers. In the same way, EV owners have benefited from direct subsidies for several years. Governments provided subsidies for EVs to promote their use and lessen reliance on fossil fuels. Subsidies serve as a method to enhance the local EV production sector. For consumers, the reduction in subsidies will increase the cost of EVs. Prices increased by 15–18% [108], making them less accessible for certain individuals. Manufacturing firms have raised issues regarding their financial health and might look into transferring costs to consumers, resulting in increased prices.

Over US$28 billion was allocated for EV incentives and tax reductions in China from 2009 to 2022 [109, 110]. With the market's development, governmental assistance and subsidies have decreased. Subsidies for EV buyers were eliminated at the close of 2022.

Table 6.11 presents the breakdown of ending Chinese government incentives.

The country sold around 3.5 million electric cars in 2022, accounting for over half of worldwide sales [111]. Sales are anticipated to consistently rise in the coming years. This reflects the development of the Chinese EV sales and the restricted influence of incentives.

The main unanswered question remains whether China's EV industry can maintain its momentum over time after the removal of subsidies.

China is recognized as an exceptionally vibrant EV market for both producers and buyers. It is considered to be the most advanced market. In 2022, the Chinese government decided to end subsidies, claiming that technology and continuous innovations drive the market instead of subsidies. The government attributed China's edge in EVs to a "developed supply chain system and competitive marketplace" [112].

Removing subsidies without a notable shift in market dynamics can happen when the law of demand and supply by itself does not sustain the crucial balance, signifying market maturity.

Figure 6.9 above indicates that sales remained steady even after the conclusion of subsidies in 2022, and there was no increase in the year before the subsidies were reduced, despite potential buyers being aware. This highlights the significance of non-subsidy influences regarding technological advancement and ownership costs in relation to ICE cars. It can be found as well increasingly a price equivalence between EVs and their ICE equivalents as previously shown. Comparatively, in 2024, China dominated global EV sales, accounting for a significant portion with 11

Table 6.11 Organization of concluding Chinese government incentives [110]

Policy	End date	Specifics
Exemption from consumption tax and vehicle & vessel tax	No indication	Automakers are granted exemptions from consumption tax and vehicle & vessel tax related to the production, subcontracted processing, and importation of electric cars
Acquisition subsidy	End of 2022	The highest subsidy is 12,600 yuan per vehicle for BEV passenger cars, and 4800 yuan for PHEVs, which includes extended-range cars
		The highest subsidy amounts to 50,400 yuan per vehicle for non-fast-charging BEV buses; 36,400 yuan for fast-charging BEV buses; and 21,300 yuan for PHEV (including extended-range) buses
		The highest subsidy for BEV trucks is 28,000 yuan per vehicle; for PHEV (including extended-range) trucks, it is 17,600 yuan
Exemption from purchase tax	End of 2027	New electric vehicles bought by 31 December 2025 are not subject to vehicle purchase tax. Electric vehicles purchased from 1 January 2026 to 31 December 2027 have their purchase tax cut by 50%
Infrastructure assistance	No indication	Arrangements for reduced electricity rates at EV charging and battery-swapping stations
		Government-directed pricing of service charges for EV charging and transition
		Costs for grid conversion related to EV charging and switching stations included in generator tariffs for electricity transmission and distribution

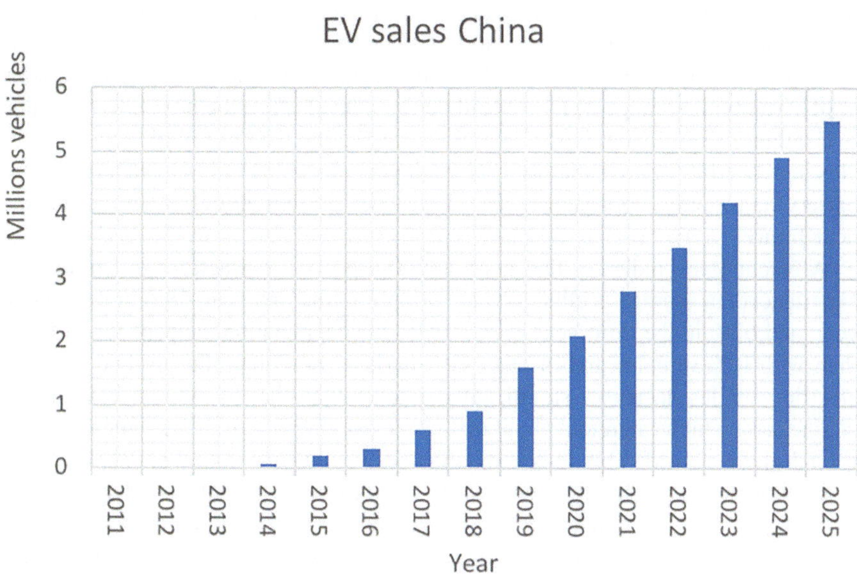

Fig. 6.9 Evolution of electric car sales in China [113]

million units sold, a 40% increase from 2023. While global EV sales reached a record 17.1 million, driven by a 25% increase year-over-year, China's sales nearly doubled the rest of the world's.

The cessation of incentives will alter the evolution of the used EV sales, as buyers will bear the actual price of new electric cars, leading to a significantly larger difference between new and used electric cars. Used EVs are likely to appear significantly more budget-friendly than new EVs, which may result in a reduced speed of value loss for EVs.

6.7.2 Second-Hand EV Market

In 2023, China saw new EV registrations reach 8.1 million, reflecting a 35% increase from 2022. In contrast, the Chinese used EV market is still quite minor, having sold only 500,000 pre-owned passenger EVs in 2020 [114], which rose to almost 800,000 by 2023. Most of these vehicles were compact, short-range models that dominated EV sales in China from 2015 to 2019 [114]. Some models were primarily aimed at securing the largest subsidy from the government rather than being fully customized to meet consumer needs. In China, the types of EVs influenced the allocation of subsidies [1]. Despite being in its early stages, the second-hand EV market in China, like in many other countries with high EV adoption, signifies an important step forward for the industry. It shows a rise in consumers' confidence in their reliability. Due to the limited transaction volume, it is difficult to fully track used EV sales at present. Changes may be expected in the years to come. Many car manufacturers are extending battery warranty coverage and offering evaluations and certifications for pre-owned EVs. This is anticipated to provide a considerable enhancement for the pre-owned car sales [114]. The typical range of EVs could begin to level off in the next few years due to continuous improvements. Certainly! The average autonomy of electric vehicles increased by approximately 13% each year since 2011 [114], but as standard battery autonomy approaches 400 km, this growth could stabilize in the near future.

A recent study indicates that over 20,000 used EV transactions in China have experienced an average decline in value of 45% in the first year, 58% in the second year, and 67% in the third year [115, 116]. This depreciation happens much faster than with ICE vehicles and in other EV markets. In China, the residual value of electric vehicles decreases significantly quicker than that of internal combustion engine cars, which can retain about 60–75% even after 3 years. The typical income for a dealer is considerably lower, staying under 3000 yuan ($423) [117]. The rapid decrease in worth of Chinese used electric vehicles stems from the rapid technological progress made by Chinese EV manufacturers, resulting in numerous and swift improvements that make older models obsolete. A favorable sign that the market for pre-owned EVs will soon enhance is that high-end EV brands like Tesla show much lower depreciation rates [114, 118, 119].

Used electric vehicles from China could discover new prospects overseas. Undoubtedly, in the past few decades, China has consistently become a significant global player in the manufacturing of both electric and internal combustion engine vehicles. Due to increased exports of new vehicles in recent years, the country surpassed Japan in 2023. In the initial quarter of 2023, China surpassed Japan to emerge as the world's top exporter of new cars, shipping out 1.07 million vehicles, with almost 40% being electric. The year 2009 marked a significant milestone as China surpassed the USA in new automobile sales [120].

When the final owner of an EV decides to replace their vehicle, it can have multiple second uses, whether being utilized locally or sent abroad, as long as the vehicle is fit for the road.

Should equal opportunities for exporting EVs and ICE vehicles arise in the future, experts forecast that 40% of vehicles in Sub-Saharan Africa will be imported EVs. While these predictions could be exaggerated, they suggest a substantial opportunity for exporting pre-owned EVs to developing countries [121].

Developed nations are progressively adopting EVs to lower carbon emissions in road transport. Considering that fleets in developing countries mainly rely on imported used vehicles, these fleets will boost their proportion of EVs by acquiring imported pre-owned EVs rather than new ones. Certainly! As electric vehicles complete their life cycle in developed countries, it is likely that they will increasingly be dispatched to developing nations for repurposing. The outcome of swift depreciation might lead to an increase in EV exports to developing countries. This requires that emerging economies have appropriate infrastructure and policies to promote demand for electric vehicles.

To enhance EV exports to least developed nations and bolster its production, China might establish new collaborations to aid these countries in formulating and advancing their electric transportation policies, particularly concerning charging infrastructure and battery replacement, akin to the trends seen with mobile phone batteries. Countries producing EVs might adopt similar approaches to grow their sales.

In fact, shipments of pre-owned EVs from major electric car markets have risen in recent years. In 2022, China sent out around 70,000 used cars, showing a notable increase from 2021, when under 20,000 vehicles were exported. Roughly 70% of these were electric vehicles. In 2023, the Ministry of Commerce released a draft policy concerning the export of used cars, which, once approved, will allow the export of used cars from all regions of China. The export of used cars from China is anticipated to increase significantly after the 2023 proposal from the Chinese Ministry of Commerce, which allows the export of used vehicles from anywhere in China, unlike the 27 cities and provinces permitted in 2019 to sell abroad any kind of secondhand cars.

6.8 Conclusion and Discussion

In the past 10 years, the EV has become an essential tool for decreasing carbon emissions in transportation on roads. Technological progress in electric cars reached a stage of maturity, yet it remains in development, particularly regarding battery advancements and the integration of AI. The market succeeded in flourishing thanks to subsidies, growing and gaining popularity as EVs offer lower operating expenses compared to ICE vehicles, even though ownership costs are higher when considering depreciation. The pre-owned vehicle market accounts for a considerable portion of the global automotive industry. Pre-owned EVs will also play a crucial role in lowering carbon emissions in road transportation.

Depreciation of EVs is a significant factor for both buyers and manufacturers, and it differs in key ways from ICE vehicles. As presented in this chapter, the breakdown of the main points that are behind the numbers showing the explanation of the depreciation of EVs:

1. Faster Initial Depreciation
   EVs have traditionally depreciated more quickly in the first few years due to:

   - Rapid technological advancements making older models obsolete
   - Concerns about battery longevity and replacement costs
   - Limited public understanding or trust in long-term EV performance

   However, this trend has been shifting in recent years.
2. Battery Life and Warranty
   Battery degradation is one of the biggest concerns affecting resale value. Modern EVs often come with warranties covering 8–10 years or 100,000 (up to 150,000 miles in some cases), which can mitigate depreciation concerns if transferred to second owners.
3. Tax Incentives and Price Drops
   Government incentives (e.g., federal tax credits, state rebates) reduce the new purchase price but don't directly affect resale prices. This can skew the depreciation curve, making a 1- or 2-year-old used EV seem disproportionately cheap.
4. Brand and Model Variability
   Some EVs, like Teslas, hold their value better than others due to strong brand perception, over-the-air software updates, and robust charging infrastructure. In contrast, older or less popular models, such as early Nissan Leafs, may depreciate rapidly.
5. Evolving Market Conditions
   The growing demand for used EVs, especially affordable ones, may contribute to stabilizing depreciation.
   Supply chain issues and high new EV prices can buoy used values.
   Regulatory pushes for zero-emission vehicles can increase used EV demand in some regions.

6. Nevertheless, there are some upcoming concerns. If battery supply costs drop substantially, new EVs may get cheaper, increasing depreciation of older ones. As well, the phasing out of incentives or saturation of the used EV market could accelerate depreciation again.
   The market for pre-owned electric vehicles is still developing, which leads to its uncertain nature. Nonetheless, some elements seem to be essential:

   - Financial incentives for new EVs result in faster depreciation of used EVs, but they assist in making EVs a standard form of transportation. Ultimately, this will benefit the used EV market as consumers develop greater trust in electric cars.
   - The integration of technology in electric vehicles, particularly AI, results in faster value loss of less recent vehicles because of the swift technological progress.
   - The advanced battery technology enhances performance, delivering better battery autonomy for more recent models, while less recent vehicles see decreased electric charge autonomy due to battery degradation over time.

At present, EVs have a shorter range than ICE vehicles, are more expensive, and depreciate faster. Luxury EVs are likely to hold their value better, while budget EVs are expected to depreciate more swiftly.

The growth of electric car sales is likely to coincide with the end of incentives, resulting in price equivalence between EVs and ICE vehicles.

This study is limited by data accessibility and demonstrates the necessity for more organized and systematic data collection to permit more accurate analysis.

The future of depreciation in EVs is being shaped by several dynamic factors, and it's evolving differently from traditional ICE vehicles. It can consequently be presented as a detailed look at what is influencing EV depreciation going forward:

1. Battery Technology Improvements:
   As battery costs fall and lifespan improves, concerns about expensive battery replacements—once a major driver of depreciation—are diminishing.
   Solid-state batteries or improved lithium-ion technologies could further stabilize EV resale values.
2. Brand and Model Differentiation:
   Tesla has shown significantly lower depreciation rates compared to other EV brands due to its strong brand, over-the-air updates, and charging network.
   Lesser-known or short-lived EV models often suffer steeper depreciation due to limited support and uncertain longevity.
3. Government Incentives and Policy Changes:
   Generous EV tax credits or rebates can distort the new vs. used EV market, making used EVs less attractive and pushing resale values down.
   Stricter emissions regulations and ICE bans may reverse this, increasing demand for used EVs and reducing depreciation.
4. Supply Chain and Production Trends:
   Global shortages or surpluses of EVs can temporarily inflate or depress resale prices.

As manufacturing scales up and supply stabilizes, depreciation patterns may become more predictable.

5. Software and Tech Relevance:
   EVs are more like smartphones on wheels—software aging, limited support for older features, and rapid tech innovation can increase depreciation.
   However, brands offering consistent software updates (like Tesla) tend to hold value better.
6. Market Demand and EV Adoption:
   As EV adoption increases and charging infrastructure improves, consumer confidence rises, helping stabilize resale values.
   Conversely, outdated or slow-charging EVs may see sharper value drops.
7. Data Transparency and Tools:
   Some platforms and EV-specific resale sites are increasingly offering accurate resale tracking, which can help reduce uncertainty and potentially flatten depreciation curves.
   In short, depreciation for EVs is becoming more nuanced, tied heavily to brand, battery tech, software relevance, and government policy. While some EVs may continue to depreciate faster than ICE cars in the short term, others, especially from leading brands, may retain value better than average over time.

References

1. Diouf B (2024) The electric vehicle transition. Environ Sci Adv 3:332
2. Patil G, Pode G, Diouf B, Pode R (2024) Sustainable decarbonization of road transport: policies, current status, and challenges of electric vehicles. Sustainability 16:8058. https://doi.org/10.3390/su16188058
3. Schloter L (2022) Empirical analysis of the depreciation of electric vehicles compared to gasoline vehicles. Transp Policy 126:268–279. https://doi.org/10.1016/j.tranpol.2022.07.021
4. Domínguez S, Pedreros S, Delgadillo D, Anzola JA (2024) Depreciation method based on perceived information asymmetry in the market for electric vehicles in Colombia. World Electr Veh J 15:511. https://doi.org/10.3390/wevj15110511
5. Woody M, Adderly SA, Bohra R, Keoleian GA (2024) Electric and gasoline vehicle total cost of ownership across US cities. J Ind Ecol 28:194–215. https://doi.org/10.1111/jiec.13463
6. Muehlegger E, Rapson DS (2022) Subsidizing low- and middle-income adoption of electric vehicles: quasi-experimental evidence from California. J Public Econ 216:104752. https://doi.org/10.1016/j.jpubeco.2022.104752
7. Burra LT, Sommer S, Vance C (2024) Free-ridership in subsidies for company- and private electric vehicles. Energy Econ 131:107333. https://doi.org/10.1016/j.eneco.2024.107333
8. Azarafshar R, Vermeulen WN (2020) Electric vehicle incentive policies in Canadian provinces. Energy Econ 91:104902. https://doi.org/10.1016/j.eneco.2020.104902
9. Costa CM, Barbosa JC, Castro H, Gonçalves R,´endez SL-M (2021) Electric vehicles: to what extent are environmentally friendly and cost effective? Comparative study by European countries. Renew Sust Energ Rev 151:111548. https://doi.org/10.1016/j.rser.2021.111548
10. Donggyun K, Choi M, Yoo N, Shin S, Lee S (2021) A new algorithm for eco-friendly path guidance focused on electric vehicles. Energy 233:121198. https://doi.org/10.1016/j.energy.2021.121198

11. Albrechtowicz P (2023) Electric vehicle impact on the environment in terms of the electric energy source—case study. Energy Rep 9:3813–3821. https://doi.org/10.1016/j.egyr.2023.02.088
12. Sheldon TL, Dua R (2019) Measuring the cost-effectiveness of electric vehicle subsidies. Energy Econ 84:104545. https://doi.org/10.1016/j.eneco.2019.104545
13. Jenn A, Lee JH, Hardman S, Tal G (2020) An in-depth examination of electric vehicle incentives: consumer heterogeneity and changing response over time. Transp Res A 132:97–109. https://doi.org/10.1016/j.tra.2019.11.004
14. Kautish P, Lavuri R, Roubaud D, Grebinevych O (2024) Electric vehicles' choice behaviour: an emerging market scenario. J Environ Manag 354:120250. https://doi.org/10.1016/j.jenvman.2024.120250
15. Li M, Lu J, Chen Z, Amine K (2018) 30 years of lithium-ion batteries. Adv Mater 30(33):e1800561. https://doi.org/10.1002/adma.201800561
16. Manthiram A (2017) An outlook on lithium ion battery technology. ACS Cent Sci 3:1063–1069. https://doi.org/10.1021/acscentsci.7b00288
17. Kim T et al (2019) Lithium-ion batteries: outlook on present, future, and hybridized technologies. J Mater Chem A 7:2942–2964. https://doi.org/10.1039/C8TA10513H
18. IEA (2022a) Global EV outlook 2022, securing supplies for an electric future. https://www.iea.org/reports/global-ev-outlook-2022
19. IEA (2022b) Evolution of li-ion battery price, 1995–2019. https://www.iea.org/data-and-statistics/charts/evolution-of-li-ion-battery-price-1995-2019
20. Suttakul P, Wongsapai W, Fongsamootr T, Mona Y, Poolsaat K (2022) Total cost of ownership of internal combustion engine and electric vehicles: a real-world comparison for the case of Thailand. Energy Rep 8:545–553. http://creativecommons.org/licenses/by-nc-nd/4.0/
21. Verma S, Dwivedi G, Verma P (2022) Life cycle assessment of electric vehicles in comparison to combustion engine vehicles: a review. Mater Today Proc 49:217–222. https://doi.org/10.1016/j.matpr.2021.01.666
22. Joshi A, Sharma R, Baral B (2022) Comparative life cycle assessment of conventional combustion engine vehicle, battery electric vehicle and fuel cell electric vehicle in Nepal. J Clean Prod 379:134407. https://doi.org/10.1016/j.jclepro.2022.134407
23. Engelmann A, Bauer I, Dolata M, Nadig M, Schwabe G (2022) Promoting less complex and more honest price negotiations in the online used car market with authenticated data. Group Decis Negot 31:419–451
24. Oktiyani R, Ela Revita MLD, Sugiarti S (2024) Depreciation calculation using the straight line method is appropriate taxation law. JIM J Ilm Mhs Pendidik Sej 9:636–643
25. Laksman E, Strömberg A-B, Patriksson M (2019) The stochastic opportunistic replacement problem, part III: improved bounding procedures. Ann Oper Res 292:711–733
26. Su C (2024) Literature review on the net present value method of project investment decision. Adv Econ Manag Polit Sci 60:60–66
27. Marzo G, Pagnozzi T (2011) Can depreciation be a variable cost? A comparison between the straight-line method and the units of production method in a lean company context. SSRN Electron J
28. Pratama BA, Zufria I, Fakhriza M (2024) Application of the production unit method to calculate accumulated depreciation of factory machinery using the Django model. Instal J Komput 16:68–77
29. Tandon A (2021) Transition finance: investigating the state of play: a stocktake of emerging approaches and financial instruments. In: OECD environment working papers, vol 179. OECD Publishing, Paris
30. Mishin YV, Mishin AY (2020) Impact of depreciation deduction calculation method on financial soundness improvement of economic agents. E3S Web Conf 208:03033
31. Liapis KJ, Kantianis DD (2015) Depreciation methods and life-cycle costing (LCC) methodology. Procedia Econ Financ 19:314–324

32. Copiello S, Bonifaci P (2018) Depreciated replacement cost: improving the method through a variant based on three cornerstones. Real Estate Manag Valuat 26:33–47
33. Kellenbrink C, Herde F, Eickemeyer SC, Kuprat T, Nyhuis P (2014) Planning the regeneration processes of complex capital goods. Procedia CIRP 24:140–145
34. Al-Saadi M, Mathes M, Käsgen J, Robert K, Mayrock M, Mierlo JV, Berecibar M (2022) Optimization and analysis of electric vehicle operation with fast-changing technologies. World Electr Veh J 13:20
35. Mahmud S, Rahman M, Kamruzzaman M, Ali MO, Emon MSA, Khatun H, Ali MR (2022) Recent advances in lithium-ion battery materials for improved electrochemical performance: a review. Results Eng 15:100472
36. Ghafari V, Bayat S, Akbari A, Yeklangi G (2023) Current and future prospects of li-ion batteries: a review. Nano Sci Technol 8:24–43
37. Koorata PK, Panchal S (2023) Thermal management of large- sized LiFePO4 pouch cell using simplified mini-channel cold plates. Appl Therm Eng 234:121286. https://doi.org/10.1016/j.applthermaleng.2023.121286
38. https://www.afr.com/wealth/investing/lithium-prices-to-keep-rising-as-demand-outpaces-supply-20230508-p5d6m3
39. Wu H, Niu M, Shao Y, Wang M, Li M, Liu X, Li Z (2024) Investigations of Li-Ion battery thermal management systems based on heat pipes: a review. ACS Omega 9:97–116. https://doi.org/10.1021/acsomega.3c08056
40. Kiani M, Omiddezyani S, Houshfar E, Miremadi SR, Mehdi Ashjaee M, Nejad AM (2020) Lithium-ion battery thermal management system with Al_2O_3/AgO/CuO nano fluids and phase change material. Appl Therm Eng 180:115840. https://doi.org/10.1016/j.applthermaleng.2020.115840
41. Kong Q, Fowler M, Entchev E, Ribberink H, McCallum R (2018) The role of charging infrastructure in electric vehicle implementation within smart grids. Energies 11:3362. https://doi.org/10.3390/en11123362
42. https://about.bnef.com/blog/electric-vehicles-accelerate-54-new-car-sales-2040/
43. Pamidimukkala A, Kermanshachi S, Rosenberger JM, Hladik G (2023) Evaluation of barriers to electric vehicle adoption: a study of technological, environmental, financial, and infrastructure factors. Transp Res Interdiscip Perspect 22:100962. https://doi.org/10.1016/j.trip.2023.100962
44. https://about.bnef.com/blog/electric-vehicle-sales-headed-for-record-year-but-growth-slowdown-puts-climate-targets-at-risk-according-to-bloombergnef-report/
45. https://www.goldmansachs.com/insights/articles/electric-vehicles-are-forecast-to-be-half-of-global-car-sales-by-2035
46. Kia Seoul, South Korea
47. https://spectrumnews1.com/ca/la-west/transportation/2023/11/07/evs-depreciate-more-than-any-other-vehicle-type%2D%2Dstudy-says
48. https://www.carparts.com/blog/electric-vehicle-depreciation-101-why-do-evs-lose-their-value-faster-than-gas-cars/?srsltid=AfmBOoqCilEFqYoy2YFDjEjssFfzZE3mzMxZudMEtxYz_J9qFwVvaMg6
49. https://www.drive.com.au/news/toyota-electric-car-resale-values
50. Chen Z, Carrel AL, Gore C, Shi W (2021) Environmental and economic impact of electric vehicle adoption in the U.S. Environ Res Lett 16:045011. https://doi.org/10.1088/1748-9326/abe2d0
51. https://choosemycar.com/resources/buying-and-selling-tips/ev-drop
52. MacInnis B, Krosnick JA, McDonald J (2023) Resistance to purchasing all-electric vehicles: evidence from a national survey. J Environ Psychol 91:102114. https://doi.org/10.1016/j.jenvp.2023.102114
53. https://www.greenenergytimes.org/2024/02/how-ev-tax-credits-affect-vehicle-depreciation/
54. https://www.govtech.com/transportation/does-electric-vehicle-depreciation-cloud-the-electric-future

55. Diouf B, Miezan E (2024) Unlocking the technology potential for universal access to clean energy in developing countries. Energies 17:1488. https://doi.org/10.3390/en17061488
56. Garikapati D, Shetiya SS (2024) Autonomous vehicles: evolution of artificial intelligence and the current industry landscape. Big Data Cogn Comput 8:42
57. Abraham KM (2020) How comparable are sodium-ion batteries to lithium-ion counterparts? ACS Energy Lett 5:3544–3547. https://doi.org/10.1021/acsenergylett.0c02181
58. Harmon JE (2024) Cathode innovation makes sodium-ion battery an attractive option for electric vehicles. Argonne National Laboratory. https://www.anl.gov/article/cathode-innovation-makes-sodiumion-battery-an-attractive-option-for-electric-vehicles
59. Rudola A, Sayers R, Wright CJ, Barker J (2023) Opportunities for moderate-range electric vehicles using sustainable sodium-ion batteries. Nat Energy 8:215–218. https://doi.org/10.1038/s41560-023-01215-w
60. Canals Casals L, Etxandi-Santolaya M, Bibiloni-Mulet PA, Corchero C, Trilla L (2022) Electric vehicle battery health expected at end of life in the upcoming years based on UK data. Batteries 8:164. https://doi.org/10.3390/batteries8100164
61. Chen W, Liang J, Yang Z, Li G (2019) A review of Lithium-Ion battery for electric vehicle applications and beyond. Energy Procedia 158:4363–4368. https://doi.org/10.1016/j.egypro.2019.01.783
62. Zheng Y et al (2013) Cell state-of-charge inconsistency estimation for LiFePO4 battery pack in hybrid EVs using mean- difference model. Appl Energy 111:571–580
63. Etxandi-Santolaya M, Mora-Pous A, Canals Casals L, Corchero C, Eichman J (2024) Quantifying the impact of battery degradation in electric vehicle driving through key performance indicators. Batteries 10:103. https://doi.org/10.3390/batteries10030103
64. Siemens White Paper. The evolution of the electric vehicle market (2020)
65. Guo Y, Shichao W, He Y-B, Kang F, Chen L, Li H, Yang Q-H (2022) Solid-state lithium batteries: safety and prospects. eScience 2:138–163. https://doi.org/10.1016/j.esci.2022.02.008
66. How long does a Tesla battery last? https://ev-lectron.com/blogs/blog/how-long-does-a-tesla-battery-last
67. https://www.in2013dollars.com/us/inflation/2010?amount=1
68. https://www.fullstackeconomics.com/p/untitled-2
69. https://caredge.com/nissan/leaf/depreciation/
70. Electric Vehicles Warranty coverage. https://www.forbes.com/sites/jimgorzelany/2022/10/31/by-the-numbers-comparing-electric-car-warranties/
71. California EV warranty. https://www.evercars.com/articles/ev-warranty-guide
72. Chen W-H, Hsieh I-YL (2023) Techno-economic analysis of lithium-ion battery price reduction considering carbon footprint based on life cycle assessment. J Clean Prod 425:139045. https://doi.org/10.1016/j.jclepro.2023.139045
73. Nissan leaf warranty UK. https://www.nissan.co.uk/owners/car-warranty.html
74. Forsythe CR, Gillingham KT, Michalek JJ, Whitefoot KS (2023) Technology advancement in driving electric vehicle adoption. PNAS Sustain Sci 120(23):e2219396120
75. https://www.mobility.ch/en/magazine/sustainability/electric-car-depreciation-what-is-the-residual-value-of-the-battery
76. https://en.wikipedia.org/wiki/Tesla_Model_S
77. https://www.carparts.com/blog/electric-vehicle-depreciation-101-why-do-evs-lose-their-value-faster-than-gas-cars/?srsltid=AfmBOoqvuRu31ICIBZibQyxKjIAxjHCMW6jNkifRHr3Syk0O8lB3Rbdc
78. https://www.tycorun.com/blogs/news/solid-state-lithium-battery
79. Manthiram A, Yu X, Wang S (2017) Lithium battery chemistries enabled by solid-state electrolytes. Nat Rev Mater 2:16103. https://doi.org/10.1038/natrevmats.2016.103
80. The Nikkei in December 2020
81. Farulla GA, Palomba V, Aloisio D, Brunaccini G, Ferraro M, Andrea Frazzica A, Sergi F (2023) Optimal design of lithium-ion battery thermal management systems based on phase change material at high current and high environmental temperature. Therm Sci Eng Prog 42:101862. https://doi.org/10.1016/j.tsep.2023.101862

82. Talele V, Moralı U, Patil M-S, Panchal S, Fraser R, Fowler M, Thorat P, Gokhale Y-P (2023) Computational modelling and statistical evaluation of thermal runaway safety regime response on lithium-ion battery with different cathodic chemistry and varying ambient condition. Int Commun Heat Mass Transfer 146:106907. https://doi.org/10.1016/j.icheatmasstransfer.2023.106907
83. Myatezh AV, Smirnov MA, Malozyomov BV, Shtang AA (2017) Performance increase of lithium-ion batteries. Adv Eng Res 133:519–524. https://doi.org/10.2991/aime-17.2017.84
84. https://global.nissannews.com/en/releases/nissan-global-ev-sales-surpass-1-million-unit-milestone
85. https://caredge.com/tesla/model-s/depreciation
86. https://medium.com/@bestpriceforcar/how-technological-obsolescence-can-affect-electric-vehicles-117cb3405db3
87. Green CD, Grinyer JR, Michaelson R (2002) A possible economic rationale for straight-line depreciation. Abacus 38:91–120. https://doi.org/10.1111/1467-6281.00099
88. Han X, Liang Y, Ai Y, Li J (2018) Economic evaluation of a PV combined energy storage charging station based on cost estimation of second-use batteries. Energy 165:326e339. https://doi.org/10.1016/j.energy.2018.09.022
89. Parker N, Breetz HL, Salon D, Conway MW, Williams J, Patterson M (2021) Who saves money buying electric vehicles? Heterogeneity in total cost of ownership. Transp Res D 96:102893. https://doi.org/10.1016/j.trd.2021.102893
90. https://www.endurancewarranty.com/learning-center/research/average-vehicle-age-increases-in-america/
91. https://static.ed.edmunds-media.com/unversioned/img/car-news/data-center/2017/feb/used-car-report/used-car-report-q4.pdf
92. https://www.valuepenguin.com/used-car-ages-study
93. https://www.carswithcords.net/2022/05/tesla-model-s-turns-10-decade-of.html
94. Huang S, Zhu Y, Huang J, Zhang E, Tao X (2024) Analysis of circular price prediction strategy for used electric vehicles. Sustainability 16:5761. https://doi.org/10.3390/su16135761
95. https://www.highradius.com/resources/Blog/double-declining-balance-depreciation/
96. https://iea.blob.core.windows.net/assets/a9e3544b-0b12-4e15-b407-65f5c8ce1b5f/GlobalEVOutlook2024.pdf
97. Klein NJ, Basu R, Smart MJ (2023) In the driver's seat: pathways to automobile ownership for lower-income households in the United States. Transp Res Interdiscip Perspect 18:100787. https://doi.org/10.1016/j.trip.2023.100787
98. https://www.nytimes.com/2024/06/03/business/electric-cars-becoming-affordable.html
99. Sordi G, Rondi A, Conti D, Casalegno A, Rabissi C (2024) Degradation of lithium-ion batteries under automotive-like conditions: aging tests, capacity loss and q-OCP interpretation. Future Batteries 3:100005. https://doi.org/10.1016/j.fub.2024.100005
100. https://thundersaidenergy.com/2024/09/05/electric-vehicles-saturation-point/
101. https://www.iseecars.com/most-popular-cars-study
102. Diouf B (2025) Modeling the used vehicle market share in the electric vehicle transition. World Electr Veh J 16:29. https://doi.org/10.3390/wevj16010029
103. Guo Z, Zhou Y (2019) Residual value analysis of plug-in vehicles in the United States. Energy Policy 125:445–455
104. Jian X (2014) Establishment and application of evaluation model of residual value for electric vehicle. Tongji University
105. Zhou Y, Santini D, Stephens T, Ward J (2017) Comparison of value retention of plugin vehicles and conventional vehicles and potential contributing factors. 95th transportation research board annual meeting
106. Zhengping T, Yun C, Yidong W, Pan M (2018) Research on the value evaluation of used pure electric car based on the replacement cost method. Materials Science and Engineering Conference
107. Schoettle B, Sivak M (2018) Resale values of electric and conventional vehicle: recent trends and influence on the decision to purchase a new vehicle. Sustain Worldwide Transp 2018

108. https://www.fastmarkets.com/insights/slowdown-in-china-ev-sales-expected/
109. https://www.reuters.com/business/autos-transportation/global-ev-sales-stay-strong-china-hits-record-despite-end-subsidies-2023-11-14/
110. https://chinadialogue.net/en/business/life-after-subsidies-for-chinas-evs/
111. https://www.canalys.com/newsroom/global-ev-sales-2022
112. https://www.cnbc.com/2024/04/08/china-says-constant-innovations-not-subsidies-behind-its-ev-edge-.html
113. https://www.businessinsider.com/china-q1-electric-vehicle-sales-double-year-over-year-2019-4
114. https://about.bnef.com/blog/nobody-wants-chinas-used-evs/
115. Sheldon TL, Dua R (2020) Effectiveness of China's plug-in electric vehicle subsidy. Energy Econ 88:104773. https://doi.org/10.1016/j.eneco.2020.104773
116. Cao X, Yang W (2019) Establishment of residual value assessment model for electric vehicle based on AHP. IOP Conf Ser Mater Sci Eng 688:033001
117. https://qz.com/interaction-goes-industrial-1851403386
118. https://www.bloomberg.com/news/newsletters/2021-05-18/hyperdrive-daily-nobody-wants-china-s-used-evs
119. https://www.japantimes.co.jp/business/2023/12/25/tech/used-evs-tough-sell-market/
120. https://thediplomat.com/2023/08/chinese-cars-gain-traction-in-latin-america-and-the-world/
121. https://www.itf-oecd.org/sites/default/files/docs/new-but-used-electric-vehicle-global-second-hand-car-trade_0.pdf

Chapter 7
Modeling the Used Vehicles Market Share in the Electric Vehicle Transition

Contents

7.1 Introduction

The used EV market is gaining in volume, as pioneers do not find the technology adequate, or they adopt newer models. A large used EV market is necessary to foster large adoption of the technology. Presently, in developed countries as well as emerging markets, a secondhand ICE vehicle is the primary method of acquisition of a car. As an illustration, in the European Union, 80% of drivers buy a used car; it reaches 90% in developing countries [1]. In the USA (United States of America), on average 70% of cars are bought used, while less than 20% of less affluent families acquire a new vehicle [2].

As EVs become a large-scale mainstream means of transport, the number of secondhand cars increases, even if among new vehicle registrations the ratio of used to new EVs remains small. There are a total of about 40 million EVs on the roads after almost 14 million new registrations in 2023. On the 14 million vehicles, the USA accounted for only 10%, Europe for 25%, and China alone for 60%. Regarding the used EV market, 400,000 vehicles changed owner in the USA, about 450,000 in Europe, and 800,000 in China alone. In 2024, in the USA, second-hand EV sales grew by about 40% compared to 2023. Even though this augmentation is noticeable,

R. Pode, B. Diouf, *Decarbonization of Road Transportation*, Green Energy and Technology, https://doi.org/10.1007/978-3-032-11388-7_7

the used EV market holds a very small proportion of the market compared to the used ICE vehicles that are fully matured: 36 million cars in the USA, about 30 million in Europe, and 20 million in China.

Nevertheless, the price of used EVs is on the continuous verge of dropping, consequently bringing them to be increasingly more competitive compared to their ICE counterparts. The main criteria for buyers to commit to the acquisition of an EV is affordability, followed by range. It is therefore understandable that affordability will stand as leverage to raise up second-hand EV market share. The reference price of USD 30,000 is already above the price of an important proportion of used EVs in the United States [3]. It is even expected that the prices fall close to USD 25,000. Such prices will make these vehicles candidates for a federal USD 4000 tax rebate, making them a more credible alternative to ICE cars. In 2023, a used Tesla, one of the top-selling EV brands in the world, could be commonly found under USD 50,000; at the beginning of 2024, the price was barely above USD 33,000. This makes them direct competitors of used SUVs or even new cars. In Europe, pre-owned EVs may be found in the market between EUR 15,000 and EUR 25,000 (USD 16,000–27,000), while plug-in hybrids are sold around EUR 30,000 (USD 32,000). Subsidies for used EVs are proposed in some European countries, such as the Netherlands (EUR 2000). It is worse noticing that incentives for new vehicles have decreased constantly since 2020, but they remain constant for used vehicles. EUR 1000 subsidies are offered in France for used EVs (EUR 1000). In 2023, second-hand EVs in China were sold on average at around CNY 75,000 (USD 11,000) [3].

Most important EV markets present the same drivers for used and new EVs, excluding the difference in subsidies, which are generally more significant for new EVs. Along with subsidies, technological progress is another important factor driving the used EV market, as well as depreciation [4, 5].

There were 8.1 million new EV registrations in China in 2023, corresponding to a 35% increase from 2022. With only 500,000 second-hand passenger EVs in 2020 and 80,000 units that changed owners in 2023, the share of the used EVs market remains relatively unimportant [6]. These were mostly compact short-range vehicles that held the most important proportion of EVs sold in China between 2015 and 2019 [6]. Subsidies in China were calculated based on EVs' range, and some models on the market were offered to get a maximum benefit from subsidies instead of just buyers' needs. It should be noticed that despite the small market share, the second-hand EV market is an important reality to be considered by the EV industry as proof of growing confidence among buyers. Such confidence may increase soon with manufacturers' extended battery warranties that may lead to major changes in the years to come. The increasing maturity of battery technology may lead to a more important used EV market share as the standard range is nearing 400 km after an average increase of 13% a year since 2011.

The outcome of a study based on 20,000 second-hand EVs sold in China shows a value diminution of 45% observed after the first year, 58% the year after, and 67% after 3 years of operation [7, 8]. EVs in China depreciate much faster than ICE cars or EVs in other important markets. ICE vehicles in China can retain 60% to 75% of their value after the first 3 years of operation. A classical dealer's margin can be considered relatively small, as it is under USD425 [9]. The fast depreciation of EVs in China more than in other markets is due to fast technological progress making older models prematurely obsolete. High-end EVs, such as Tesla models, hold their value better [6, 10–12]; this raises hopes for the future of EVs.

Even if the secondhand EV market in China remains small, there is an important potential for exports. China progressively gained a global leader position in the automobile sector in EVs but also in ICE vehicles as well. Chinese exports of new vehicles increased in the past years. With 1.07 million vehicles exported, China overtook Japan and became the world leader in 2023; almost 40% of those vehicles were EVs. As a turning point, in 2009 China passed the USA in new car sales [13]. China's exports are expanding to second-hand vehicles, as the export of second-hand vehicles became legal in 2019. This development in the automobile in China holds great potential for the export of second-hand vehicles in markets such as Sub-Saharan Africa or developing countries in general [14].

EVs are the main tool to decarbonize road transport around the world. Developing countries that so far rely on imported used vehicles will most likely build their future fleet based on used EVs from developed countries, as it is currently done for ICE cars that are given a second life.

Developing countries that need charging infrastructures for their EV fleet may need assistance from partners as main importers. China exported about 20,000 used vehicles in 2021 and jumped to 70,000 used cars in 2022, which made up 70% of EVs.

The Chinese Ministry of Commerce issued a change of policy in 2023 by approving second-hand car exports from every country round in comparison. Comparatively, in 2019, exports were allowed only in 27 localities (cities and provinces).

Many factors are subject to influence the adoption of the second-hand EV instead of a new one. In first place there is affordability, but it is a double-edged factor, as the lower cost of EVs is becoming a reality, but as long as subsidies make new EVs more affordable, the choice of a used EV over a new one will not be straightforward. Subsidies negatively affect depreciation, so it is also for the range decrease that is related to battery technology and fast technological progress that makes older EV models prematurely obsolete, contributing therefore to their depreciation.

As it is now for the mature ICE car market, a major point in the EV transition is to predict how long it will take before new and used EVs reach an equilibrium in their market shares. The objective in this chapter is to contribute to bridging this gap in the EV literature by proposing a model to describe the dynamic until equilibrium in the evolution of the second-hand market share vs. new EVs in the EV transition. In its hypothesis, the model considers an S-curve for both new and used EV adoption curves. The S-curve parameters for new EV adoption are based on collected data, and a hypothetical ratio between new and used EVs is taken as an input variable, while the output gives the time it will take to reach equilibrium between new and second-hand EVs in the EV transition. Such an approach is new in the literature in the field of electric transport. The two types of S-curves differ in their conditions to pick up and progress with a given speed of adoption. It will therefore be included in the variation in the parameters in the S-curve for new EVs compared to used EVs to describe the time-independent speed of adoption of each one of these two qualities of vehicles. The evolution of the adoption of second-hand EVs may be a symptom of the fast depreciation of EVs that can correspond to slower adoption of used EVs. Fast depreciation vs. high demand of used EVs will be considered in the equation of the S-curve.

The methodology in this research is based on data collection to extract the parameters of the S-curve that describes the adoption of new EVs. The S-curves for used EVs are based on parameters selected depending on two major different situations: high demand of used EVs or low demand that may correspond to fast depreciation of EVs.

7.2 The S-Curve and the EV Transition

The way a technological innovation is adopted in a society follows a non-linear path known as an S-curve (Fig. 7.1); the first stage is slow and the second step is fast before the last slow stage corresponding to saturation. The shape of an S-curve is attributed to system feedback: economies of scale, social diffusion, and technological reinforcement as well.

As an innovation in a market, the adoption of the EV evolves following an S-curve (Fig. 7.1), with an introduction phase and a rapid growth before saturation as a last step [15–17].

The slope of the curve describes how fast the technology is accepted after an initial introduction phase; it will depend on the context of each market before full adoption is eventually achieved. The pickup is slow before an exponential growth,

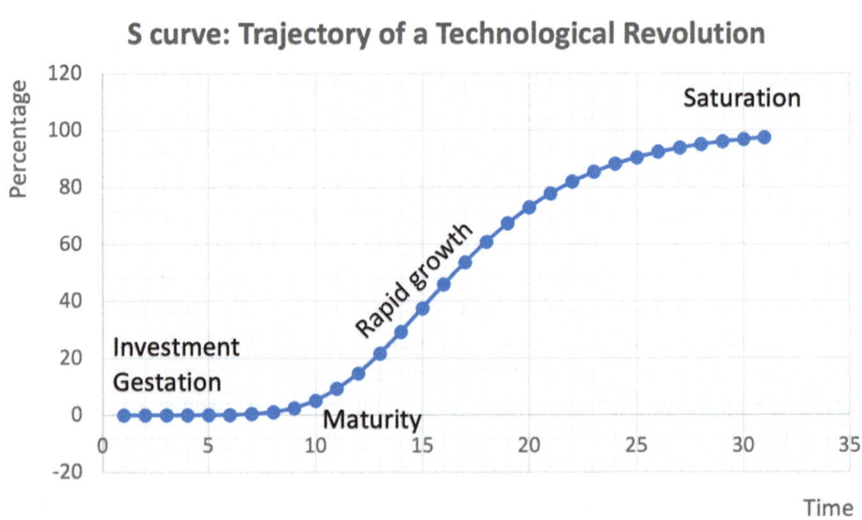

Fig. 7.1 S-curve as a typical trajectory of a technological transition

and a slowdown before saturation is eventually reached for a full transition, as in any S-curve characterizing a market penetration.

The major increase in new registrations for all countries occurs at an inflection point corresponding to maturity on Fig. 7.1, as if there was a form of confidence from buyers after pioneers opened the game. There is then an exponential growth till a certain level before a slowdown to convince the most reticent buyers. Once efficiency is proven and a solid base of trust is built, the majority of buyers will invest in the accepted innovation. The last category of buyers steps at a slow pace to complete the full EV transition. Among all the buyers, some can afford new products; others will wait before they can have access to preowned products.

Figure 7.2 shows the evolution of the ratio of EVs in new registrations in markets with some of the most meaningful EV penetration: Norway, Sweden, Denmark, the Netherlands, Germany, the United Kingdom, France, Portugal, and the United States of America [18]. These curves show the early stages and pickup of an S-curve.

An S-curve model can provide a description of the EV penetration in these markets. The rate R(t) of market penetration of EVs as a function of time is given by Eq. (7.1) below, where each market can be well represented by two variables α and k to be found by matching the curve R(t) and the actual data [18]. The couple of variables, α and k, will be specific to each market and found when graph R(t) matches the graph obtained from actual data.

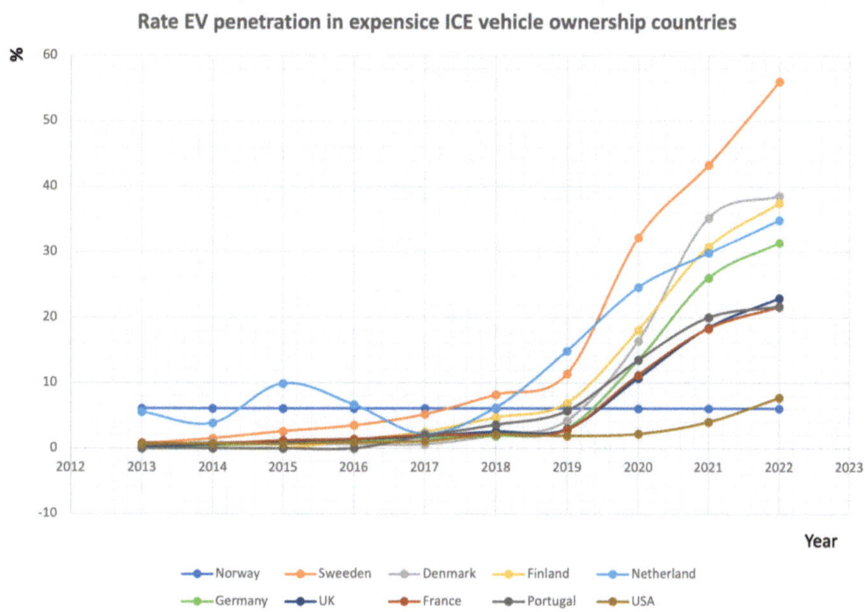

Fig. 7.2 The rate of EV penetration in expensive ICE vehicle ownership countries from 2013 to 2022. (Reprinted from [18])

$$R(t) = \frac{100}{\left[1 + \exp(-\alpha t)\right]^k} \quad \textit{where}\, \alpha > 0\, \textit{and}\, k > 0\, \textit{and where}\, t\, \textit{is the time}, \tag{7.1}$$

The meaning and values of α and k are presented in detail in Reference [15]. Figure 7.3 shows Norway's EV transition actual data and simulation. Here are the values of α and k for the country with the highest EV penetration rate, Norway: $\alpha = 0.29$ and $k = 11$ [19].

The above curve shows that Norway will reach 100% EV transition by 2030, corresponding to 100% new registrations made of EVs.

While the global EV penetration rate and its characteristic variables, which are extracted from Fig. 7.4, are $\alpha = 0.12$ and $k = 11$ [20].

The above Fig. 7.4 predicts a global full EV transition around 2055.

Figure 7.5 above shows the effect of the change in the value of coefficient α for a given value of k. As α increases, the pickup time gets shorter and the slope is sharper, meaning a faster adoption of the technology after pickup. Smaller values of α give curves with a smaller slope, equivalent to a slower adoption of the technological innovation.

Figure 7.6 above shows the effect of variation of coefficient k for a given value of α. As variable k changes, the slope remains constant, but the pickup time increases, meaning a later maturity of the technological innovation. The pickup time increases with the value of k, as shown in Fig. 7.6 above.

There is an important question in the unicity of the couple (α, k) for the S-curve to match the actual data that describe the EV penetration rate in each market. It can be proven that the couple (α, k) is unique for a given market.

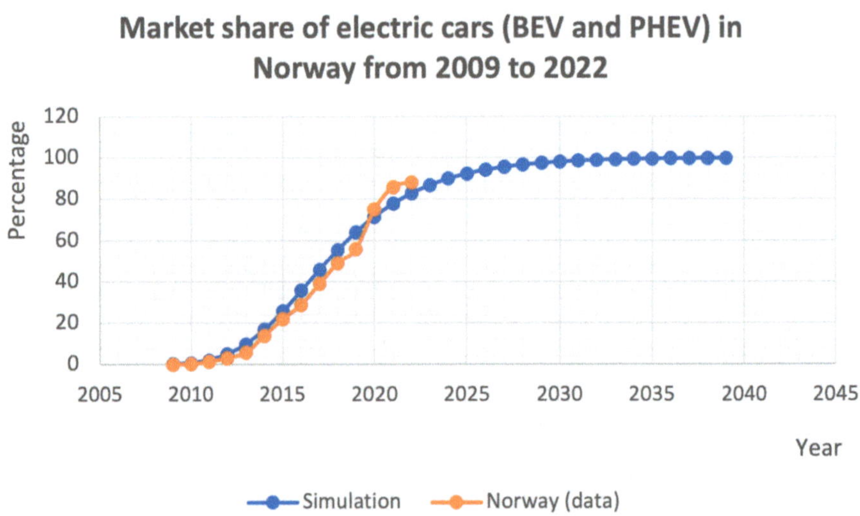

Fig. 7.3 Norway EV Transition statistics and simulation. (Reprinted from [5])

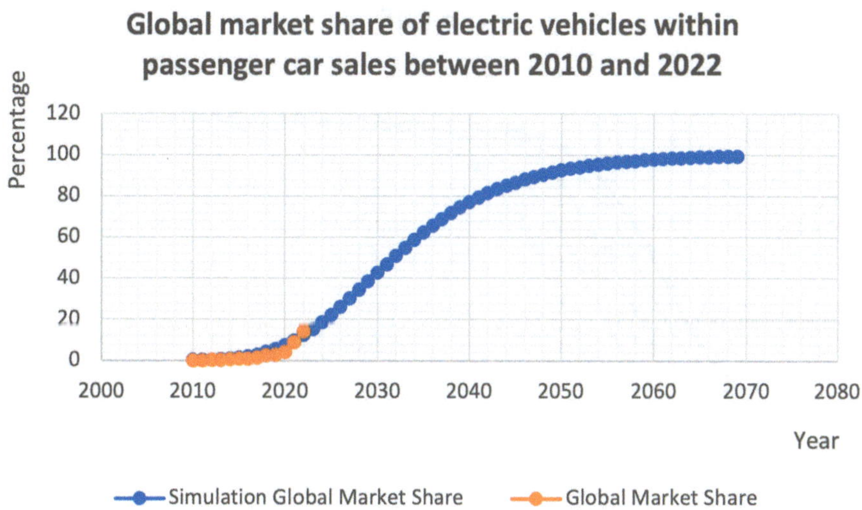

Fig. 7.4 Global EV Transition statistics and simulation. (Reprinted from [21])

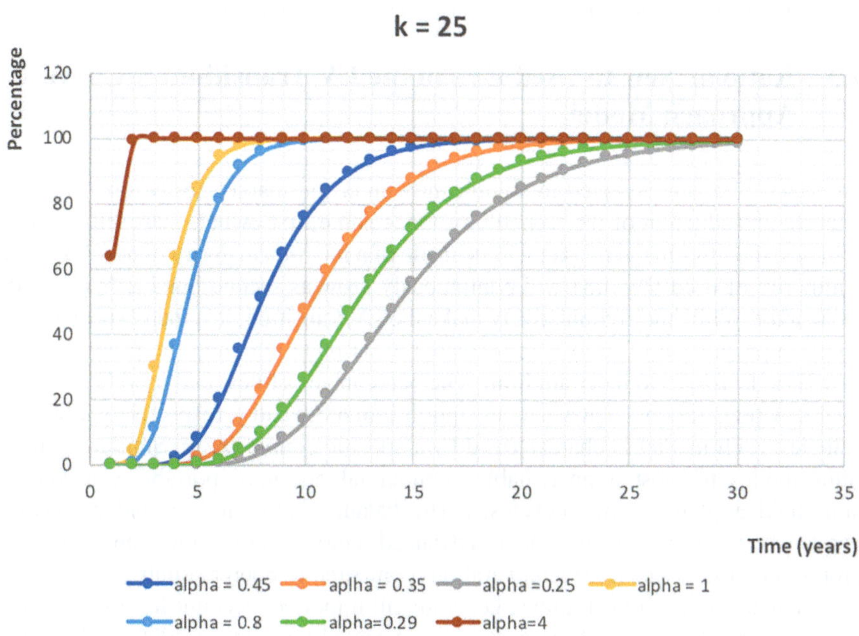

Fig. 7.5 Evolution of the curve R(t) for a given value of k = 25 and different values of α

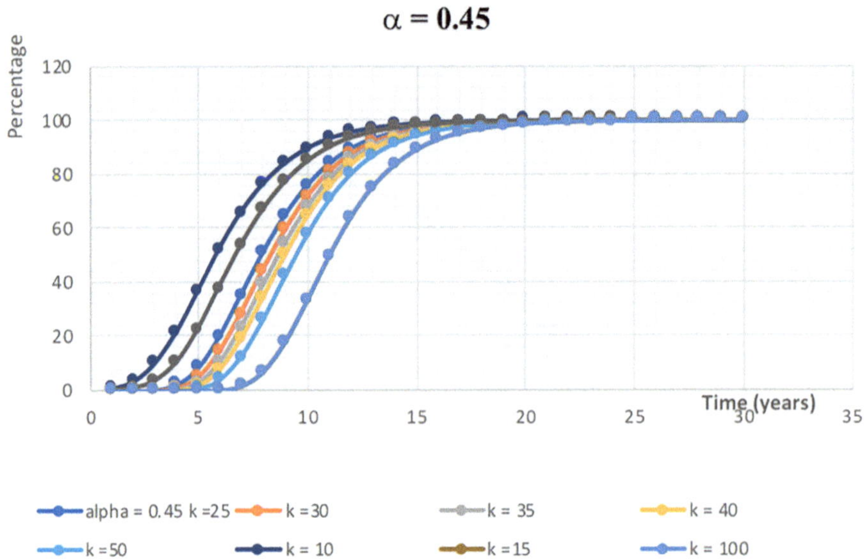

Fig. 7.6 α = 0.45 with different values of k, larger values of k corresponding to a later pickup of the technology, and the slopes remain almost the same as long as α is constant

7.3 Ratio of New to Used EVs in the EV Transition: S-Curve Approach Model

Preowned EVs are expected to significantly enter the market not before 10 years after the introduction of the technology. Once some first owners start selling their vehicles, the offer in the market will be made of a mix of a majority of new EVs and a minority of used EVs that will continuously grow as second-hand EVs arrive in the market. While they cumulatively make the EV transition, it is then expected that after the proportion of used EVs increases to its maximum and the proportion of new EVs decreases to a certain limit. The percentages of new and used EVs in the total new registrations to arrive to an equilibrium are x and y, respectively. For ICE vehicles, each market is characterized by a specific couple (x, y). Similar equilibrium applies to most nonperishable commercial products: portable electronics, household appliances, motorcycles… The balance between new and preowned goods will be market dependent. In advanced nations can be found an important proportion of new cars in new registrations compared to poorer countries where the new vehicles in new registrations, generally all imported, account for a very small ratio. Nevertheless, around the world it is expected that almost 100% of EVs are purchased new for the first decade.

Used EVs are considered a new technology in this model, exactly as they are for new EVs. The hypothesis is therefore that in the market, the adoption of the second-hand EVs will follow the same course as new EVs. In the early stage of adoption, only a few preowned EVs are available for sale, resulting in only a few potential buyers who may be hesitant to adopt them for many different reasons. These reasons include the cost of acquisition relative to new EVs, especially when subsidies significantly benefit new EV purchases; concerns regarding battery reliability and range; cost of maintenance; issues of obsolescence; and, importantly, warranty considerations. As the number of EVs increases, there will be more second-hand EVs available, and more consumers will buy second-hand EVs before a last step when more reticent buyers will acquire a used EV. EVs' fast depreciation will play a major role in this process. Ultimately, there will be an equilibrium in the ratio between new and used EVs in the market.

The EV transition is described by an S-curve [15–17] for both new and used EVs as a hypothesis, with the following equations:

R(t) is for the total penetration rate of EVs, u(t) is the penetration rate of used EVs, and n(t) is for new EVs,

$$R(t) = \frac{100}{\left[1+\exp(-\alpha t)\right]^{k}}, \quad \text{where } \alpha > 0 \text{ and } k > 0 \tag{7.2}$$

The penetration rate of used EVs u(t) is so that:

$$u(t) = \frac{y}{\left[1+\exp(-\beta t)\right]^{m}}, \quad \text{where } \beta > 0 \text{ and } m > 0, t > n \text{ years} \tag{7.3}$$

$$\mathrm{u}(\mathrm{t}) = 0, \quad \text{if } \mathrm{t} \le \mathrm{n} \text{ years} \tag{7.4}$$

The proportion of new EVs:

$$n(t) = R(t), \quad t \le n\ years \tag{7.5}$$

$$\mathrm{n}(\mathrm{t}) = \mathrm{R}(\mathrm{f}) - \mathrm{u}(\mathrm{t}) \quad \text{if } \mathrm{t} > \mathrm{n} \text{ years} \tag{7.6}$$

where the variables α and β, as well as k and m, determine the first inflection of the curves as well as the slopes. In Fig. 7.7 below, the hypothesis is $\alpha = \beta$ and k = m, it corresponds to a case of high demand of second-hand EVs resulting from a low depreciation of EVs.

The output of the model, its prediction, is the number of years necessary to find an equilibrium in the ratio of new to second-hand EVs in new registrations. The input data will be the parameters α, k, β, and m, as well as the expected ratio at equilibrium. The final ratio will depend on policies not presented in this manuscript.

Figure 7.7 above is obtained for the global market based on the hypothesis α = 0.25 and k = 11, α = β and k = m [15], and supposing that at saturation new registrations are made at 70% of used EVs and 30% of new EVs, this is the current ratio for ICE vehicles in the USA as a reference market [2]. With such a hypothesis, based on Fig. 7.7 for α, k, β, and m, the market reaches equilibrium more than 30 years after initial adoption of EV technology.

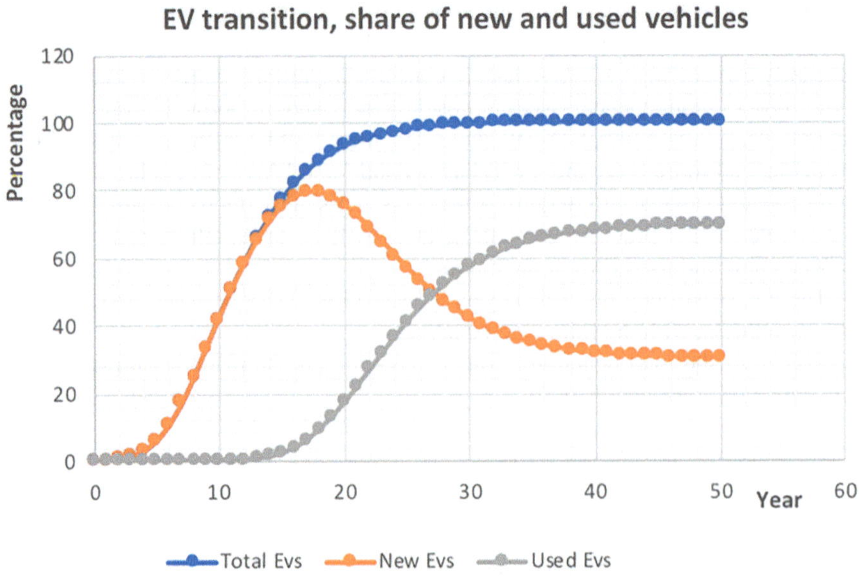

Fig. 7.7 Market share between new and used items, $\alpha = 0.25$ and $k = 11$, $\alpha = \beta$ and $k = m$

Based on Fig. 7.7, globally, 26 years after the adoption of EVs, the proportion of new vehicles and used vehicles will be the same in new registrations: 50% each. The proportion of new EVs will continue decreasing, and that of used EVs will increase until equilibrium for 14 more years, making it 40 years after the introduction of EV powertrain technology.

In the scenario of high demand of used products, it is possible to have $k = m$; the second-hand goods are introduced in the market very early after the technology is adopted. It is the case when exceptional production does not match demand, and used products can be almost as expensive as new products or even more expensive because the waiting list may be too long and there is a sort of long-term shortage of the product on the market; hypothetically there is no depreciation, and there may even be a gain in value.

If the product faces rapid depreciation, the supply is much more important than the demand; the condition will be $m > k$ and $\beta < \alpha$ (Fig. 7.8) as the demand for used products will be much lower than for new products, and it corresponds to a delayed pickup and a slower market progress of used vehicles compared to new vehicles.

Variables: $k = 11$, $m = 11$, $\alpha = 0.25$, $\beta = 0.1$

In the case of rapid depreciation, as in Fig. 7.8 above, the proportion of new EVs will remain more important than that of used EVs in new registrations long after the full adoption of the technology.

***USA market*: $\alpha = 0.13$ and $k = 11.1$**

The USA market is characterized by $\alpha = 0.13$ and $k = 11.1$ based on actual data and S-curve based simulation as presented in Figs. 7.9 and 7.10.

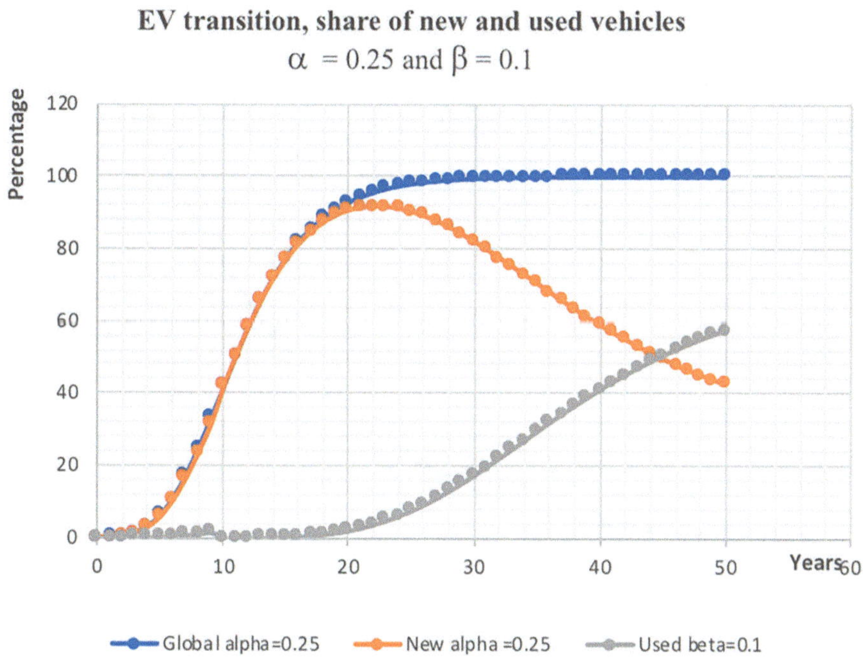

Fig. 7.8 Market share between new and used items, $k = m = 11$, $\alpha = 0.25$, and $\beta = 0.1$

In 2023, a total of 1.4 million new EVs were sold in the USA versus 400,000 used EVs sold. This gives a total of 1.8 million new EV registrations in 2023, corresponding to a proportion of 22.22% of used EVs.

Figure 7.9 above shows the progress in the EV adoption in the US market the first 10 years.

Figure 7.10 above shows the prediction of the EV adoption in the US market based on the actual data for the first 10 years as represented in Fig. 7.9 and a simulation based on the S-curve approach.

Presently, due to different factors, EVs depreciate faster than any other powertrain technology. In 2023, EVs lost on average almost 50% of their initial value after the first 5 years [22]. The Chevrolet Bolt, with 51.1% loss in value, and the Nissan Leaf, with a 50.8% loss after only 3 years of operation, are among the fastest depreciating EV models [23].

7.3.1 β and the Depreciation Rate

Figure 7.11 above shows the comparison between the total EV penetration in the US market and the share of the used market in new EV registration for different values of β, from fast adoption of second-hand EVs corresponding to $\alpha = \beta = 0.13$ to

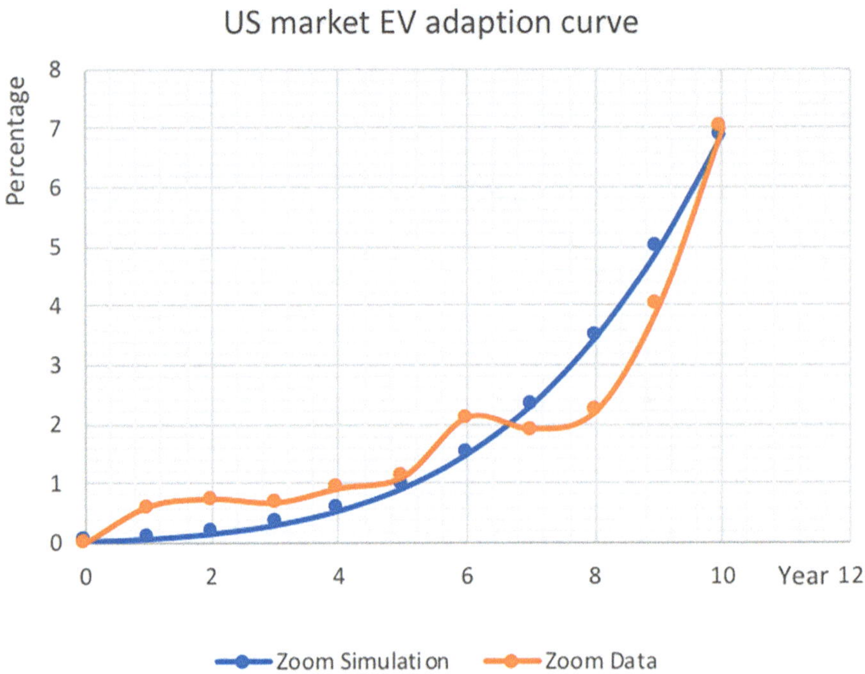

Fig. 7.9 US market EV adoption ratio in new registrations between 2012 and 2022

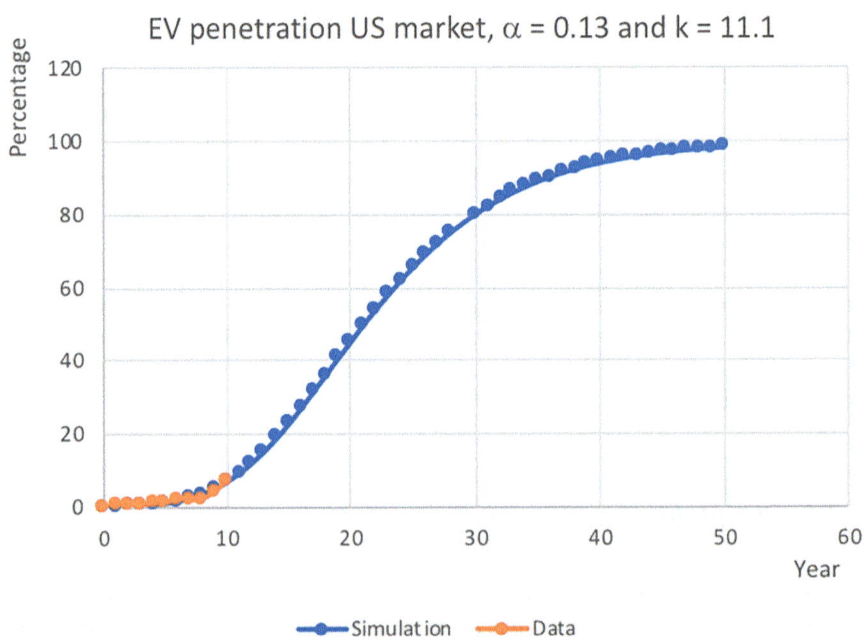

Fig. 7.10 Prediction of the US market EV transition

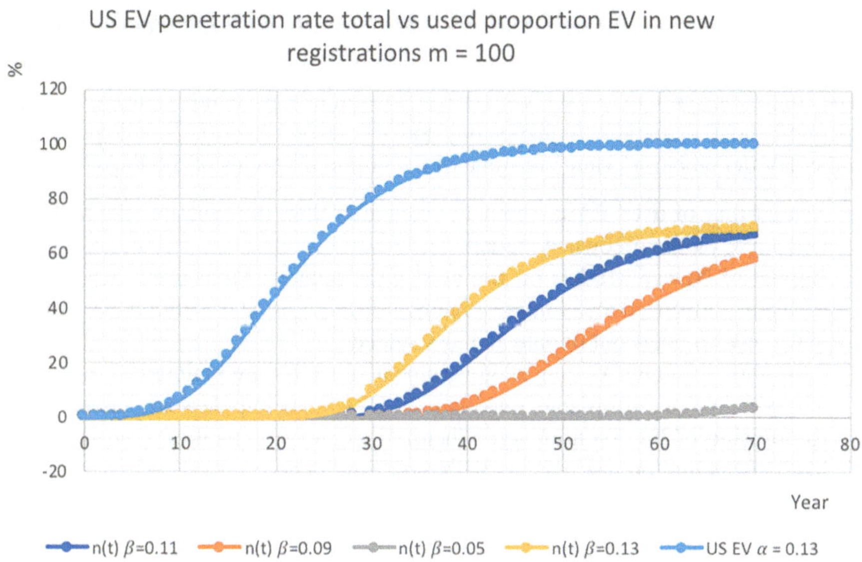

Fig. 7.11 US market share between new and used items, m = 100, α = 0.13, and β = 0.05, 0.09, 0.11, and 0.13

β = 0.05 matching with slow adoption of used EVs. The values of β may decrease as the depreciation rate increases.

7.3.2 Different Factors Affecting EV Depreciation

The depreciation of EVs is mainly function of different points that can be listed below [24]:

- **Impact of technology**
- **Brand Reputation**
- **Government Incentives**
- **Mileage and Age**
- **Battery Health and Maintenance**
- **Battery cost and lifespan**
- **Consumer appreciation**
- **Market Share**
- **Availability of charging Infrastructure**

7.3.3 Repercussion of Used EVs Market on EVs Depreciation

The continuous growth of the EV market will result in a more important share of the second-hand EV market. A higher demand for second-hand EVs will positively affect EV depreciation. Here are some of the important points of the impact.

- **Availability of used EVs:** About 10 years after their introduction, the first EVs will arrive on the preowned market.
- **Impact of Newer Models on the used EVs market:** More recent EV models with more advanced technological integration, such as AI or better battery range, make older models out of date and less valued.
- **Economic aspects:** As long as government subsidies are maintained, new EVs will remain more affordable, making used EVs financially less attractive.
- **Reputation of EV Technology:** When EV technology matures, consumers are likely to acquire pre-owned cars with more confidence.

7.4 Results and Discussion

Depreciation of EVs, particularly due to progress in technology and subsidies, has an important effect on the dynamic in the second-hand market. In fact, there is no doubt that EVs are operational and practical; the progress in battery technology plays a major role in the widespread adoption of EVs [25, 26]. Batteries with higher energy density, fast and efficient charging, and longer life make EVs practical and viable commercial goods and open opportunities for a secondhand market. But EVs are more than just an electric motor with a battery. Technology, including battery development, plays a major role in EVs' depreciation rate [26] and therefore on the second-hand EVs market. Fundamental additions come from software integration and artificial intelligence, allowing EVs to offer a revolution in driving experience and sustainability. Innovative software, data, and systems used by a computer to instruct operations are an integral part of EV functionality development. The software offers multiple advantages compared to traditional mechanical systems; it allows updates and improvements without the need for physical modifications.

The software plays the role of backbone of EV operation. The software oversees management and operation of the vehicle. Data collection is a full part of the system as it is reintegrated to evaluate the vehicle's performance and possible improvements.

There are five major types of axes in EV software application and development, all related to vehicle control and management.

- Vehicle design and materials: It oversees the performance and adequate functionality of key components of the vehicle: the battery, electric motors, brakes, lights…
- Maintenance: Management, predictive maintenance, and safety
- Monitoring: EV issues and EV driver interaction
- Navigation: particularly to reduce range anxiety with the location of public charging stations

- Vehicle to grid: interconnectivity between vehicle and grid for more sustainability, allowing car owners to go off-grid.

In addition to battery technology, each of these five points above, when it reaches maturity, may have an important effect on depreciation rates and consequently on the second-hand EV market share progression and future equilibrium in the proportion of used EVs vs. new EVs in new registrations.

The model in this chapter to predict the time needed to reach equilibrium in the market shares between new and used EVs based on the dynamic of adoption of the two types of vehicles is realistic. Nevertheless, the parameters β and m in the S-curve, in Eq. (7.3), describing used EV adoption, are static or time-independent. Such a point could be opened to debates, as the time needed before equilibrium is reached can be so long that the parameters β and m in the equation describing u(t) can get their value changed. The variation of β and m will depend on the evolution of depreciation that affects the adoption of used EVs. The rate of adoption of used EVs itself may affect the depreciation of EVs, generating a form of wave that can ultimately lead to higher or lower depreciation rates.

The market for used EVs also affects the depreciation of electric vehicles. With the expansion of the EV market, the proportion of the secondhand EV market will also become more prominent, and these pre-owned vehicles will experience improved depreciation rates. Here are some key aspects concerning the effect of the second-hand EV market on the depreciation of pre-owned EVs.

- Disponibility of used EVs: following over a decade of use, the initial batch of EVs will enter the used car market. This market will continue to expand. The growing availability of used EVs could result in a saturation where supply exceeds demand, causing a quicker depreciation rate.
- Impact of newer models on the used EV market: More recent EV models with more advanced technology, such as integration of AI or better battery range, make older models out of date and less valued by buyers.
- Economic factors: Government incentives for new EVs increase their affordability, rendering used EVs less appealing to buyers unless they are priced substantially lower.
- Consumer awareness and trust in EV technology: as EV technology advances and matures, consumers will be more inclined to invest in used EVs, leading to increased value.

The EV transition will be made of new EVs for those who can afford them and used EVs for the most important proportion of the population in different countries. Nevertheless, the adoption of used EVs may change as technology progresses or subsidies [17–20, 23, 24] vary. A completely matured technology will generate depreciation rates less dependent on obsolescence.

The model presented in this investigation is based on a static approach of the S-curve with parameters β and m. Considering the above comments, a more accurate approach of β and m could be taken time-dependent, and β(t) and m(t) could be used to predict the amount of time needed to reach equilibrium in each market.

$$u(t) = \frac{y}{\left[1 + \exp\left(-\beta(t)t\right)\right]^{m(t)}} \tag{7.7}$$

β(t) is particularly important in the speed of adoption of used EVs if the S-curve is adopted as a fit description for the adoption of EVs in the electric mobility transition.

Equation (7.7) above is the S-curve equation for the secondhand EV adoption based on dynamic parameters m(t) and β(t) that inform, respectively, about the time-dependence of the pickup time of the secondhand EV market as well as the time-dependence of the speed of adoption after pickup.

7.5 Conclusion

The model, presented in this study, which is based on the S-curve, describes the adoption of an innovation within a market context. Both new EVs and pre-owned EVs are considered innovations, with used EVs serving as options for consumers with lower budgets and being introduced into the market with a certain delay compared to new EVs. This model examines the coexistence of new and used EVs in the context of new registrations. It further predicts the number of years required to achieve market equilibrium after a continuous increase in the proportion of used EVs and a corresponding decrease in the proportion of new EVs, spanning from the initial introduction of EV technology to its full market maturity.

Secondhand products may be highly demanded or face a low demand. Fast depreciation corresponds to low demand of used EVs, but as subsidies are cut or gradually decrease and battery quality increases along with the cost drop, EVs will keep their better value, and they will very likely push up the demand. In this model it is supposed that the final balance between new and used EVs in a given market will be similar to that of ICE vehicles. Such a hypothesis may need more attention due to the specific reality of EVs. Increased gathering of current data will improve the forecast of the eventual balance between new and used EVs in the marketplace and the shift towards full EV adoption. The variables k, α, m and β are characteristic of a market. High demand for used EVs will lead to a shorter period before equilibrium is reached, in contrast to low demand.

This research is limited to a standard situation but could be extended to different circumstances that could affect the evolution of an innovation in a given market. For example, safety concerns could be included with the battery, as in the early days of the EV, or ICE vehicle manufacturers switching to EVs could bring to the market new products and a new dynamic in customers' choices. In that regard, the results could be more accurate with time-dependent parameters in the S-curve equations, particularly for used EVs.

Understanding the dynamics and ratio between new and used EVs is fundamental regarding the future of waste management as battery recycling or reconditioning and the EV industry in general.

The hypothesis of the ratio of new to used vehicles being considered to be the same for ICE vehicles and EVs as well can be discussed further due to the faster depreciation of EVs now strongly affecting the used EVs market.

This research shows the importance of systematic data collection as well to better understand the statistical behavior and dynamics in the EV transition and even in the adoption of other technologies.

References

1. https://iea.blob.core.windows.net/assets/a9e3544b-0b12-4e15-b407-65f5c8ce1b5f/GlobalEVOutlook2024.pdf
2. Klein NJ, Basu R, Smart MJ (2023) In the driver's seat: pathways to automobile ownership for lower-income households in the United States. Transport Res Interdiscip Perspect 18:100787. https://doi.org/10.1016/j.trip.2023.100787
3. https://www.nytimes.com/2024/06/03/business/electric-cars-becoming-affordable.html
4. Sordi G, Rondi A, Conti D, Casalegno A, Rabissi C (2024) Degradation of lithium-ion batteries under automotive-like conditions: aging tests, capacity loss and q-OCP interpretation. Future Batteries 3:100005. https://doi.org/10.1016/j.fub.2024.100005
5. Etxandi-Santolaya M, Mora-Pous A, Casals LC, Corchero C, Eichman J (2024) Quantifying the impact of battery degradation in electric vehicle driving through key performance indicators. Batteries 10:103. https://doi.org/10.3390/batteries10030103
6. https://about.bnef.com/blog/nobody-wants-chinas-used-evs/
7. Sheldon TL, Dua R (2020) Effectiveness of China's plug-in electric vehicle subsidy. Energy Econ 88:104773. https://doi.org/10.1016/j.eneco.2020.104773
8. Cao X, Yang W (2019) Establishment of residual value assessment model for electric vehicle based on AHP. IOP Conf Ser Mater Sci Eng 688:033001
9. https://qz.com/interaction-goes-industrial-1851403386
10. Lebeau P, Macharis C, Van Mierlo J (2019) How to improve the total cost of ownership of electric vehicles: an analysis of the light commercial vehicle segment. World Elect Vehicle J 10:90. https://doi.org/10.3390/wevj10040090
11. https://www.bloomberg.com/news/newsletters/2021-05-18/hyperdrive-daily-nobody-wants-china-s-used-evs
12. https://www.japantimes.co.jp/business/2023/12/25/tech/used-evs-tough-sell-market/
13. https://thediplomat.com/2023/08/chinese-cars-gain-traction-in-latin-america-and-the-world/
14. https://www.itf-oecd.org/sites/default/files/docs/new-but-used-electric-vehicle-global-second-hand-car-trade_0.pdf
15. Diouf B (2024) The Electric Vehicle Transition. Environ Sci Adv 3:332
16. Patil G, Pode G, Diouf B, Pode R (2024) Sustainable decarbonization of road transport: policies, current status, and challenges of electric vehicles. Sustainability 16:8058. https://doi.org/10.3390/su16188058
17. Muehlegger E, Rapson DS (2022) Subsidizing low- and middle-income adoption of electric vehicles: Quasi-experimental evidence from California. J Public Econ 216:104752. https://doi.org/10.1016/j.jpubeco.2022.104752
18. Burra LT, Sommer S, Vance C (2024) Free-ridership in subsidies for company- and private electric vehicles. Energy Econ 131:107333. https://doi.org/10.1016/j.eneco.2024.107333
19. Azarafshar R, Vermeulen WN (2020) Electric vehicle incentive policies in Canadian provinces. Energy Econ 91:104902. https://doi.org/10.1016/j.eneco.2020.104902
20. Costa CM, Barbosa JC, Castro H, Gonçalves R, Lanceros-Mendez S (2021) Electric vehicles: to what extent are environmentally friendly and cost effective? Comparative study by European countries. Renew Sustain Energy Rev 151:111548. https://doi.org/10.1016/j.rser.2021.111548

21. Ku D, Choi M, Yoo N, Shin S, Lee S (2021) A new algorithm for eco-friendly path guidance focused on electric vehicles. Energy 233:121198. https://doi.org/10.1016/j.energy.2021.121198
22. Albrechtowicz P (2023) Electric vehicle impact on the environment in terms of the electric energy source—case study. Energy Rep 9:3813–3821. https://doi.org/10.1016/j.egyr.2023.02.088
23. Sheldon TL, Dua R (2019) Measuring the cost-effectiveness of electric vehicle subsidies. Energy Econ 84:104545. https://doi.org/10.1016/j.eneco.2019.104545
24. Jenn A, Lee JH, Hardman S, Tal G (2020) An in-depth examination of electric vehicle incentives: consumer heterogeneity and changing response over time. Transport Res Part A 132:97–109. https://doi.org/10.1016/j.tra.2019.11.004
25. Pamidimukkala A, Kermanshachi S, Rosenberger JM, Hladik G (2024) Barriers and motivators to the adoption of electric vehicles: a global review. Green Energy Intell Transport 3:100153
26. Gautam P, Pode G, Pode R, Ayetor GK, Diouf B (2024) Depreciation in the electric vehicle transition: sustainability of the second-hand electric vehicle market. Vehicles 6:2044–2074. https://doi.org/10.3390/vehicles6040101

Chapter 8
The Second-Hand Market in the Electric Vehicle Transition

Contents

8.1 Introduction

EVs have progressed into a dependable technology; certain countries aim for 100% of new vehicle registrations to be made of EVs [1, 2]. Presently, EVs and ICE vehicles have comparable performances even if EVs show shorter ranges and faster depreciation rates. Nevertheless, according to the literature review [3–5], it appears that faster depreciation of EVs, compared to ICE vehicles, has limited impact on the vehicle purchasing decision-making process. However, upon closer examination,

R. Pode, B. Diouf, *Decarbonization of Road Transportation*, Green Energy and Technology, https://doi.org/10.1007/978-3-032-11388-7_8

purchasers indirectly factor in the depreciation of EVs among their primary motivations or apprehensions about buying an EV. In fact, in priority they take into account aspects like price, range, and charging network accessibility. Other consumer preferences and context can have major influences on the market evolution. It can be about technology, brand reputation, oil prices, or other economic conditions. For example, when gas prices rise sharply or environmental regulations tighten significantly, there can be an important surge in the demand for EVs, resulting in higher prices and slower depreciation. Conversely, if incentives decrease or consumer focus turns to the newest technology cars, the interest of potential buyers in EVs might wane, resulting in a rapid decline in value.

Another important element leading to the decline in the value of electric cars concerns the after-warranty coverage. It is primarily due to the potential need for battery replacement. These elements are the primary variables influencing the pace of depreciation of EVs. This can be comprehended by examining how the loss of value of EVs affects drivers' choices to purchase an electric car. Therefore, it is essential to investigate the reality of EVs and the depreciation of popular models. Manufacturers and governments promote EVs via decarbonization strategies for road transport because of reduced ownership expenses, ignoring the variable depreciation that occurs.

Governmental structures provided subsidies for EVs to encourage their uptake [6–8] and lessen reliance on oil simultaneously. EVs are eco-friendly compared to ICE cars [9–11], and they generate zero emissions. Conversely, subsidies may serve as a means to enhance the local EV manufacturing sector.

Subsidies for acquisitions have led to a distortion in the market [12–14]. The anticipated removal of subsidies is likely to inject new energy into the pre-owned market because of a greater cost disparity between new and used electric cars. However, focusing solely on operation and not on acquisition will probably see barely a minimal impact. This will stay valid unless there is significant advancement in energy storage that makes autonomy decrease less sensitive. Battery presents the double role of driving EV market and being a main player in EV depreciation. In fact, recognized for their impressive energy density and compactness, Li-ion batteries have outperformed earlier battery technologies regarding power generation, charge sustainability, and diminished environmental harm [15–17]. In the last almost 30 years, Li-ion batteries have seen an annual increase in efficiency of 5–10% and a decrease in cost of 15–18% [18, 19]. This rapid progress has resulted in an extraordinary rise in the worldwide total of EV fleet, which expanded more than 1000-fold from 2010 to 2024 [18]. Forecasts from the International Energy Agency (IEA) indicate a possible increase of 12–21 times in EV stocks from 2021 to 2030 [18]. Even with this hopeful outlook, numerous obstacles still need to be tackled, such as safety concerns; problems regarding mineral availability and their social and policy consequences; environmental challenges tied to mineral extraction; infrastructure challenges; and disposal challenges. All these may contribute directly or indirectly to the adoption but also the depreciation of EVs.

At the moment, the future impact of subsidy reduction is not clear even if the existence of certain companies may be threatened. As subsidies come to be removed, a few EV manufacturing companies might struggle to remain competitive. Producers with sufficient influence might take in the subsidy reduction, whereas lesser competitors could encounter significant challenges. Consequently, just a handful of big corporations might control the market down the line. Nonetheless, various factors may still propel the expansion of the EV market, such as increasing fuel costs, governmental regulations, and advancements in technology. Should these factors develop positively, the EV market could keep expanding swiftly in the years ahead without financial aid.

The EV market after subsidies could introduce a new dynamic in the pre-owned electric car market around the world.

The decline in value of electric cars affects both owners and manufacturers. Currently, ICE vehicles maintain the same appreciation more effectively compared to EVs [3, 20–22]; this may change as the EV market reaches maturity, particularly in technology. Tesla cars are a good illustration, as they depreciate less rapidly primarily because of the brand's strong reputation and established cutting-edge technology.

As technology advances and consumer demand rises, it is anticipated that EVs will hold their value more effectively, resulting in a reduced pace in loss of value and a bigger used car market.

The decline in value of electric cars is a real issue during an evaluation period of 4–5 years after the purchase of a new vehicle. The swift loss of value might postpone the time when EVs reach the used car market.

The research on depreciation concentrated primarily on calculation techniques related to different types of products. The existing literature on depreciation methods has identified a variety of approaches that allow public and private organizations to compute asset depreciation more precisely and in accordance with their unique requirements. These techniques have developed throughout the years, adjusting to various settings. The straight-line method is among the easiest and most widely utilized techniques. Alternative methods encompass component depreciation, which analyzes distinct parts of the asset individually, recoverable amount, revalued cost, and variable depreciation [23, 24]. The declining balance method speeds up depreciation during the initial years of the asset's lifespan. The double declining balance method is notably aggressive, whereas the declining balance paired with straight-line depreciation permits a shift to linear depreciation in the subsequent years. Additional variations consist of variable declining balance, stepped declining balance, adjusted declining balance, and component declining balance [25, 26]. The units of production method for depreciation relies on the asset's actual usage, like operating hours or the number of units created. This method has variations such as adjusted production units depreciation and component production units depreciation [27, 28]. The sum-of-years digits method is an alternative accelerated depreciation approach that allocates a greater expense during the initial years. Variants consist of the altered sum-of-years digits, reverse sum-of-years digits, component-based sum-of-years digits, and productivity-modified sum-of-years digits [29, 30]. The general accelerated depreciation method encompasses methods like the modified accelerated cost recovery system. Other options include alternative accelerated depreciation and customized accelerated depreciation, designed to meet particular requirements [31]. There exist various less frequent yet equally beneficial methods and approaches in particular situations. Included are the replacement value method, which determines depreciation based on the expense of replacing the asset instead of its initial value; the parabolic-Kuentzle method, which modifies depreciation following a parabolic curve, suitable for assets that experience non-linear wear; the Ross method, applied in cases where the asset's depreciation occurs irregularly throughout its useful life; and accelerated depreciation for intangible assets, which emphasizes the swift reduction in value of assets like patents or copyrights [32–34].

The role of second-hand EVs in the EV transition is becoming increasingly important, especially as more countries push for decarbonization of transport. Lower upfront costs make second-hand EVs more accessible to middle- and low-income households. Pre-owned EVs help democratize clean transport, not just keeping it in the hands of wealthier early adopters. The faster market penetration increases the overall share of EVs on the road more quickly. A robust secondhand market helps reduce the average vehicle age and phase out older, polluting vehicles. Extended vehicle lifespan encourages circular economy principles by extending the usable life of batteries and vehicles. Supports sustainability by reducing waste and maximizing the value extracted from manufacturing emissions.

Nevertheless, battery health with battery degradation over time affects range and resale value.

A healthy second-hand EV market indicates a mature EV ecosystem; buyers, sellers, service networks, and regulations are all in place. There are major challenges, as battery replacement cost can be high for older EVs, affecting resale value. Misinformation or lack of consumer knowledge about EVs may deter used EV purchases. There is currently a limited supply of used EVs, but this will improve as more new EVs age. Fleet retirements from companies and ride-share services will significantly expand the second-hand EV pool. Smart reconditioning (battery swaps, software updates) will help older EVs remain viable. Second-hand EVs will likely become the gateway for EV adoption in developing countries.

This chapter explores the interplay between the different factors affecting the value of EVs and the second-hand market. The main factors that affect EV depreciation are subsidies, progress in technology, including batteries, and battery deterioration. The battery technology affects EVs in different aspects: performance that makes EVs a practical reality and makes older models obsolete, cost that affects the price of new EVs and deterioration that impacts depreciation. Clearly, the speed of depreciation of EVs does not benefit a significant second-hand market, especially in the initial phases of an innovative industry like EVs. It is fair to anticipate that once the elements causing the depreciation of EVs are managed, especially subsidies and improved battery technology, and price parity is reached between EVs and ICE vehicles, the used electric car market will expand significantly.

The approach to sustainability in this manuscript is the ability to maintain or support over time subsidies of all sorts before reaching maturity of markets and the price parity between new EVs and new ICE vehicles. Particularly in the context of dependence on natural resources like fossil fuels and environmental issues like global warming. It emphasizes the need to use financial resources to foster technological progress in a way that meets current needs to reduce pollution from the transport industry without limiting the possibility of developing an important future pre-owned market.

This study attempts at understanding the second-hand EV market in relation to technological progress and depreciation. This involves analyzing how rapid advances in EV technology affect the resale value, adoption trends, and overall market behavior of used EVs. The study of depreciation of EVs in this manuscript attempts at understanding how their value declines over time, particularly in the first 3–5 years, compared to their initial purchase price. The goal of this study includes:

- The drivers of the EV market are cost of ownership, technology, and subsidies.
- Analysis of resale value trends across different EV makes and models in different countries.
- Comparison of depreciation rates of EVs versus ICE vehicles.
- Assessment of the impact of governments' policies or market conditions on resale values.
- Identification of other major factors affecting EV depreciation, such as progress in battery technology and battery degradation, swift progress in technology such as AI and brand perception.
- The evolution of the second-hand EV market

Regarding the monetary analysis, the depreciation in this manuscript is generally presented within a country based on a given currency. Even though the depreciation rates can be compared between two different countries or economic zones.

8.2 EV Market Drivers from Consumers' Perspective

8.2.1 The Battery

The battery is the first factor that made EVs a practical reality and continuously drives their uptake. Advancements in energy storage technology will be determinant in the environmental effects and sustainability of the EV transition. Among different possible technologies, Li-ion based chemistry seems to have a bright future even with advancements in alternative technologies like sodium-ion (Na-ion) [35–37].

A value between 100% and 80–70% is an acceptable state of health (SOH) for batteries [38]. When the capacity drops below 80–70%, it raises some issues, and the battery must be replaced for reliable EV functioning. The cost of battery replacement remains exorbitant.

Certain technical and economic issues in EV batteries involve the tradeoff between cost and the limited availability of certain elements like cobalt. Lithium Iron Phosphate (LFP) [1, 38–40] technology, which is free from cobalt, is commonly utilized in EVs, yet it still has a lower energy density compared to Lithium Nickel Manganese Cobalt (LNMC) technology [1, 38, 41].

At present, a typical EV is observed to have a yearly average loss of 2.5% of the battery capacity in EVs annually, which means that in 7–10 years, their capacity declines to 70–80%, necessitating a battery replacement [38, 42]. Consequently, due to the SOH, a decade-old EV will not be appealing to buyers.

Almost 60% of the value of an EV was from the battery in 2015; by 2020, this dropped to 33%, and it is projected to reach 20% by the end of 2025 [43]. Additionally, enhanced battery longevity will alleviate consumer concerns regarding replacement expenses and after-warranty support. Customer concerns include not just battery expenses but also quicker charging, extended range, and increased energy density for lighter automobiles.

Another consequence of the drop regarding battery prices is the upcoming price equivalence between EVs and ICE cars. It would be a perfect situation for reduction of incentives [43].

Progress in Li-ion battery technology has been determinant in the uptake of electric cars. Improvements in battery technology have been essential for the adoption of electric vehicles. Major progress was achieved in safety, performance, and range by using more sustainable materials, enhanced durability, and batteries with higher energy density.

Technological progress is still advancing, as significant achievements remain attainable in energy density for batteries that are lighter and offer longer range, especially aiming towards generalized solid-state batteries [44]. In spite of all the significant accomplishments that have established the EV as the leading option for decarbonizing road transportation, the battery appears to be the primary instrument that enables EVs to become a sustainable reality, supported by an adequate charging network. However, the primary factor that distinguishes the EV driving experience from that of ICE vehicles extends beyond just the powertrain: it is probably the technology that introduces the most significant innovation within the EV experience, and this reality becomes increasingly evident over time with the growing adoption of DNN (deep neural networks), ML (machine learning), and AI (artificial intelligence). Such programs progressively establish autonomous driving as a norm in the EV experience, along with enhanced battery management, increased safety, improved driver-vehicle interaction, and soon greater automobile internet, similar to the evolution in mobile communication (2G through 5G). Adopting a smartphone from a basic 2G phone allows greater possibilities than phone calls. Swift advancements in technology appear to be the key influence in the electric mobility business; they significantly contribute as well to the quick decline in EVs' value.

Lithium batteries using solid electrolytes without a separator, or solid-state Li-ion batteries, are considered the batteries of the future, offering improved energy density and greater safety [45, 46].

Lithium will very likely continue to hold an important role in the electric car industry. Innovative battery technology is being developed rapidly by car manufacturers, which includes solid-state batteries. The improved storage capacity may improve autonomy by up to 50%. A prototype was created by Toyota for a vehicle that utilizes a solid-state battery, with manufacturing expected to begin in 2024 [47].

Energy storage in electric transportation demands an extended lifespan, lower degradation rates, reduced volume, and less susceptibility to external factors. A significant level of investigation is underway to improve battery quality and efficiency via experiments, theoretical methods, or simulations [48–53]. To achieve such an objective, more research is needed in materials and design for higher efficiency, security, and better control of external factors [48–53].

8.2.2 Cost of Ownership

An automobile's global cost of ownership comprises the buying price, any grants, and operational costs, including loss of value over time; this parameter is often omitted or entirely disregarded in comparisons between EVs and ICE cars.

The reduced ownership costs of EVs compared to ICE vehicles are frequently highlighted as a key factor in countries with significant EV adoption [1]. Reduced operational costs are accompanied by subsidies and rapid technological advancements to enhance the shift to EVs. The latter two elements contribute to the widespread acceptance of EVs and the quick depreciation of EVs. High-tech EVs, such as Tesla models, featuring advanced technology and AI integration, experience a lower depreciation rate due to higher market demand compared to more affordable EVs, exemplified here by the Nissan Leaf. These simpler cars experience a steeper rate of depreciation. There is continuous progress in battery technology and performance, featuring greater storage density that enables extended range at reduced costs as battery prices consistently decline [54, 55]. The growing technological incorporation in the automotive sector, especially in EVs, is AI; AI aids in machine learning, safety, efficiency, battery management, and driver assistance. The expense of ownership is a key factor in the EV sector, especially in nations where EV adoption is at its peak, as shown in Table 8.1. The cost of ownership primarily relates to the price of electricity in comparison to gasoline. Nations that have the cheapest electricity and the most expensive gasoline are likely to see the greatest EV adoption rate, especially in urban regions. In Table 8.1 it is indicated that the average EV driver is a person earning an important revenue, who lives in a country where gasoline is significantly more expensive than electricity, and resides in an urban area not densely populated and where electric vehicles have a certain level of acceptance.

Norway, Iceland, and Sweden therefore exhibit the most important levels of EV adoption (Table 8.1).

In 2022, EV sales accounted for 14% of the worldwide car market [41], and various studies estimate that by 2040, EV sales will comprise 54% of new vehicles available on the global market, with forecasts indicating approximately 70 million EVs sold by that year [43, 44].

The main consideration for car buyers when opting for an EV is affordability, followed by the vehicle's range and the availability of public charging stations [1]. From the perspective of the owners, numerous other criteria can be articulated in terms of cost-effectiveness. This suggests a potentially thriving second-hand EV

Table 8.1 Factors influencing the EV market excluding subsidies in the top 20 markets

Country	% of e-cars in new registrations in 2023 [56]	Number of cars per 1000 [57]	kWh per capita [58]	1 L/1 kWh [1]	GDP per Capita [59]	Density pop/km^2 [60]	Urban Pop (%) [61]
Norway	86	550	28,000	16.2	89,154	43	84
Iceland	72	779	51,920	15.4	68,728	4	94
Sweden	43	473	16,290	5.1	61,029	60	88
Denmark	35	472	59,000	3.6	68,008	352	88
Finland	31	660	14,880	4.9	53,655	43	86
Netherlands	30	501	6710	4.0	57,768	520	93
Germany	26	578	5640	3.5	51,204	605	78
Switzerland	22	546	8820	8.6	91,992	545	74
Portugal	20	549	4440	6.1	24,567	289	67
United Kingdom	19	480	4070	4.4	46,510	715	84
France	19	570	8380	9.1	43,659	303	82
Belgium	18	507	6380	3.4	51,247	985	98
China	16	194	7100	15.0	12,556	385	64
Italy	9	682	4460	3.4	35,657	509	72
Spain	8	553	5850	4.6	30,103	243	81
Greece	7	550	5700	7.3	20,193	205	80
Canada	7	707	15,780	10.9	51,988	11	82
South Korea	6	500	12,030	12.2	34,998	518	81
United States	5	850	12,710	5.6	70,249	91	83
New Zealand	4	889	8530	8.4	48,781	49	87

market if the cars are priced competitively against new models and the battery preserves a considerable part of its initial capacity to satisfy range requirements. Subsidies have a major influence on buying costs, thereby impacting the gap between new and pre-owned EVs.

The two top countries in the list of countries with the highest EV penetration rates, Norway and Iceland, occupy the second and fourth places in GDP per capita as presented in Table 8.1. Norway and Iceland also have some of the highest ratios of gasoline cost over electricity.

It is an indication that in the absence of subsidies, the operational expenses of an EV would be significantly more advantageous in these countries than in any of the other nations on the list.

Ultimately, the markets that exhibit the greatest EV adoption are those where owning ICE vehicles is costlier and in nations with more substantial subsidies.

Many of the other factors are closely linked to batteries, including vehicle cost, EV performance, safety, and/or longevity. Despite the competitive maturity of EV technology, further research is still required due to market demands. As time goes on, more research is needed to develop batteries with enhanced performance that

Table 8.2 Electric vs. ICE Kia Soul cost of ownership for the first 5 years of operation

Kia Soul	Electric	ICE
Purchase cost	$43,095	$20,490
Fuel	$0.133/kWh	$1.18/L
Average Energy consumption	17 Wh/100 km	8.6 L/100 km
Total cost of Energy 5 years	$1130.5	$5074
Accumulated traveled	50,000 km	50,000 km
5 years Value Loss	52%	42%
Cost of ownership after 3 years	$23,539.9	$13,679.8

can compete better regarding cost, energy density, longevity, and simpler material recycling.

The non-subsidy parameters listed in Table 8.1 above clearly apply to both new EVs and used EVs. The primary distinction between owners of new EVs and those of used EVs is probably related to budget.

The hidden factor of the reduced ownership expense concerning electric cars is their steeper depreciation rate, which is often overlooked in comparison with ICE cars considering cost of ownership is considerably more important for an EV than for an ICE vehicle of the similar brand and model.

For instance, in 2023, the price of an electric Kia Soul is about USD 43,095 USD the cost, whereas a gasoline Kia Soul costs USD 20,490 [62]. Table 8.2 shows the difference in cost of ownership for the first 3 years, between an electric Kia Soul and an ICE Kia Soul while the vehicles operated under the manufacturer's warranty coverage.

Table 8.2 juxtaposes EV and ICE Kia Souls, considering the vehicles' depreciation. An in-depth evaluation of the swift decrease in EV values should lead to an extended ownership period for EV buyers to maintain reasonable ownership expenses.

Evidently, the depreciation rate of electric vehicles will differ by country depending on various influencing factors. For instance, attraction for pre-owned EVs will rise, and the rate of depreciation will fall as subsidies decrease. Subsidies have an important impact on the time required to achieve parity in the cost of ownership between EVs and ICE vehicles.

Attaining equal cost of ownership between electric and internal combustion engine vehicles will offer a notable financial advantage to support the shift towards full EV adoption alongside a more dynamic used EV marketplace.

8.2.3 Technology

Technology is the primary force accelerating the growth and transformation of the EV market. Technology is the primary driver of the EV transition, enabling cleaner, more efficient, and economically viable alternatives to ICE vehicles. Technology is not just enabling EVs; it is redefining how engineers design, build, drive, and interact with vehicles. It is accelerating a systemic shift from fossil fuels to electrified, smart, and sustainable transportation. Lithium-ion batteries are becoming more efficient, storing more energy in smaller, lighter packs, increasing EV range. Solid-state batteries promise even greater safety and performance [63]. Battery prices have fallen dramatically over the last decade, making EVs more affordable (Fig. 8.1). Cost per kilowatt-hour (kWh) has dropped from over $1000 in 2010 to under $150 in 2024, making EVs more accessible [64].

The energy density of Li-ion batteries has tripled-plus since the early 1990s. Future gains (lab-scale solid-state, anode-free designs) may double these figures, but for now, commercial limits hover in the 300–350 Wh/kg range with ~500 Wh/L volumetric energy density for premium cells in 2025. The evolution of energy density in lithium-ion batteries from 1992 to 2030 reflects major advancements in materials science, battery design, and manufacturing [65].

Table 8.3 shows the evolution of the battery density of Li-ion batteries from the early 1990s.

Table 8.4 shows the technical progress while comparing the 2011 Nissan Leaf and the 2018 Nissan Leaf. It is particularly striking to see the difference in battery size 24 kWh in 2011 versus 40 kWh in 2018 and the price difference $37,550 versus $36,200 for the 2018 model. As after inflation adjustment the 2011 model with lower specifications is more expensive than the 2018 model, this is almost essentially due to the battery price drop from about $800/kWh in 2011 to about $200/kWh in 2018 (Fig. 8.1). This comparison is particularly relevant due to the sharp price drop between 2011 and 2018. The battery price drop is much slower after 2018.

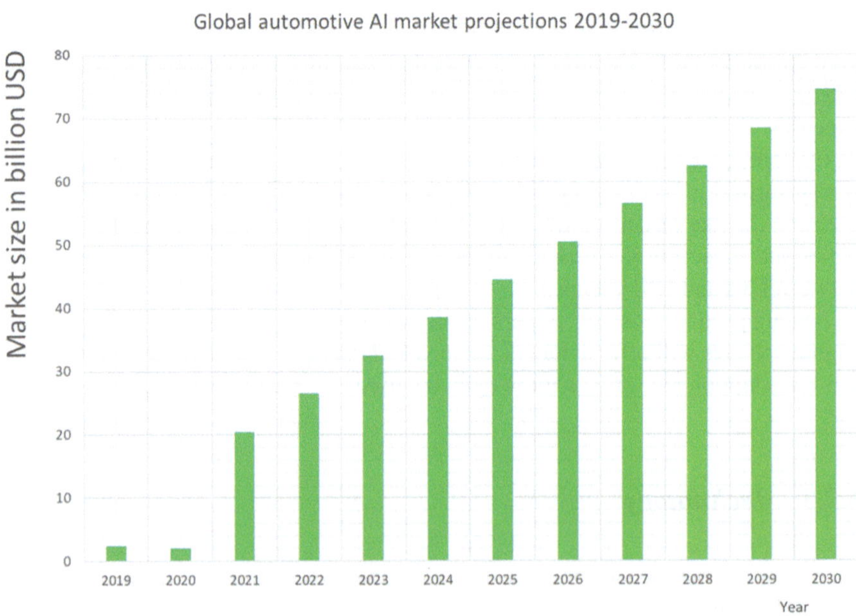

Fig. 8.1 Global automotive AI market projections 2019–2030 [67]

Table 8.3 Evolution Li-ion battery density of energy 1992 to 2025 and projection to 2030 [65]

Year	Average energy density (Wh/kg)	Notes
1992	~90–100	Early commercial Li-ion (Sony) using $LiCoO_2$ cathodes
2000	~120–140	Gradual improvements in cathode/anode materials
2010	~180–200	Improved electrolytes, better electrode packing
2020	~250–270	Nickel Manganese Cobalt (NMC), Nickel Cobalt Aluminum (NCA), cathodes, advanced manufacturing
2023	~280–300	High-Ni NMC/NCA, silicon-anode blends emerging
2025	~320–350	More silicon in anodes, better electrolytes
2030 Forecast	~400–500+	Potential commercial use of solid-state batteries

Table 8.4 Comparison between Nissan Leaf 2011 and Nissan Leaf 2018 [66]

Nissan Leaf 2018  SL	VS.	Nissan Leaf 2011 SL 
110 kW	Motor	80 kW
147	Horsepower	107
236 lb-ft	Torque	207 lb-ft
8.8 s	Acceleration 0–80	10 s
40 kWh	Battery size	24 kWh
0.28	Drag coefficient	0.29
150 miles (EPA)	Driving range	73 miles (EPA)
3508 lbs	Curb weight	3375 lbs
30 cu.ft.	Max cargo cap.	24 cu.ft.
106'	Wheelbase	106'
MacPherson Strut Front, Tension Beam Rear	Suspension	MacPherson Strut Front, Tension Beam Rear
5	Seating capacity	5
$36,200	Base MSRP	$34,540
N/A	Inflation adj. MSRP	$37,550 (2018)

Developments in fast-charging technologies (like solid-state batteries and ultra-fast chargers) reduce charging times from hours to minutes. Smart grid integration allows optimized charging during off-peak hours, reducing stress on electricity networks. Wireless and bidirectional charging (vehicle-to-grid, V2G) is emerging to enhance convenience and grid resilience. Modern electric drivetrains are more efficient and require less maintenance than ICEs (over 85% efficiency vs. ~25–30%). The Regenerative Braking technology captures kinetic energy and converts it into electricity, extending vehicle range. Over-the-air (OTA) updates, autonomous driving, and vehicle-to-grid (V2G) communication are transforming EVs into smart mobility platforms. AI and machine learning enable better vehicle diagnostics, reducing downtime and improving safety. Rapid growth in public and home charging stations makes EV ownership more convenient. Regenerative braking and advanced thermal management improve overall energy use. Software updates (like Tesla's OTA updates) enhance features and fix issues post-sale. Lightweight materials (e.g., aluminum, carbon fiber) help reduce energy consumption. Automation and robotics increase manufacturing efficiency and lower costs. Recyclable and sustainable materials are being integrated into battery and car production. In wireless charging, inductive charging pads and dynamic charging roads are in development for hands-free energy replenishment. In recycling technologies, innovations in battery recycling and second-life use reduce environmental impact. Automakers are adopting sustainable practices, like using renewable energy and eco-friendly materials. There is a more important integration of renewable energy. EVs can act as energy storage units for smart grids, aiding in the integration of solar and wind power. Vehicle-to-grid (V2G) and vehicle-to-home (V2H) technologies allow EVs to supply energy back to homes or the grid during peak demand. Technological strides in sensors and AI are making autonomous EVs more feasible.

AI supports machine learning, safety, efficiency, battery control, driver engagement, intelligent driving … AI is increasingly essential to vehicle technology, and additionally, its incorporation in EVs is accelerating quickly, as illustrated in Fig. 8.1.

Algorithms inspired by evolutionary technologies are increasingly being explored and applied in the field of autonomous driving to enhance decision-making, control systems, and adaptation in complex environments. These algorithms are often bio-inspired and fall under the broader categories of evolutionary computation and genetic programming.

These include Genetic Algorithms (GA), Genetic Programming (GP), Evolution Strategies (ES), and Differential Evolution (DE). EAs are used to generate optimal or near-optimal driving paths while avoiding obstacles, considering dynamic constraints, and maintaining safety. GAs can evolve rule-based or neural network-based driving policies by simulating various driving scenarios. Balancing speed, fuel efficiency, comfort, and safety under different driving conditions. Evolves artificial neural networks (ANNs) using genetic or evolutionary methods. Notable techniques include:

- NEAT (Neuro-Evolution of Augmenting Topologies)
- CMA-ES (Covariance Matrix Adaptation Evolution Strategy)

Applications include:

- Train neural networks that map raw sensory inputs (camera, LiDAR, radar) directly to control commands.
- Neuro-evolution helps design controllers that adapt to new or unseen driving conditions, such as inclement weather or road anomalies.
- Evolved models can improve upon human demonstrations by optimizing performance over time.

Evolutionary algorithms are particularly suited for training in simulated environments, where many scenarios can be tested, mutated, and evolved over generations. Tools like CARLA or LGSVL simulators allow rapid iteration of evolved driving strategies.

Genetic programming can evolve symbolic representations of decision trees or logic structures that combine multi-sensor inputs for robust decision-making.

Evolution strategies help fine-tune fusion parameters under varying environmental conditions.

EAs support online adaptation by evolving driving strategies in real-time or semi-real-time, especially for:

- Changing traffic patterns
- Novel urban layouts
- Hardware degradation or sensor failures

Good for non-differentiable, noisy, or multi-modal optimization problems. Can explore novel solutions that might be hard to discover via gradient-based learning. Flexibility in evolving both control logic and model architectures. Evolutionary methods require large numbers of simulations. Evolved solutions in simulation may not always generalize well to real-world environments. Some evolved behaviors might be hard to interpret or verify. Uber ATG and NVIDIA have experimented with evolutionary methods in reinforcement learning setups for driving policy optimization. Similarly, DeepMind has used neuro-evolution (like NEAT) to train control policies in complex simulations, which can be translated to driving domains.

Ride-sharing and subscription models are reshaping car ownership patterns, favoring electric fleets. AI and machine learning improve battery management, predictive maintenance, and route optimization. Connected vehicles enable real-time data exchange for traffic, charging availability, and software enhancements. EVs are foundational for autonomous vehicle (AV) development due to their simpler mechanical systems and digital control. Mobility-as-a-service models often prefer EVs due to lower total cost of ownership.

8.2.4 Subsidies

Subsidies have indeed been a major driver of the EV transition globally [68]. Subsidies reduce upfront costs. EVs generally have higher sticker prices than comparable internal combustion engine vehicles due to costly battery packs. Government subsidies, like tax credits, rebates, and grants, help reduce this initial price gap, making EVs more attractive to consumers. In the U.S., for example, federal tax credits of up to $7500 have significantly boosted EV adoption. The EU and China offer both purchase subsidies and incentives like tax exemptions and license plate privileges. Subsidies encourage industry investment. In fact, subsidies and government incentives also spur automakers to ramp up EV production, invest in R&D, and localize battery manufacturing, key steps in reducing long-term costs and boosting innovation. Public funding helps roll out EV charging networks, which are critical for reducing "range anxiety" and enabling mass adoption. Infrastructure subsidies lower risk for private investors and accelerate deployment.

Subsidies have an important role in driving market demand. Fleet procurement programs and corporate incentives push demand for EVs in commercial and government fleets, helping scale production and reduce costs through economies of scale

Subsidies support as well equitable access to EV technology. In some regions, EV subsidies are targeted toward low-income buyers to ensure a just transition and broader societal benefit

Nevertheless, subsidies may have some major negative effects. Over-reliance on subsidies may distort markets and delay cost competitiveness. Gradual phase-outs are essential to encourage self-sustaining market dynamics. Policies like carbon pricing, fuel economy standards, and ZEV mandates can complement or even outperform subsidies in certain contexts

Subsidies are not the only driver of the EV transition, but they have been essential in catalyzing early adoption, stimulating industry shifts, and enabling infrastructure development. As EVs reach price parity and mature technologically, the role of subsidies may evolve, but for now, they remain a cornerstone of global EV policy.

8.2.5 Consumer Trust and Adoption

Brand reputation plays a critical role in the transition to electric vehicles (EVs), influencing consumer adoption, investor confidence, and policy support. Established automakers (e.g., Tesla, BMW, and Toyota) benefit from trust in their engineering, safety, and after-sales service. This reassures consumers when trying relatively new EV technologies. Brands known for innovation (e.g., Tesla, Rivian) are more likely to be perceived as leaders in EV development, making their models more appealing despite potentially higher prices. A strong brand reputation can offset concerns about EV reliability, battery life, or charging infrastructure by implying that the brand stands behind its product. Automakers like Polestar and Tesla have cultivated a distinct EV identity. This focused branding helps create emotional appeal and loyalty, especially among early adopters and eco-conscious consumers. Premium brands leverage reputation to sell high-end EVs (Porsche Taycan), while others use trust in affordability (Nissan with the Leaf) to target mainstream segments. A strong reputation makes it easier to attract investors and secure funding for Research and Development (R&D) and manufacturing capacity expansions. Trusted brands are more likely to be chosen as partners by battery suppliers, tech firms, and charging infrastructure companies. Brands with strong reputations may have better relationships with regulators, enabling them to shape EV policies or secure incentives and subsidies. A positive brand image encourages firms to adopt environmentally responsible practices, minimizing regulatory and reputational risks as sustainability becomes a public expectation. Well-regarded brands are better positioned to recover from EV-related issues (battery recalls, fires, or range discrepancies) because of pre-existing consumer goodwill.

Tesla's reputation for innovation has created a loyal customer base and investor enthusiasm, accelerating EV adoption despite limited model variety and customer service issues. For Volkswagen, post-Dieselgate, VW has made EVs central to its brand rebuild, and its commitment to electrification (via the ID series) is helping shift public perception. For Ford, the Mustang Mach-E and F-150 Lightning leverage Ford's strong American legacy to build trust in EVs among traditional truck and muscle car buyers.

8.3 The Global Pre-owned EV Market

Depreciation shapes the second-hand EV market. Depreciation, while normally a price-lowering force, ironically fuels both high demand and low supply in the second-hand EV market. It discourages owners from selling and attracts price-sensitive buyers, resulting in a squeeze that boosts prices and intensifies competition for used models. Depreciation plays a unique role in shaping the second-hand EV market, especially when demand is high but supply is low. Depreciation slows down supply. EVs depreciate faster than internal ICE vehicles in their first few years due to rapid battery technology improvements, government incentives that lower new EV prices, and consumer uncertainty about battery life and replacement cost. Because of this steep early depreciation, existing EV owners may hesitate to sell their vehicles; they would take a significant loss. This hesitation leads to fewer EVs entering the second-hand market, worsening the supply shortage. Depreciation Drives Demand in the Used Market. While new EVs are still relatively expensive, depreciated second-hand EVs become more affordable, making them attractive to budget-conscious buyers. First-time EV adopters want to test the waters. This makes demand for used EVs stronger, especially in regions where fuel prices are high, urban policies favor EVs (e.g., low-emission zones), and charging infrastructure is growing. Price imbalance emerges. The fast depreciation should mean lower prices and high supply, but instead low supply keeps second-hand prices high, sometimes even stabilizing or appreciating values for popular models (e.g., Teslas, Nissan Leafs). Buyers compete over a small pool of EVs, pushing up resale value and distorting typical depreciation patterns. There is a feedback loop effect. High demand and limited supply keep prices artificially high for used EVs. This in turn slows further depreciation, as owners realize their EVs hold value. Depreciation encourages more people to buy new EVs (with subsidies) while waiting for resale values to hold. Further delays existing owners from selling, tightening supply even more.

Leasing companies play a crucial role in shaping the depreciation rate of EVs, primarily because they are key stakeholders in the used vehicle market and manage large EV fleets. Their role affects both how depreciation is calculated and how market trends evolve.

1. Setting Residual Values

Leasing companies determine residual values, the expected market value of a vehicle at the end of a lease. This figure is foundational to calculating depreciation, defined as:

Depreciation = Initial Purchase Price − Residual Value

For EVs, leasing companies consider:

- Battery degradation risks
- Technological obsolescence (e.g., newer models with better range)
- Government subsidies and their impact on new EV prices
- Secondary market demand for used EVs

2. Data-Driven Forecasting

Leasing firms collect vast data from fleet performance, maintenance, and resale values. This data informs:

- Battery longevity and performance trends
- Market appetite for specific EV models
- Real-world total cost of ownership (TCO)
- These insights help refine depreciation curves more accurately than traditional methods.

3. Market Influence

As major fleet operators and sellers of used vehicles, leasing companies influence market pricing, which in turn affects depreciation norms. They may flood or restrict the used EV market, affecting supply-demand balance. They often negotiate buy-back guarantees or manufacturer partnerships, which can stabilize depreciation rates.

4. Risk Management and Pricing

Leasing companies are risk-averse. Because EVs are newer tech, they tend to apply conservative residual values, especially early in the adoption curve. This leads to higher calculated depreciation until more predictable performance data emerges.

5. External Factors Considered by Lessors

When setting depreciation-related values for EVs, leasing companies must factor in:

- Charging infrastructure expansion
- Regulatory pressures (ICE bans)
- Technological improvements in battery capacity and cost
- Manufacturer warranties and incentives

These elements are reflected in their models, affecting how depreciation is projected.

Table 8.5 summarizes the role of Leasing companies in the development of second-hand EVs.

Table 8.5 Role of leasing companies in the depreciation of EVs

Role of leasing companies	Impact on depreciation rate
Set residual values	Directly affects depreciation calculations
Analyze real-World data	Improves accuracy of depreciation forecasting
Influence secondary market	Affects resale values, shaping depreciation trends
Hedge risk conservatively	Can inflate depreciation rates initially
Respond to external signals	Adjust projections based on policy & tech changes

8.3.1 The Vehicles of the Second-Hand Market

The top sources of second-hand EVs are:

Fleet and leasing companies [69]:

These companies regularly cycle vehicles every 2–4 years. Examples: corporate fleets, rental companies, or government EVs. Their contribution is often the largest single source of used EVs. These vehicles are typically well-maintained and newer: Nissan Leaf, Tesla Model 3, Hyundai Kona EV.

Private owners (individual resales):

Private sellers upgrade or change cars regularly. The market share varies by region, often 20–40% of the second-hand EV market [70]. These vehicles present a wide variety of ages and conditions. Motivation includes upgrades to newer models with longer range.

Car Dealerships and Trade-ins:

People trade in their EVs when buying a new car. Dealers recondition and resell EVs or auction them off. There may often be overlaps. In fact, these cars may originally come from private owners or fleets.

Rental Car Companies:

Rental agencies increasingly rotate their EV stock into the second-hand market. The recent trend is that Hertz and others added EVs but also began offloading them in 2023–2024. The condition of these vehicles is moderate mileage, used heavily but maintained.

Manufacturer buybacks/Certified Pre-Owned (CPO) Programs:

Automakers offer CPO vehicles, which are inspected and warrantied. There is a limited volume but rising in importance as OEMs build out used EV programs. Examples: Tesla CPO, BMW i-series programs.

In nations that import used EVs, the origin often reflects where surplus supply exists: Japan: a major exporter of used EVs (Nissan Leaf) to countries like New Zealand, the UK, and parts of Africa. USA: Exports some used Teslas and other EVs, especially into Canada or Latin America, Europe: EVs from Germany, the UK, and the Netherlands often flow to Eastern Europe and Nordic countries. The origin of second-hand EVs in the market can be ranked by importance as volume and influence, based on factors like market size, early adoption, government incentives, and fleet turnover. Here is a breakdown of the top countries of origin of second-hand EVs:

United States:
The importance of this origin is from the large EV market with high early adoption in states like California. The key sources from this market are off-lease Teslas, Nissan Leafs, Chevy Bolts, and Ford EVs, and fleet turnover from companies and ride-hailing services. There is an important impact, as it is a major source of affordable used EVs exported to other markets.
China:
The importance of this origin is due to the fact that China is the world's largest EV market. The vehicles are from the high volume of low-cost EVs (BYD, Wuling), short vehicle ownership cycles, and government-subsidized fleets now retiring older models. These used EVs have an impact on growing exports to developing countries and internal rural markets.
European Union (especially Germany, the Netherlands, Norway, and the UK):
The importance of this origin comes from aggressive EV policies, early adoption, and short leasing cycles. The key sources are Tesla, Renault, BMW, VW, and Hyundai EVs from corporate and personal leases. These vehicles have an important impact as they drive second-hand EV markets in Eastern Europe and Africa.

Secondary Origin:

Japan:
The second-hand EVs are mostly Nissan Leaf and Mitsubishi i-MiEV. This source is particularly relevant as Japan is an early EV producer with consistent export of used cars, including EVs, to Southeast Asia, Australia, and Africa. The impact of Japanese used EVs is limited, but the export volume of used EVs is growing.
South Korea:
The used EVs are mostly Hyundai Kona EV, Kia Soul EV, Kia EV6, and Ioniq. These vehicles increase domestic adoption with export potential. There is also an impact as an emerging contributor to the regional used EV supply.

There are export trends to increase toward:

- Africa: importing second-hand EVs mainly from Europe and Japan.
- Southeast Asia: Growing demand, especially from middle-income buyers.
- Eastern Europe: A significant destination for Western Europe's used EVs.

The leasing market acts as a feeder, stabilizer, and trust-builder for the second-hand EV market. It promotes EV adoption, ensures a consistent and quality-controlled supply of used EVs, and improves affordability and transparency, which are key to growing the second-hand EV ecosystem. The leasing market plays a crucial role in the development and sustainability of the second-hand EV market. The leasing market plays an important role in the vehicle supply for the second-hand market. There is a high turnover of leased EVs; leasing contracts typically last 2–4 years. At the end of these terms, a steady stream of relatively new, well-maintained EVs enters the secondhand market. Many leasing companies serve corporate fleets. As more companies electrify their fleets through leasing, this increases the future supply of used EVs. Leasing companies influence residual value as they closely monitor and project future resale values, which helps stabilize and even boost confidence in used EV values. With more EVs being leased, better data is collected on depreciation trends, making second-hand pricing more accurate and transparent. The leasing market allows better consumer accessibility. Lower upfront cost of leasing encourages EV uptake. In fact, many consumers lease rather than buy EVs due to high initial costs. This helps boost EV adoption, indirectly feeding into the used market. The leasing market makes used EVs more affordable. After leases expire, EVs are resold at lower prices, making them more accessible to a broader demographic. Leased EVs often come with battery health reports. Leasing companies maintain records of battery condition and service history, which builds buyer confidence in used EV purchases. Leasing absorbs some of the perceived risk around EV ownership, such as battery degradation and technology obsolescence, which increases initial uptake and feeds supply to the second-hand market. Leasing companies often resell off-lease EVs with extended warranties or as certified pre-owned vehicles, making them more attractive. Leasing ensures that EVs are used efficiently and then reintroduced to the market, supporting sustainability goals, therefore maximizing vehicle lifespan. Leased EVs may be refurbished, or their batteries repurposed for energy storage before resale.

The second-hand EV market is poised for continued growth, supported by increasing consumer interest, technological advancements, and a growing supply of used vehicles. However, addressing challenges related to depreciation and consumer confidence will be crucial for sustained market development.

The second-hand EV market is experiencing significant growth and transformation in 2025, driven by increased consumer demand, falling prices, and a surge in supply from lease returns.

Market growth and demand are marked by record sales: in the UK, used EV transactions rose by 57.4% in 2024, reaching 188,382 units and capturing a 2.5% market share, up from 1.7% in 2023 [71]. In the U.S., used EV sales increased by 62.6% year-over-year in 2024, with 287,175 units sold. Tesla still dominates the market. Used Tesla sales surged by 27% in April 2025 compared to the previous month, giving Tesla a dominant 47% share of the used EV market [72].

Pricing trends and affordability are to be considered. Depreciation has an important impact on the used EV market. EVs are experiencing faster depreciation compared to ICE vehicles. In the UK, EVs retain only 49% of their value after 2 years, compared to 70% for petrol and diesel vehicles [73]. Price parity may be achieved in near future. In the UK, one in three used EVs are now priced under £20,000, making them more accessible to a broader audience [74]. The battery cost will change the market landscape. Advancements in battery technology have led to a significant drop in prices, with forecasts predicting battery costs to fall to $80/kw-h by 2026, down from $115/kWh in 2024 [75].

Supply dynamics and lease returns will change the market structure. A wave of off-lease EVs is entering the secondhand market. In 2025, approximately 123,000 leased EVs will return to the market, with projections of 329,000 in 2026 and 650,000 by 2027 [76].

There is a need for higher inventory levels. In the U.S., the supply of used EVs has tightened, decreasing by 36% year-over-year, indicating strong demand [76]. There are challenges and other considerations to take into account. Rapid depreciation of EVs poses challenges for owners and leasing companies, potentially leading to higher leasing costs and reduced resale values.

There is a changing consumer confidence. In fact, while battery replacement costs are decreasing, concerns about battery longevity and replacement remain a consideration for potential buyers.

8.3.2 Challenges

The used EV market is expanding, yet it faces several significant challenges that hinder its growth and consumer adoption. Battery health is a paramount concern for used EV buyers. Over time, EV batteries degrade, leading to reduced driving range and performance. This degradation not only affects the vehicle's usability but also its resale value. For instance, in Europe, the second-hand EV market grapples with issues related to battery health status, making it a critical factor for potential buyers. Used EVs tend to depreciate faster than their ICE counterparts. This accelerated depreciation is influenced by factors such as rapid technological advancements, leading to newer models with better features, and aggressive pricing strategies by manufacturers. For example, EVs have been reported to retain only 49% of their value after 2 years, a significant drop compared to the 70% retention rate for petrol and diesel vehicles [3]. The availability and accessibility of charging stations remain a significant barrier. In regions like Andhra Pradesh, India, the scarcity of charging infrastructure has been a major deterrent to EV adoption, with only 601 charging stations statewide, equating to 1/205 km of road [77]. This limitation is even more pronounced for used EV owners who may not have access to home charging solutions. The variety of used EV models available in the market is relatively limited compared to ICE vehicles. This scarcity restricts consumer choice, making it challenging for buyers to find a used EV that fits their specific needs and preferences. The limited availability is partly due to the still-growing new EV market, which naturally leads to fewer used models being available. While EVs generally require less maintenance than ICE vehicles, concerns about the cost and availability of specialized repair services persist. Battery replacements, in particular, can be expensive, and not all service centers are equipped to handle EV-specific issues. This uncertainty can deter potential buyers from considering used EVs. Government incentives and support for EVs vary widely by region. In some areas, used EV buyers may not benefit from subsidies or tax incentives, making new EVs with available incentives more financially attractive. This disparity can skew consumer decisions away from the used EV market. The rapid pace of technological advancement in the EV sector means that older models can quickly become outdated. Features such as battery efficiency, range, and infotainment systems improve with each new model, making used EVs less appealing to tech-savvy consumers. There exists a knowledge gap among consumers regarding the benefits and limitations of used EVs. Misconceptions about battery life, charging infrastructure, and maintenance can lead to hesitation in purchasing used EVs. Enhanced consumer education is essential to bridge this gap and foster confidence in the used EV market.

Addressing these challenges requires concerted efforts from manufacturers, policymakers, and stakeholders to enhance battery technologies, expand charging infrastructure, and provide consistent incentives. Such measures will be pivotal in bolstering consumer confidence and accelerating the growth of the used EV market.

EV Battery replacement is one of the main challenges used EV buyers may have to face. When considering EVs, two significant cost factors over time are battery replacement and depreciation. For a long-term plan for an EV, battery replacement could become relevant. But if the vehicle is to be sold or traded within 5–7 years, depreciation will almost always have a greater financial impact.

8.3.2.1 Battery Replacement

EV batteries are among the most expensive components, often costing \$5000–\$20,000 depending on the make and model [78]. Battery replacement is a long-term cost, often avoided if the car is sold before major degradation occurs. The cost makes battery replacement economically unviable for older vehicles with low residual value. EV battery replacement can range from \$4000 to over \$20,000, depending on the vehicle and battery capacity [78]. For example: ~\$6000–\$8000 for Nissan Leaf [79], ~\$12,000–\$20,000+ for Tesla Model S/X [80]. This cost can deter consumers from buying used EVs or keeping older ones on the road. Most modern EV batteries last 8–15 years or 100,000–300,000 miles. Most EVs come with an 8-year/100,000-mile warranty for the battery [81]. Actual battery failures are rare within warranty; degradation (loss of range) is more common than total failure.

Figure 8.2 shows the battery pack of a BMW i3 to be replaced.

Furthermore, in addition to their high economic cost, battery replacement in EVs presents a range of challenges spanning technical and environmental dimensions. The key issues are:

1. Lack of Standardization
   EV batteries vary significantly across manufacturers in terms of design, size, voltage, and integration. This lack of standardization complicates mass production of replacement units and battery-swapping infrastructure. Lack of standard battery sizes and architectures across manufacturers hinders interchangeability and mass reuse. This fragmentation complicates refurbishment and second-life applications.
2. Battery Degradation Uncertainty
   It is hard to predict exactly how a battery will degrade over time. Degradation depends on temperature, charging habits, driving patterns, and chemistry, which complicates decisions on when replacement is necessary.

Fig. 8.2 BMW i3 battery replacement

3. Limited Infrastructure for Battery Swapping
   Battery swapping, quickly exchanging a depleted battery for a charged one, is technically viable but not widely adopted. Requires standardized battery designs and heavy infrastructure investment. Unlike refueling or even charging infrastructure, battery swapping stations are rare and generally limited to fleet vehicles in places like China. Swapping requires precise alignment and compatibility, which again is hindered by non-standardized designs; it is less feasible globally due to lack of coordination.
4. Supply Chain and Raw Material Constraints
   Lithium, cobalt, nickel, and other key materials face supply bottlenecks. Recycling and second-life use are still developing, which limits circularity and sustainability in battery replacement cycles.
5. Technical Complexity
   Many batteries are integrated into the vehicle's structure (especially in newer EVs), making them difficult and costly to access and replace. Replacement often requires certified technicians and specialized tools.
6. Environmental Impact
   Improper disposal or recycling of used batteries can cause environmental harm. Recycling processes are energy-intensive and not yet widespread or efficient enough. Disposing of or recycling lithium-ion batteries is complex and resource-intensive. Inadequate recycling infrastructure can lead to e-waste or environ-

mental pollution. The battery supply chain also involves ethical concerns (e.g., cobalt mining).

7. Warranty and Service Restrictions
   Many manufacturers restrict battery replacement or repair to authorized service centers, limiting consumer choice and increasing costs.
8. Software Lock-In and Compatibility
   Some EVs may require software reprogramming or pairing of new batteries, further complicating third-party replacement options.

- Possible Solutions & InnovationsBattery leasing models to reduce upfront cost.
- Improved battery chemistries with longer life and lower degradation.
- Second-life applications for used EV batteries (e.g., grid storage).
- Investment in recycling technologies and circular economy initiatives.
- Industry-wide standards for battery formats and diagnostics.

Depreciation is typically a larger and more immediate cost than battery replacement, especially in the first 3–5 years. EVs depreciate faster than gas cars historically (due to tech obsolescence, tax credits lowering new EV prices, range anxiety, and charging infrastructure concerns). Average depreciation: Some EVs lose 50–60% of their value in 3 years, compared to 35–50% for ICE vehicles. Tesla models depreciate more slowly due to brand strength and over-the-air updates. Factors accelerating depreciation are rapid EV tech advancement, newer models with longer ranges and faster charging, government incentives on new EVs.

Table 8.6 shows a. comparison between the role of battery replacement and depreciation in the EV market.

8.3.3 *Technology*

Technology has had a significant impact on the second-hand EV market, both improving its value proposition and shaping consumer perceptions. Advanced diagnostic tools allow precise evaluation of EV battery health, which is the most critical (and expensive) component in an EV. Platforms like Recurrent and OEM-specific tools, such as Tesla's battery degradation reports, provide transparency to buyers. This reduces uncertainty and risk, making used EVs more appealing. Many EVs receive over-the-air (OTA) software updates, meaning that even used models can

Table 8.6 Comparison of cost of battery and depreciation in EVs

Factor	Battery replacement	Depreciation
Timing	8–15 years, long-term	Immediate, especially in first 3–5 years
Cost range	$4000–$20,000+	30–60% value loss in 3–5 years
Predictability	Can be estimated by age/mileage	Influenced by market & tech trends
Mitigation	Warranty, proper care	Choose EVs with strong resale (e.g., Tesla)
Impact	Large, but less common	More significant in short-term ownership

gain new features or performance enhancements long after the initial purchase. This extends the technological relevance of older models and boosts resale value. Integrated AI-based telematics systems help forecast component wear and suggest preventive maintenance, which is helpful for both sellers and buyers. Buyers are more confident purchasing used EVs with detailed usage and maintenance logs enabled by technology. Platforms like Carvana, AutoTrader, and EV-specific marketplaces use AI-driven tools to match buyers with suitable used EVs.

These platforms often include battery health reports, financing options, and warranty comparisons, streamlining the purchase process. The expansion of charging networks and smart home chargers makes owning older EVs more practical, even if their range is lower than new models. Second-hand EVs become more usable and attractive, especially in urban or suburban environments. Machine learning models are being used by insurers, dealerships, and fleet managers to predict depreciation more accurately. This helps buyers make informed decisions and encourages leasing companies to offload EVs into the second-hand market with confidence. Companies are using AI and robotics to refurbish EVs, especially fleets, more efficiently, which improves the quality and safety of used vehicles. Battery repurposing, as for energy storage, also helps manage end-of-life batteries sustainably, reducing environmental concerns. Battery replacement cost is key. Although diagnostics help, tech cannot yet solve the high cost of replacing worn-out batteries. Compatibility and software lock-in have major importance. Some manufacturers limit features or updates for used vehicles, which can reduce long-term appeal.

Technology has made the secondhand EV market more transparent, trustworthy, and efficient. It helps mitigate traditional concerns around battery degradation and vehicle longevity, opening the market to more mainstream buyers. As these tools become more accessible and standardized, the second-hand EV market is likely to grow significantly.

8.3.4 Subsidies

Subsidies for second-hand EVs can significantly influence the EV market, consumer behavior, and environmental outcomes. Subsidies reduce the upfront cost of used EVs, making them accessible to middle- and lower-income households. Helps democratize EV adoption, as new EV subsidies primarily benefit wealthier buyers who can afford higher-priced vehicles. Encourages owners of new EVs to sell earlier, feeding more EVs into the second-hand market. More consumers consider used EVs as a viable option, increasing overall EV penetration. By propping up demand, subsidies help stabilize or increase residual values of EVs, which can reduce total cost of ownership concerns for both new and used buyers. More EVs on the road, regardless of whether they are new or used, displace ICE vehicles, reducing emissions. Incentivizing the use of second-hand EVs increases the total mileage and environmental benefit per EV produced. Broader EV ownership encourages demand for charging infrastructure and promotes general EV awareness. Works best when combined with support for public charging networks, battery health diagnostics, and EV education. Supporting second-hand EVs can often deliver similar environmental benefits at lower subsidy costs than new EV programs. Must address challenges like battery degradation, odometer fraud, and warranty limitations in used vehicles. France and Germany have piloted subsidies for used EVs with success in increasing adoption. California offers grants and financing assistance specifically for low-income buyers purchasing used EVs.

8.3.5 The Market

An analytical overview of the global pre-owned car market evolution from 2025 to 2034 can be based on current trends, historical data, and forward-looking projections [82].

The global pre-owned car market evolution (2025–2034) can be observed at different points:

- Market size and growth trajectory

  2025 Market value: ~$1.5 trillion globally.
  Dominated by: USA, China, Europe.
  Projected global growth (2025–2034): 6–8%.
  2034 Market projected value: ~$3.5 trillion.

- Key trends shaping the market

  Electrification of used car fleets.
  Rise of used electric vehicles (EVs) as more new EVs enter the market.
  EV depreciation trends influence affordability.
  A secondary market for EVs becomes essential in mass adoption strategies.

In the United States the typical age of vehicles reached 12.6 years on average as a new record lifespan of [83]. Typically, a new vehicle is sold to a second owner after 4–5 years of usage. A study shows that a typical secondhand car buyer is looking for a vehicle with an average age of about 6.5 years [84].

In 2025, the global used-car market is valued at more than USD 1.7 trillion; it is projected to reach almost USD 3.5 trillion in 2034 (Table 8.7), as represented in Fig. 8.3. Such progress corresponds to a growth rate between 6% and 8% [85]. Figure 8.4 shows the pre-owned market evolution. North America is the largest market for pre-owned cars. North America accounts for almost one-third of the

Table 8.7 Global pre-owned cars market evolution (2025–2034) [82]

Metric	2025 estimate	2034 projection
Global market size	~$1.5 trillion	~$2.7–3.5 trillion
Used EV market share (%)	~5	20–30

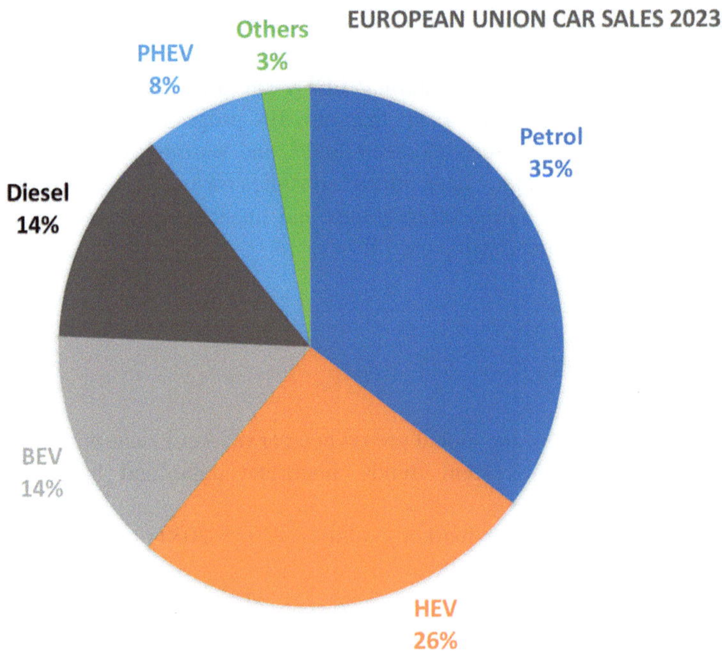

Fig. 8.3 European Union car market by fuel in 2023. (Source: Statzon) [88]

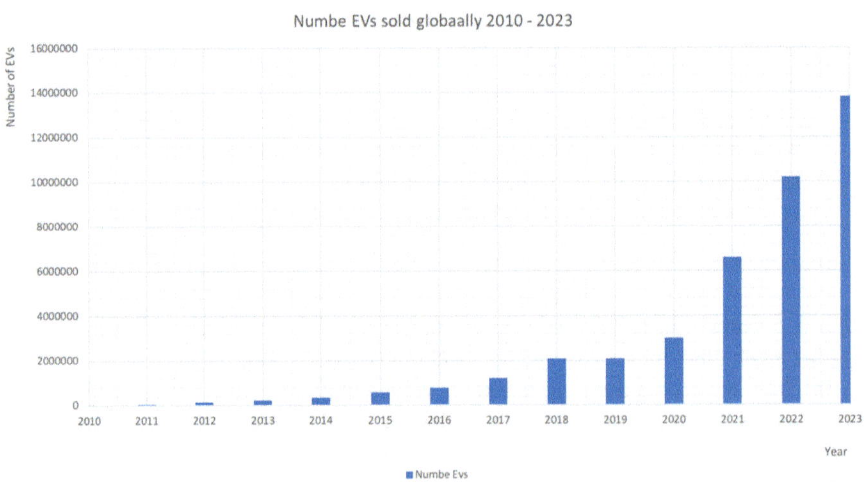

Fig. 8.4 Global electric vehicle market penetration [89]

global market, while Europe follows; China, the Middle East, and Africa constitute the least significant markets, holding merely 5% of global sales [86]. At the same time, the United States represented under 10% of all new electric vehicle registrations globally in 2022 [87].

A brand-new car is typically beyond the financial reach of many individuals. Households with high incomes are more prone to owning new cars compared to households with low incomes. In Norway, where the EV penetration rate is the highest, a 2019 study indicates that 31% of new car registrations were made by the wealthiest 10% of families, whereas the lowest income 10% of households accounted for merely 0.6% of registrations [88]. This difference holds greater significance when looking at the EV market. In 2019, Norway saw that 37% of newly registered EVs were from the wealthiest 10% of households, while only 0.7% were from the bottom 10% of households [88]. Less expensive, used EVs can meet the demands of a wider range of consumers globally.

A key factor consistently influencing the market for used vehicles is the rising demand for EVs and hybrid electric vehicles (HEVs). The spectrum of the used vehicles market is illustrated in Fig. 8.3 below, highlighting that fuel powered vehicles dominate the market sales but there is an important progress of EV of all sort.

Figure 8.3 shows that even though the EV market is continually expanding, sales for used EVs are still quite minimal when compared to conventional ICE vehicles. The growing practicality and cost-effectiveness of used EVs could partially change the market landscape in the years ahead. Additionally, the widespread appeal of premium variants of HEVs and plug-in hybrid electric vehicles (PHEVs) has enhanced the reputation of EVs. The dynamics of pre-owned electric car sales might exhibit distinct patterns once the market achieves complete maturity.

The second-hand EV sales are currently expanding as early owners change vehicles. Pre-owned EV sales play a vital role in promoting widespread adoption, particularly as new EVs are still costly and used models become more accessible. For ICE vehicles, purchasing a used car is usually the main way to obtain a vehicle in developing as well as developed nations, and it is expected to be analogous with electric cars. Studies show that 80% of European Union residents purchase used cars, with this proportion rising to around 90% in low- and middle-income countries [88]. Roughly 70% of vehicles sold in the U.S. are used cars, and just 17% of less affluent families purchase a new vehicle [88].

As major EV markets develop, a growing volume of used EVs is entering the secondary market. However, the size of the used EV market is still relatively small when compared to new EV registrations. Approximately 14 million new electric vehicles were recorded globally in 2023, bringing the total number on the roads to 40 million. In 2023, approximately 60% of new electric vehicle registrations took place in China, about 25% in Europe, and 10% in the United States, representing nearly 95% of worldwide electric vehicle sales. On the other hand, projections suggest that in 2023, the pre-owned EV market reached around 800,000 units in China, 400,000 in the United States, and more than 450,000 when combining France, Germany, Italy, Spain, the Netherlands, and the United Kingdom. In the United States, the sale of pre-owned EVs is expected to increase by 40% in 2024 versus 2023. Despite progress, the second-hand EV market remains small in comparison to the well-established second-hand ICE markets, which have had decades to grow: 30 million in France, Germany, Italy, Spain, the Netherlands, and the UK together; almost 20 million in China; and 36 million in the USA.

Still, used EVs are progressively becoming more budget-friendly, allowing them to rival pre-owned ICE counterparts soon. In the United States, for instance, a significant share of used EVs are currently listed at under USD 30,000 [89]. Additionally, there are projections that the average price will decrease to around USD 25,000, the threshold for used EVs to qualify for the second-hand vehicle reduction of USD 4000 at the federal level, thereby allowing them to be competitive compared to top-selling fuel-powered vehicles, new or used. The cost of a pre-owned Tesla in the U.S. was more than USD 50,000 in early 2023 and decreased to slightly over USD 33,000 in 2024. This allowed Tesla to be competitive compared to both used SUVs and numerous new vehicles. In Europe, used EVs are available for prices ranging from EUR 15,000 to EUR 25,000 (USD 16,000–27,000), while the price is approximately EUR 30,000 (USD 32,000) for pre-owned plug-in hybrid vehicles. Several European nations provide financial incentives for used EVs, including the Netherlands (EUR 2000), where since 2020 the support for new vehicles continues decreasing gradually, in contrast to the stable assistance for pre-owned vehicles, and France (EUR 1000). In 2023, in China, the average cost of used EVs was approximately CNY 75,000 (USD 11,000) [89].

When examining countries that have the greatest electric car adoption, similar factors will account for the trends in the sales of second-hand EVs. The primary distinction lies in direct subsidies, which can be crucial in the decision to adopt a new EV. There ought to be a significant distinction between markets supported by subsidies and those after subsidies. Subsidies aren't the sole constraint influencing the used EV market; rapid technological advancements, alongside the loss of value of electric car parts like motors and energy storage, also affect the outcome of the worldwide used electric car sales.

The rise and growth of electric vehicles in the global EV sales and their contribution to the electrification of road transport will likely follow an S-curve pattern (Fig. 8.4) [1]. It pertains to an early stage where uptake is slow because consumers do not have enough understanding and confidence in investing in used electric vehicles. With the increasing popularity of EVs driven by subsidies, consumers will become more confident in EV technology and will be more likely to buy used EVs as they learn what to anticipate. This will continue until an equilibrium is established between new and pre-owned EVs in the market annually. The proportion of new vehicles to used ones in the market differs by nation. In the United States, in 2022, new cars represented about 30% of new registrations, whereas used cars represented approximately 70% [89].

The S-curve is described by the subsequent Eq. (8.1):

$$R(t) = \frac{100}{\left[1+\exp(-\alpha t)\right]^k}, \quad \text{where } \alpha > 0 \text{ and } k > 0, \tag{8.1}$$

where R(t) represents the market penetration rate of pre-owned electric vehicles, as t tends to infinity, R(t) tends towards saturation. In the above curve, $\alpha = 0.25$ and $k = 11$ as mentioned in Reference [1], with α and k obtained through an empirical method [1]. The details of the S-curve are presented in Reference [1]. An important question that needs a forecasted response is the trend that will affect the proportion of used EVs during the shift to electric mobility, both globally and within specific nations. The rate at which EVs depreciate and technological progress will have a major impact. Pre-owned EVs are expected to markedly enter the market, since EVs were launched 10 years ago to enter the international road transport. It is expected that the ratio of new to pre-owned EV registrations will level off at x% for new EVs and y% for used EVs. A similar equilibrium is noted in most non-perishable consumer items: mobile phones, laptops, bicycles... This equilibrium between innovation and the transfer of ownership will depend on market conditions. Countries with developed economies that generally have good access to finances are likely to show a larger proportion of new vehicles in total new registrations, while in the least advanced nations, almost all new registrations consist of pre-owned imported vehicles. It is recognized that worldwide, in the first decade, almost all EV registrations will comprise new EVs.

In recent years, the resale values of electric vehicles have been increasing. In Europe, the resale value of electric vehicles sold after a year has steadily increased from 2017 to 2022, surpassing all other types of vehicles and exceeding 70% by mid-2022 [88]. The resale value of electric cars after 3 years was below 40% in 2017 but has been closing the gap with other powertrains, reaching around 55% by mid-2022 [88]. This result arises from multiple factors, including increasing prices of new EVs and technological advancements that allow vehicles and batteries to retain their value longer; these factors decrease depreciation and enhance demand for used EVs. Similar patterns were observed in China. With existing electrification rates, 33 million homes in the EU will gain access to used electric vehicles from now until 2035. However, if the leasing industry speeds up its adoption of EVs, this figure would increase by 56%, reaching 51 million, according to a report by Transport & Environment (T&E) [90]. These 18 million extra households would be reducing their expenditures on vehicle costs.

The leasing industry serves as a significant pathway for vehicles in the pre-owned car market. Every year in the EU, four out of ten cars entering the used car market originate from the leasing industry. If leasing firms transition to electric vehicles more quickly, they will hasten the speed at which the pre-owned market becomes eco-friendly. T&E urges Europe's leading seven leasing firms to pledge to eliminate fossil fuel cars and to lease only battery electric vehicles (BEVs) starting in 2028. The entire industry must accomplish this by 2030.

Currently, nearly 80% of EU residents purchase their vehicles used. Approximately 90% of individuals in low and middle income brackets purchase vehicles from the used car market. The proportion remains considerable (62%) even within affluent groups.

A used battery EV has significantly reduced ownership expenses compared to a pre-owned petrol car, according to research conducted by consumer organization BEUC. Families can save nearly €6000 in 7 years by choosing electric over petrol cars [91]. According to T&E, a consistent flow of electric vehicles in the used car market is crucial for European households aiming to cut expenses. The significant arrival of budget-friendly electric vehicles in the used car market is achievable… and leasing firms have control over this. If they transition to green quicker, European households will also go green at a lower expense. It's no exaggeration to claim that leasing firms can speed up and make electric vehicles accessible for the 80%. Stef Cornelis concludes that their responsibility is environmental and social.

Eight million pre-owned electric vehicles priced at €10,000 are on the streets.

The study indicates that if the leasing industry takes the initiative instead of merely responding to the market regarding the adoption of battery electric vehicles, it could result in an additional 18 million used BEVs available by 2035. Crucially, eight million of these vehicles would be priced at €10,000 or under, which represents the typical amount low- and middle-income families spend on a second-hand car.

However, T&E research indicates that despite assertions of green leadership, leasing companies are not progressing rapidly enough in the shift to electric vehicles. For instance, none of Europe's major leasing firms has established a timeline to eliminate fossil fuel cars, and the adoption of battery electric vehicles simply matches the general market trend. A T&E undercover probe in France and Germany reveals that leasing company sales personnel do not consistently assist their clients in transitioning to electric vehicles.

The matter of pre-owned EVs can become more complex when we consider the possibility of future coexistence between electric cars and fuel cars. Certainly, as demonstrated, energy transition technologies are often anticipated to follow S-curves: rapidly rising, followed by achieving full market acceptance. However, it is very likely that EVs will continue to exist alongside ICE vehicles, making it essential to evaluate the future balance between the two technologies worldwide. Some analysts predict that electric vehicles could attain a saturation level of 15–30% of sales by the years 2025–2030 [88]. In 2023, electric vehicle sales, presently at 15% of worldwide vehicle sales, will keep progressing, particularly due to specific policies. Consequently, what effects would the limited EV potential have on energy consumption and the ability to provide and source materials? The expense remains a major barrier to the acceptance of electric vehicles. Consequently, without coercive measures in place, income will limit EV uptake, resulting in a level of saturation in the EV market. Data on income distributions will determine the barrier for EV adoption. Perspectives on climate present another hurdle to the uptake of EVs. The saturation of the electric vehicle market would greatly impact energy and materials markets due to the substantial demand. The need for EVs and ICE vehicles will be tied to oil demand, electricity usage for EVs, lithium needs for batteries, and the impacts on new energy resources and investments.

Tesla vehicles maintain their value well for a variety of reasons. Tesla cars get remote updates for the duration of their operational life. This technical decision minimizes value loss from obsolescence since upgrades to newer models will also benefit older cars. Moreover, another element is the uniqueness that buyers have started to link with Tesla. The company additionally manages a network of superchargers. Tesla, similar to other producers, doesn't depend as heavily on consumer

Table 8.8 Top electric car brands market share in Europe 2023 [88]

Vehicles	Sales (%)	Remarks
Tesla	17.6	Reputation & Tech
Volkswagen	9.9	Price
BMW	6.4	Price
Mercedes Benz	5.1	Symbol
Audi	5	Brand name
MG	5	Symbol

incentives. Tesla enjoys significant popularity and acclaim among electric vehicle brands, acknowledged as the leading seller of new EVs, showcasing the most cutting-edge technology and undergoing the least depreciation, a trend that is similarly observed in the used EV market. Pre-owned EVs that see the biggest price drops are typically older models, often thought to have outdated technology, particularly the Nissan Leaf and Chevrolet Bolt EV. In 2023, Tesla dominates the USA's second-hand electric vehicle market, with the Model 3 representing 34.9% of sales.

Models aged 1–5 years face the sharpest decline in value. A report by STATEON [88] identified as one of the leading dealers in Europe as a reference market for 2023, details the sales share of models aged 1–5 years:

Table 8.8 shows the top-selling used EVs will feature the Teslas and Volkswagen. A comprehensive model of the used EV market is provided in reference [88].

8.3.6 Emerging Markets

The market for used electric vehicles (EVs) in developing countries is an emerging and complex space influenced by economic, infrastructural, policy, and technological factors.

Cost-sensitive buyers in developing countries are turning to used EVs as a more affordable alternative to new ones. Urban areas with worsening pollution and rising fuel costs are driving interest in EVs. Lower prices compared to new EVs are the main drivers of the market. Incentives/subsidies in some countries for electric mobility (India, Kenya, Nigeria). Importation of used EVs from developed countries (Japan, UK, US) due to early EV adoption there. Growth in ride-sharing and delivery services needs affordable fleet vehicles.

There are some major challenges:

- Battery degradation: Shorter range and performance anxiety.
- Limited charging infrastructure: Few charging stations outside major cities.
- Lack of policy clarity: Some governments have inconsistent or underdeveloped EV regulations.
- Technical knowledge gap: Fewer trained mechanics and EV service centers.
- Import regulations: High tariffs or restrictions on used EVs in some nations.

Table 8.9 Key markets due to large population

Country	Highlights
India	Rapid urbanization, strong government push for EVs, growing used EV market
Nigeria	Interest in low-cost mobility, but infrastructure challenges persist
Kenya	Nairobi emerging as EV hub; strong support for 2- and 3-wheel EVs
South Africa	Some uptake, but limited by high costs and weak charging network
Philippines	Growth in used e-trikes and electric jeepneys

EVs in developing countries open new opportunities:

- Local refurbishment businesses: Repair and upgrade imported used EVs for resale.
- Battery second-life markets: Repurposing old EV batteries for energy storage.
- Government incentives: When paired with infrastructure investment, these can rapidly grow the market.
- Fleet electrification: Taxis, logistics, and ride-hailing services adopting used EVs for cost savings.

- Market TrendsChina's used EV exports are increasing to countries in Africa and Southeast Asia.
- Battery leasing models are being explored to reduce upfront costs.
- Micro-EV and two/three-wheeler segments are growing fastest.

Table 8.9 shows key markets for second-hand EVs due to large population.

8.4 Demand and Supply in the Second-Hand EV Market

The second-hand EV market is still very dynamic and can be characterized in many different ways in relation to demand and supply depending on how the question is analyzed.

8.4.1 High Demand, Low Supply of Second-Hand EVs

Demand is higher than supply due to growing consumer interest, lower running costs, and environmental benefits. Supply is limited because the market is still maturing, and vehicles are not circulating fast enough yet. The high demand and low supply in the second-hand EV market can be explained by a combination of economic, technological, and behavioral factors.

Demand is high because used EVs are significantly cheaper than new ones, making them accessible to a broader range of consumers. More people want eco-friendly transportation, and buying a used EV is both green and cost-effective. As fuel prices rise, many switch to EVs to save money on fuel, boosting demand, especially for affordable used models. Some governments offer tax breaks or access to low-emission zones for EVs, encouraging even second-hand purchases [2]. Improved charging infrastructure with more charging stations will translate in more confidence in owning any EV, new or used.

Supply is low because the EV market is relatively young. Only in the last 5–7 years have EVs been sold at scale, so fewer used vehicles are available. There is a longer ownership cycles with EVs. In fact, EV owners tend to hold onto their vehicles longer, often due to satisfaction or slower battery degradation than expected. Leasing companies, car-sharing fleets, and rental companies are not retiring EVs as quickly as they do with ICE cars.

EVs depreciate fast, but used EVs may hold their value better in the future due to demand, making owners less likely to sell them quickly. Supply chain issues (especially for batteries) over the past few years slowed new EV production, which in turn delays their entry into the used market.

8.4.2 Low Demand, High Supply of Second-Hand EVs

In the secondhand EV market, a potential situation of low demand and high supply can be explained by a mix of economic, technological, and psychological factors. Demand can be low because EV batteries lose capacity over time, reducing range and performance. Buyers fear costly battery replacements after purchase. There are rapid tech advancements. New EVs come with longer range, faster charging, and better features. Buyers prefer newer models, making older ones less appealing. Limited trust and awareness impact demand. Many buyers are unfamiliar with how EVs age compared to gas cars. Concerns about reliability and unknown maintenance costs reduce confidence. In some regions, public and home charging is not widely accessible. Buyers avoid EVs if charging is not convenient or affordable. Governments often provide bigger subsidies for new EVs than for used ones. This skews consumer interest toward new rather than second-hand EVs.

Supply is high when a large portion of EVs are leased, especially fleet or company cars. When leases end, often after 2–3 years, vehicles enter the used market in bulk. EV drivers want to upgrade for better tech. EV owners tend to upgrade quickly to newer models with better range or features. This leads to more second-hand EVs being sold. When there is important fleet turnover, companies or car-sharing services replace EVs frequently. These vehicles then enter the used market, increasing supply.

The export and policy shifts can be game changers. Some countries import a lot of used EVs from more developed markets. Policy changes like tax breaks expiring can cause a sudden sell-off of older EVs. This can create a market imbalance. Too many older EVs and not enough buyers willing to take the risk. Prices drop, but that doesn't always solve the core trust and battery concerns.

8.4.3 Demand Equals Supply of Second-Hand EVs

This situation may occur when the EV market reaches full maturity, as it has presently for ICE vehicles. The demand equals supply in the secondhand EV market when price and other conditions are such that everyone who wants to buy or sell can do so. This is a dynamic balance, not fixed; it changes with tech, policy, price, and consumer behavior.

In the secondhand EV market, demand equals supply when the number of people willing and able to buy used EVs matches the number of people willing and able to sell them at a given price. This does not always happen, but it is an ideal equilibrium point in economics. Here are key reasons why demand may equal supply in this market. Market equilibrium can be achieved when the price of used EVs adjusts naturally; it balances the number of buyers and sellers. If supply exceeds demand, prices fall; if demand exceeds supply, prices rise until they meet in the middle.

Equality between demand and supply in used EVs can come from increasing market maturity. As more new EVs are sold, the number of used EVs increases. At the same time, more consumers are willing to buy second-hand EVs due to lower prices compared to new EVs, better understanding of EV technology, and improved charging infrastructure.

This parallel growth in both the supply of more used EVs and demand, more EV-interested buyers, can push the market toward equilibrium.

Policy and incentives will play a role. In some regions, government incentives or tax benefits for owning EVs or trading old cars in can stimulate both sides: sellers are motivated to upgrade and list their old EVs, and buyers are attracted by discounts or low operating costs.

This keeps supply and demand closely aligned.

Improved battery technology & confidence is another important factor. As EVs become more reliable and batteries degrade more slowly than feared, buyer confidence in used EVs increases, keeping demand strong and predictable.

Online marketplaces and transparency play an important role. Platforms like Autotrader, Carvana, or Tesla's own used vehicle portals help match buyers and sellers efficiently, helping markets reach equilibrium faster.

8.4.4 Possible Future

The used EV market is on a sharp growth trajectory, driven by lease returns, declining battery costs, and evolving consumer demand. While depreciation creates challenges, certified pre-owned programs, tax incentives, and sustainability initiatives will shape a more mature and trustworthy marketplace. Over the next decade, expect a vibrant used EV ecosystem supported by innovation in battery lifecycle management and policy support.

As millions of lease-returned EVs flood the market, expected at ~123,000 in 2025, rising to ~650,000 by 2027, the used EV stock is set to grow sharply [76].

Already seeing this: in the U.S., used EV sales surged ~62.6% in Q4 2024 versus 2023, accounting for nearly 2% of the used-car market. Rapid depreciation, where 3-year-old EVs can lose 60–75% of value (vs. 45–50% for ICE vehicles), except for some Teslas, means bargain pricing for buyers, though trickier times for sellers [76].

CarMax noted used-EV prices fell ~40% between 2022 and early 2025, while interest doubled, highlighting their growing appeal [92]. Falling prices are driving younger and more price-sensitive buyers into the used EV space. Key segments influencing demand include 1–3-year-old EVs (still under warranty), a broadening range of battery capacities, and increasing certified pre-owned EV programs. EV battery replacement costs are dropping; the cost may soon be lower than typical ICE engine work by 2030. Certified pre-owned (CPO) programs offering battery guarantees and inspections will drive buyer confidence. U.S. used-EV tax credits (up to $4000) are currently boosting demand; buyers are racing to close deals before potential policy reversals. Regional incentives and stricter emissions rules will continue to shape market dynamics. North America & Europe: Primary growth markets with expanding charging infrastructure, rising demand for CPO EVs, and strong policy backing. Asia Pacific & LatAm: The fastest EV fleet growth in the new-vehicle space will soon feed larger used inventories. Battery "second life" applications using spent EV batteries for home or grid storage are becoming common, enhancing sustainability and residual value. Upcoming recycling innovations and regulation will increasingly impact long-term market value. A major question to ask is the meaning for stakeholders. For buyers, there is a great bargain now, especially on 1–3-year-old EVs with warranties and decent battery life. For dealers there is a need to find better ways to assess battery health, certify EVs, and build trust. For manufacturers & policymakers, there must be a focus on battery longevity, second-life use cases, and recycling infrastructure. For investors, there is an expected growth in used EV marketplaces, battery repurposing, and certification services.

Table 8.10 shows the outlook summary.

Table 8.10 Outlook summary

Timeline	Supply	Pricing	Demand drivers
2025–2027	Over one million lease returns	Competitive, bargain-driven pricing	Younger buyers, tax credit urgency
2028–2032	High volume with >650 k returns/year	Market stabilizes; battery costs drop	Certified pre-owned reliability
2030+	Saturated market; maturity	Price premiums for battery health, warranties	Strong second-life/battery recycling support

8.5 Discussion and conclusion

In the past 10 years, the EV has become an essential tool for decreasing carbon emissions in transportation on roads. Technological progress in electric cars reached a stage of maturity, yet it remains in development, particularly regarding battery advancements and the integration of AI. The market succeeded in flourishing thanks to subsidies, growing and gaining popularity as EVs offer lower operating expenses compared to ICE vehicles, even though ownership costs are higher when considering depreciation. The pre-owned vehicle market accounts for a considerable portion of the global automotive industry. Pre-owned EVs will also play a crucial role in lowering carbon emissions in road transportation.

Depreciation of EVs is a significant factor for both buyers and manufacturers, and it differs in key ways from ICE vehicles. As presented in the manuscript, the breakdown of the main points that are behind the numbers shows the explanation of the depreciation of EVs:

1. Faster Initial Depreciation
   EVs have traditionally depreciated more quickly in the first few years due to:
   - Rapid technological advancements making older models obsolete
   - Concerns about battery longevity and replacement costs
   - Limited public understanding or trust in long-term EV performance

However, this trend has been shifting in recent years.

2. Battery Life and Warranty
   Battery degradation is one of the biggest concerns affecting resale value. Modern EVs often come with warranties covering 8–10 years or 100,000 (up to 150,000 miles in some cases), which can mitigate depreciation concerns if transferred to second owners.
3. Tax Incentives and Price Drops
   Government incentives (e.g., federal tax credits, state rebates) reduce the new purchase price but don't directly affect resale prices. This can skew the depreciation curve, making a 1- or 2-year-old used EV seem disproportionately cheap.
4. Brand and Model Variability
   Some EVs, like Teslas, hold their value better than others due to strong brand perception, over-the-air software updates, and robust charging infrastructure. In contrast, older or less popular models, such as early Nissan Leafs, may depreciate rapidly.
5. Evolving Market Conditions
   The growing demand for used EVs, especially affordable ones, may contribute to stabilizing depreciation.
   Supply chain issues and high new EV prices can buoy used values.
   Regulatory pushes for zero-emission vehicles can increase used EV demand in some regions.

6. Nevertheless, there are some upcoming concerns. If battery supply costs drop substantially, new EVs may get cheaper, increasing depreciation of older ones. As well, the phasing out of incentives or saturation of the used EV market could accelerate depreciation again.
   The market for pre-owned electric vehicles is still developing, which leads to its uncertain nature. Nonetheless, some elements seem to be essential:
   - Financial incentives for new EVs result in faster depreciation of used EVs, but they assist in making EVs a standard form of transportation. Ultimately, this will benefit the used EV market as consumers develop greater trust in electric cars.
   - The integration of technology in electric vehicles, particularly AI, results in faster value loss of less recent vehicles because of the swift technological progress.
   - The advanced battery technology enhances performance, delivering better battery autonomy for more recent models, while less recent vehicles see decreased electric charge autonomy due to battery degradation over time.

At present, EVs have a shorter range than ICE vehicles, are more expensive, and depreciate faster. Luxury EVs are likely to hold their value better, while budget EVs are expected to depreciate more swiftly.

The growth of electric car sales is likely to coincide with the end of incentives, resulting in price equivalence between EVs and ICE vehicles.

This study is limited by data accessibility and demonstrates the necessity for more organized and systematic data collection to permit more accurate analysis.

The future of depreciation in EVs is being shaped by several dynamic factors, and it's evolving differently from traditional ICE vehicles. It can consequently be presented as a detailed look at what is influencing EV depreciation going forward:

1. Battery Technology Improvements
   As battery costs fall and lifespans improve, concerns about expensive battery replacements, once a major driver of depreciation, are diminishing.
   Solid-state batteries or improved lithium-ion technologies could further stabilize EV resale values.
2. Brand and Model Differentiation
   Tesla has shown significantly lower depreciation rates compared to other EV brands due to its strong brand, over-the-air updates, and charging network.
   Lesser-known or short-lived EV models often suffer steeper depreciation due to limited support and uncertain longevity.
3. Government Incentives and Policy Changes
   Generous EV tax credits or rebates can distort the new versus used EV market, making used EVs less attractive and pushing resale values down.
   Stricter emissions regulations and ICE bans may reverse this, increasing demand for used EVs and reducing depreciation.
4. Supply Chain and Production Trends

Global shortages or surpluses of EVs can temporarily inflate or depress resale prices.
As manufacturing scales up and supply stabilizes, depreciation patterns may become more predictable.

5. Software and Tech Relevance.
   EVs are more like smartphones on wheels—software aging, limited support for older features, and rapid tech innovation can increase depreciation.
   However, brands offering consistent software updates (like Tesla) tend to hold value better.
6. Market Demand and EV Adoption
   As EV adoption increases and charging infrastructure improves, consumer confidence rises, helping stabilize resale values.
   Conversely, outdated or slow-charging EVs may see sharper value drops.
7. Data Transparency and Tools
   Some platforms and EV-specific resale sites are increasingly offering accurate resale tracking, which can help reduce uncertainty and potentially flatten depreciation curves.
   In short, depreciation for EVs is becoming more nuanced, tied heavily to brand, battery tech, software relevance, and government policy. While some EVs may continue to depreciate faster than ICE cars in the short term, others, especially from leading brands, may retain value better than average over time.

References

1. Diouf B (2024) The electric vehicle transition. Environ Sci Adv 3:332
2. Patil G, Pode G, Diouf B, Pode R (2024) Sustainable decarbonization of road transport: policies, current status, and challenges of electric vehicles. Sustainability 16:8058. https://doi.org/10.3390/su16188058
3. Schloter L (2022) Empirical analysis of the depreciation of electric vehicles compared to gasoline vehicles. Transp Policy 126:268–279. https://doi.org/10.1016/j.tranpol.2022.07.021
4. Domínguez S, Pedreros S, Delgadillo D, Anzola JA (2024) Depreciation method based on perceived information asymmetry in the market for electric vehicles in Colombia. World Electr Veh J 15:511. https://doi.org/10.3390/wevj15110511
5. Woody M, Adderly SA, Bohra R, Keoleian GA (2024) Electric and gasoline vehicle total cost of ownership across US cities. J Ind Ecol 28:194–215. https://doi.org/10.1111/jiec.13463
6. Muehlegger E, Rapson DS (2022) Subsidizing low- and middle-income adoption of electric vehicles: quasi-experimental evidence from California. J Public Econ 216:104752. https://doi.org/10.1016/j.jpubeco.2022.104752
7. Burra LT, Sommer S, Vance C (2024) Free-ridership in subsidies for company- and private electric vehicles. Energy Econ 131:107333. https://doi.org/10.1016/j.eneco.2024.107333
8. Azarafshar R, Vermeulen WN (2020) Electric vehicle incentive policies in Canadian provinces. Energy Econ 91:104902. https://doi.org/10.1016/j.eneco.2020.104902
9. Costa CM, Barbosa JC, Castro H, Gonçalves R,́endez SL-M (2021) Electric vehicles: to what extent are environmentally friendly and cost effective? Comparative study by European countries. Renew Sust Energ Rev 151:111548. https://doi.org/10.1016/j.rser.2021.111548

10. Donggyun K, Choi M, Yoo N, Shin S, Lee S (2021) A new algorithm for eco-friendly path guidance focused on electric vehicles. Energy 233:121198. https://doi.org/10.1016/j.energy.2021.121198
11. Albrechtowicz P (2023) Electric vehicle impact on the environment in terms of the electric energy source—case study. Energy Rep 9:3813–3821. https://doi.org/10.1016/j.egyr.2023.02.088
12. Sheldon TL, Dua R (2019) Measuring the cost-effectiveness of electric vehicle subsidies. Energy Econ 84:104545. https://doi.org/10.1016/j.eneco.2019.104545
13. Jenn A, Lee JH, Hardman S, Tal G (2020) An in-depth examination of electric vehicle incentives: consumer heterogeneity and changing response over time. Transp Res A 132:97–109. https://doi.org/10.1016/j.tra.2019.11.004
14. Kautish P, Lavuri R, Roubaud D, Grebinevych O (2024) Electric vehicles' choice behaviour: an emerging market scenario. J Environ Manag 354:120250. https://doi.org/10.1016/j.jenvman.2024.120250
15. Li M, Lu J, Chen Z, Amine K (2018) 30 years of lithium-ion batteries. Adv Mater 30(33):1800561. https://doi.org/10.1002/adma.201800561
16. Manthiram A (2017) An outlook on lithium ion battery technology. ACS Cent Sci 3:1063–1069. https://doi.org/10.1021/acscentsci.7b00288
17. Kim T et al (2019) Lithium-ion batteries: outlook on present, future, and hybridized technologies. J Mater Chem A 7:2942–2964. https://doi.org/10.1039/C8TA10513H
18. IEA (2022a) Global EV outlook 2022. Securing supplies for an electric future. https://www.iea.org/reports/global-ev-outlook-2022
19. IEA (2022b) Evolution of Li-ion battery price, 1995–2019. https://www.iea.org/data-and-statistics/charts/evolution-of-li-ion-battery-price-1995-2019
20. Suttakul P, Wongsapai W, Fongsamootr T, Mona Y, Poolsaat K (2022) Total cost of ownership of internal combustion engine and electric vehicles: a real-world comparison for the case of Thailand. Energy Rep 8:545–553. http://creativecommons.org/licenses/by-nc-nd/4.0/
21. Verma S, Dwivedi G, Verma P (2022) Life cycle assessment of electric vehicles in comparison to combustion engine vehicles: a review. Mater Today Proc 49:217–222. https://doi.org/10.1016/j.matpr.2021.01.666
22. Joshi A, Sharma R, Baral B (2022) Comparative life cycle assessment of conventional combustion engine vehicle, battery electric vehicle and fuel cell electric vehicle in Nepal. J Clean Prod 379:134407. https://doi.org/10.1016/j.jclepro.2022.134407
23. Engelmann A, Bauer I, Dolata M, Nadig M, Schwabe G (2022) Promoting less complex and more honest price negotiations in the online used car market with authenticated data. Group Decis Negot 31:419–451
24. Oktiyani R, Ela Revita MLD, Sugiarti S (2024) Depreciation calculation using the straight line method is appropriate taxation law. JIM J Ilm Mhs Pendidik Sej 9:636–643
25. Laksman E, Strömberg A-B, Patriksson M (2019) The stochastic opportunistic replacement problem, part III: improved bounding procedures. Ann Oper Res 292:711–733
26. Su C (2024) Literature review on the net present value method of project investment decision. Adv Econ Manag Polit Sci 60:60–66
27. Marzo G, Pagnozzi T (2011) Can depreciation be a variable cost? A comparison between the straight-line method and the units of production method in a lean company context. SSRN Electron J
28. Pratama BA, Zufria I, Fakhriza M (2024) Application of the production unit method to calculate accumulated depreciation of factory machinery using the Django model. Instal J Komput 16:68–77
29. Tandon A (2021) Transition finance: investigating the state of play: a stocktake of emerging approaches and financial instruments. In: OECD environment working papers, vol 179. OECD Publishing, Paris
30. Mishin YV, Mishin AY (2020) Impact of depreciation deduction calculation method on financial soundness improvement of economic agents. E3S Web Conf 208:03033

31. Liapis KJ, Kantianis DD (2015) Depreciation methods and life-cycle costing (LCC) methodology. Procedia Econ Financ 19:314–324
32. Copiello S, Bonifaci P (2018) Depreciated replacement cost: improving the method through a variant based on three cornerstones. Real Estate Manag Valuat 26:33–47
33. Kellenbrink C, Herde F, Eickemeyer SC, Kuprat T, Nyhuis P (2014) Planning the regeneration processes of complex capital goods. Procedia CIRP 24:140–145
34. Al-Saadi M, Mathes M, Käsgen J, Robert K, Mayrock M, Mierlo JV, Berecibar M (2022) Optimization and analysis of electric vehicle operation with fast-changing technologies. World Electr Veh J 13:20
35. Abraham KM (2020) How comparable are sodium-ion batteries to lithium-ion counterparts? ACS Energy Lett 5:3544–3547. https://doi.org/10.1021/acsenergylett.0c02181
36. Harmon JE (2024) Cathode innovation makes sodium-ion battery an attractive option for electric vehicles. Aragonne National Laboratory. https://www.anl.gov/article/cathode-innovation-makes-sodiumion-battery-an-attractive-option-for-electric-vehicles
37. Rudola A, Sayers R, Wright CJ, Barker J (2023) Opportunities for moderate-range electric vehicles using sustainable sodium-ion batteries. Nat Energy 8:215–218. https://doi.org/10.1038/s41560-023-01215-w
38. Canals Casals L, Etxandi-Santolaya M, Bibiloni-Mulet PA, Corchero C, Trilla L (2022) Electric vehicle battery health expected at end of life in the upcoming years based on UK data. Batteries 8:164. https://doi.org/10.3390/batteries8100164
39. Chen W, Liang J, Yang Z, Li G (2019) A review of lithium-ion battery for electric vehicle applications and beyond. Energy Procedia 158:4363–4368. https://doi.org/10.1016/j.egypro.2019.01.783
40. Zheng Y et al (2013) Cell state-of-charge inconsistency estimation for LiFePO4 battery pack in hybrid EVs using mean-difference model. Appl Energy 111:571–580
41. Pamidimukkala A, Kermanshachi S, Rosenberger JM, Hladik G (2023) Evaluation of barriers to electric vehicle adoption: a study of technological, environmental, financial, and infrastructure factors. Transp Res Interdiscip Perspect 22:100962. https://doi.org/10.1016/j.trip.2023.100962
42. Etxandi-Santolaya M, Mora-Pous A, Canals Casals L, Corchero C, Eichman J (2024) Quantifying the impact of battery degradation in electric vehicle driving through key performance indicators. Batteries 10:103. https://doi.org/10.3390/batteries10030103
43. Siemens White Paper. The evolution of the electric vehicle market (2020)
44. Guo Y, Shichao W, He Y-B, Kang F, Chen L, Li H, Yang Q-H (2022) Solid-state lithium batteries: safety and prospects. eScience 2:138–163. https://doi.org/10.1016/j.esci.2022.02.008
45. https://www.tycorun.com/blogs/news/solid-state-lithium-battery
46. Manthiram A, Yu X, Wang S (2017) Lithium battery chemistries enabled by solid-state electrolytes. Nat Rev Mater 2:16103. https://doi.org/10.1038/natrevmats.2016.103
47. The Nikkei in December 2020
48. Koorata PK, Panchal S (2023) Thermal management of large- sized LiFePO4 pouch cell using simplified mini-channel cold plates. Appl Therm Eng 234:121286. https://doi.org/10.1016/j.applthermaleng.2023.121286
49. Farulla GA, Palomba V, Aloisio D, Brunaccini G, Ferraro M, Andrea Frazzica A, Sergi F (2023) Optimal design of lithium-ion battery thermal management systems based on phase change material at high current and high environmental temperature. Therm Sci Eng Prog 42:101862. https://doi.org/10.1016/j.tsep.2023.101862
50. Wu H, Niu M, Shao Y, Wang M, Li M, Liu X, Li Z (2024) Investigations of Li-Ion battery thermal management systems based on heat pipes: a review. ACS Omega 9:97–116. https://doi.org/10.1021/acsomega.3c08056
51. Kiani M, Omiddezyani S, Houshfar E, Miremadi SR, Mehdi Ashjaee M, Nejad AM (2020) Lithium-ion battery thermal management system with Al_2O_3/AgO/CuO nanofluids and phase change material. Appl Therm Eng 180:115840. https://doi.org/10.1016/j.applthermaleng.2020.115840

52. Talele V, Moralı U, Patil M-S, Panchal S, Fraser R, Fowler M, Thorat P, Gokhale Y-P (2023) Computational modelling and statistical evaluation of thermal runaway safety regime response on lithium-ion battery with different cathodic chemistry and varying ambient condition. Int Commun Heat Mass Transfer 146:106907. https://doi.org/10.1016/j.icheatmasstransfer.2023.106907
53. Myatezh AV, Smirnov MA, Malozyomov BV, Shtang AA (2017) Performance increase of lithium-ion batteries. Adv Eng Res 133:519–524. https://doi.org/10.2991/aime-17.2017.84
54. Mahmud S, Rahman M et al (2022) Recent advances in lithium-ion battery materials for improved electrochemical performance: a review. Results Eng 15:100472
55. Ghafari V, Bayat S, Akbari A, Yeklangi G (2023) Current and future prospects of Li-ion batteries: a review. Nano Sci Technol 8:24–43
56. IEA (2022) Global EV outlook
57. Global Petrol Prices. https://www.globalpetrolprices.com/gasoline_prices/
58. https://ourworldindata.org/grapher/per-capita-electricity-generation
59. World Bank. https://data.worldbank.org/indicator/NY.GDP.PCAP.CD
60. https://wisevoter.com/country-rankings/renewable-energy-by-country/
61. https://en.wikipedia.org/wiki/List_of_countries_by_renewable_electricity_production
62. Kia Seoul, South Korea
63. Yildiz E, Serpelloni M, Salvadori A, Cabras L (2024) A comparative review of models for all-solid-state Li-Ion batteries. Batteries 10:150. https://doi.org/10.3390/batteries10050150
64. Diouf B, Miezan E (2024) Unlocking the technology potential for universal access to clean energy in developing countries. Energies 17:1488. https://doi.org/10.3390/en17061488
65. https://physicsworld00.com/a/lithium-ion-batteries-break-energy-density-record/
66. https://www.autoevolution.com/news/by-the-numbers-2011-nissan-leaf-vs-2018-nissan-leaf-comparison-122189.html
67. Garikapati D, Shetiya SS (2024) Autonomous vehicles: evolution of artificial intelligence and the current industry landscape. Big Data Cogn Comput 8:42
68. Sheldon TL, Dua R (2024) The dynamic role of subsidies in promoting global electric vehicle sales. Transp Res A Policy Pract 187:104173. https://doi.org/10.1016/j.tra.2024.104173
69. Baklouti A, Schutz J, Dellagi S, Chelbi A (2022) Selling or leasing used vehicles considering their energetic type, the potential demand for leasing, and the expected maintenance costs. Energy Rep 8:1125–1135. https://doi.org/10.1016/j.egyr.2022.07.074
70. Mashrur SM, Mohamed M (2025) Uncovering factors affecting consumers decisions for pre-owned electric vehicles. Transp Res Part D Transp Environ 139:104555
71. https://www.fleetnews.co.uk/news/record-rise-in-demand-for-used-electric-vehicles-and-market-share
72. https://www.businessinsider.com/tesla-used-car-sales-increase-average-sale-price-2025-5
73. Niri AJ, Poelzer GA, Zhang SE, Rosenkranz J, Pettersson M, Ghorbani Y (2024) Sustainability challenges throughout the electric vehicle battery value chain. Renew Sust Energ Rev 191:114176. https://doi.org/10.1016/j.rser.2023.114176
74. LaMonaca S, Ryan L (2022) The state of play in electric vehicle charging services—a review of infrastructure provision, players, and policies. Renew Sust Energ Rev 154:111733. https://doi.org/10.1016/j.rser.2021.111733
75. Orangi S, Manjong N, Clos DP, Usai L, Burheim OS, Strømman AH (2024) Historical and prospective lithium-ion battery cost trajectories from a bottom-up production modeling perspective. J Energy Storage 76:109800. https://doi.org/10.1016/j.est.2023.109800
76. https://www.recurrentauto.com/research/used-electric-vehicle-buying-report
77. https://cleanmobilityshift.com/policy-regulation/road-to-electric-decoding-andhra-pradeshs-strategy-to-boost-ev-adoption/
78. https://www.cars.com/articles/how-much-do-electric-car-batteries-cost-to-replace-465308/
79. https://www.mrmoneymustache.com/the-nissan-leaf-experiment/
80. https://www.carboeurope.org/tesla-battery-replacement-cost/

81. Nguyen-Tien V, Zhang C, Strobl E, Elliott RJR (2025) The closing longevity gap between battery electric vehicles and internal combustion vehicles in Great Britain. Nat Energy 10:354–364
82. https://www.businessresearchinsights.com/market-reports/used-car-market-118118
83. https://www.spglobal.com/mobility/en/research-analysis/average-age-vehicles-united-states-2024.html
84. https://www.cars.com/articles/2024-could-be-a-buyers-market-if-youre-buying-new-report-478345/
85. https://www.news.market.us/used-car-market-news/
86. https://wardsintelligence.informa.com/wi967636/global-vehicle-sales-top-92-million-units-in-2023-december-volume-up-11
87. https://www.iea.org/reports/global-ev-outlook-2024/trends-in-electric-cars
88. https://statzon.com/insights/e-mobility-europe-an-overview-of-europes-latest-electric-vehicles-data
89. Diouf B (2025) Modeling the used vehicle market share in the electric vehicle transition. World Electr Veh J 16:29. https://doi.org/10.3390/wevj16010029
90. Buhmann KM, Criado JR (2023) Consumers' preferences for electric vehicles: the role of status and reputation. Transp Res Part D Transp Environ 114:103530. https://doi.org/10.1016/j.trd.2022.103530
91. https://www.transportenvironment.org/articles/leasing-sector-could-flood-second-hand-market-with-electric-cars
92. https://www.forbes.com/sites/michaelharley/2025/04/24/carmaxs-ev-trends-report-shows-ev-searches-nearly-doubled-in-the-past-three-years/

Chapter 9
Is the Grid Ready for the Electric Vehicle Transition?

Contents

9.1 Introduction

EVs of every kind are currently replacing 1.5 million barrels of oil per day, which is approximately 3% of overall demand for road fuel [1]. The competition has begun to reduce carbon emissions from road transportation and encourage drivers to transition to EVs in many countries. Nonetheless, EVs represent the future of transportation, provided sufficient energy can be produced to sustain them all. In 2020, there were 16.5 million EVs on roads globally [2, 3]; this number is projected to increase to 300 million by 2030 [2, 3]. With these increasing figures, in nations with the greatest EV adoption, grids might face difficulties without investment and changes in charging behaviors. As an example, it is forecasted that maximum net electricity demand may rise by as much as 50% with complete electrification of the fleet in the US. For the specific Western US area, analysts show that when the penetration reaches 30% to 40% of vehicles on the road, the grid will undergo considerable strain without substantial investments and modifications in charging behaviors. Due

R. Pode, B. Diouf, *Decarbonization of Road Transportation*, Green Energy and Technology, https://doi.org/10.1007/978-3-032-11388-7_9

to the time required to charge EVs with standard chargers, drivers often leave their vehicles to charge in the evening and overnight, increasing strain on the grid. Generally, electric energy demand reaches its highest point between 5–9 pm when individuals come home from work and utilize household electric appliances. This suggests that if vehicle owners charged their cars during the day, particularly with the potential of solar energy, it could reduce expenses and assist the grid as the number of EVs rises to achieve sustainability targets. It is feasible to sync EV charging times with high renewable energy generation and apply dynamic pricing or incentives for EV users to charge with renewable energy sources. The transformation of road transportation with the rise of EVs, with projections indicating that 42% of worldwide light-duty vehicle sales will be electric by 2030 [2, 3], is a significant challenge in the grid's readiness and resilience for the transition to EVs. Grid resilience, which is an essential factor in maintaining the stability and dependability of an electrical grid, includes the ability to adjust to unexpected obstacles and swiftly bounce back from difficulties, whether these arise from human-induced problems, like the surge from massive EV charging, or from natural calamities. Resilience in the case of massive EV adoption is especially gauged by a grid's ability to keep delivering essential loads, even when confronted with high-impact occurrences, as from simultaneous massive demand from EV fleet charging [4]. Attaining grid resilience requires a multifaceted strategy, which encompasses bolstering essential infrastructure to endure severe conditions and deploying sophisticated control algorithms. The surge in electricity demand will place more pressure on the grid. In order to match the rapid increase in EV adoption, utilities must respond promptly to ready the grid for this emerging load type. However, conventional utility business models and regulations, which are effective for reactive planning regarding new loads as they would be for new residential complexes, currently do not permit the proactive grid investments needed for EVs. The uncertainty surrounding the precise charging requirements for EVs, such as when, where, and how much, stands in stark contrast to regulatory demands for prudent investments aimed at minimizing costs for ratepayers. Significant challenges are present in certain areas at the local distribution transformer or feeder level, whereas other places benefit from a considerable amount of unused capacity and will not need grid upgrades. Given that grid hosting capacities vary considerably across different locations, even within the same distribution network, planning for the rollout of EV charging presents a highly site-specific challenge. Microgrids, as a potential local solution, represent a significant source of resilience within the larger grid [5, 6]. They provide islanding capabilities, enabling them to function autonomously during power surges or outages, and they ensure a steady flow of renewable energy.

Figure 9.1 above shows the general structure of a microgrid. Microgrids can operate independently, integrate into a larger community network, or act as black start systems, improving the reliability of power systems. Microgrids can provide an extra level of resilience, and a systematic framework aids in comprehending and managing grid conditions during and following surge occurrences. This comprehensive strategy is essential for maintaining a constant supply of electricity, even during challenging circumstances [5, 6].

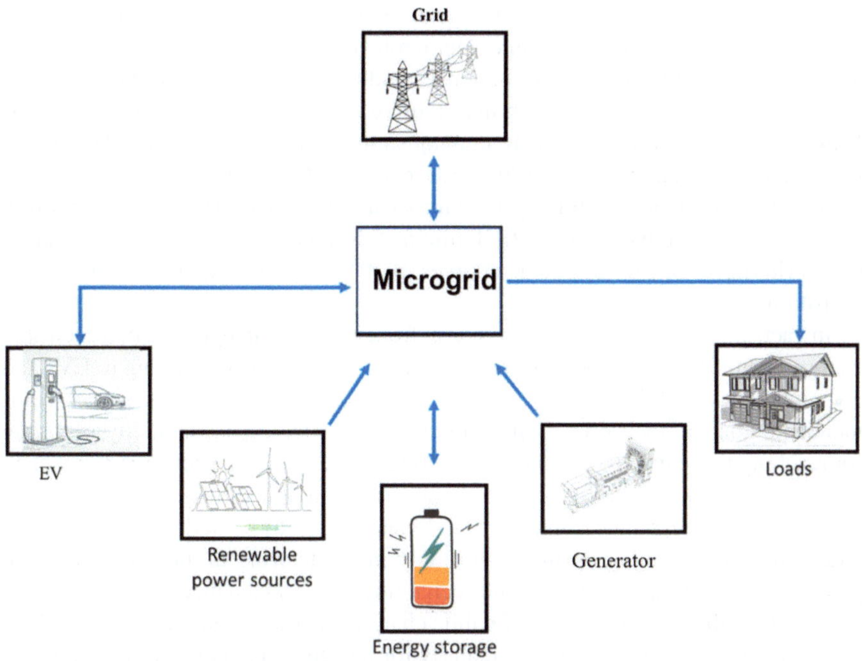

Fig. 9.1 A general structure of a microgrid

Another innovative strategy in EV charging could emerge from time-of-use pricing. This strategy will promote charging during off-peak times by providing reduced electricity rates at those hours. Increased attention towards smart charging and the application of algorithms and communication technologies to plan charging during the best times. In that regard, it might be required to have a better organized Vehicle-to-Grid (V2G) Technology [7, 8] enabling EVs to supply energy back to the grid during high demand periods. The grid's resilience will be necessary due to the extra load from EVs, which can cause localized grid strain, particularly in regions with outdated infrastructure.

In fact, in a completely electrified fleet, the demand on the grid from road transport could potentially exceed current systems capacity during peak times. Although it is improbable for EV charging times to align throughout a power system, it remains feasible that numerous EVs or the entire local fleet may be charging simultaneously within a distribution transformer or zone substation area, potentially leading to complications.

Distribution grids require the capability to simulate future charging demands and effects across different scenarios. This encompasses comprehending EV penetration by area, charging habits, and the capacity to impact these through flexibility management. It may include existing and prospective rooftop solar photovoltaic penetration, along with storage accessibility for a strong and rapid development of renewable energy potential. It involves forecasting the future effects on grid infrastructure, taking into account storage capacity and needs, and evaluating investment necessities well ahead of time. Numerous networks worldwide acknowledge the necessity of addressing these challenges and are eager to advance. Nevertheless, the existing regulatory framework may require modifications in several instances. Numerous regions of the world are making significant investments in solar energy. For instance, in Europe, the REPowerEU plan [9, 10] outlines that the European Union's new solar energy strategy aims for solar panels on every residential rooftop across the continent by 2029 [9, 10]. Utilizing this daytime energy will be crucial as the world transitions to renewable sources, particularly in adequation with the EV transition.

In fact, a realistic solution may come from a higher integration of renewable energy, even if it will not be achieved without challenges. Transitioning to EVs and a simultaneous exploitation of renewable energy to tackle climate change are commendable goals in their own right. Reducing fossil fuel use substantially is crucial for reaching those goals. However, attempting to execute such transitions on a massive scale within a short period is fraught with difficulties, risks, and unexpected outcomes that necessitate committed and clear vision to be addressed effectively and realistically. Scaling up means not only manufacturing millions of EVs each year but also offering assistance with recharging and maintenance.

Presently, there is an increase of public charging stations along with the EV fleet. They are an adequate solution, particularly for fast charging and densely populated areas such as major cities, where home charging is not an available solution. Public charging infrastructures represent a viable business model that allows the development of EVs by reducing range anxiety.

The transition to EVs offers a chance as well to enhance the incorporation of renewable energy within the transportation industry. Figure 9.2 illustrates an example of a public charging station with the integration of photovoltaic solar panels.

By synchronizing EV charging with times of elevated renewable energy production, such as photovoltaic solar, it can be guaranteed that EVs will run on clean energy, further lowering their carbon impact and improving transportation sustainability.

Simultaneously transforming the energy and transportation sectors will involve a wide range of both known and unexpected factors that will interact intricately in complex and unpredictable ways. As EVs and renewable energy expand, the challenges and solutions will involve progressively larger populations and areas. Each proposed solution is expected to create further obstacles. Moreover, increasing scale threatens individuals' entrenched beliefs, ways of life, and livelihoods, many of which could be altered or made obsolete. Adjusting to technological shifts is tough, but adapting to social shifts is even harder.

The challenge in massive adoption of EVs is not necessarily energy but more power capacity. This challenge is even greater when considering the fast progress of EV adoption as shown by the S-curve describing the adoption of EVs in different countries, particularly those with higher EV penetration rates [11, 12].

Despite these challenges, some countries are almost having 100% of new registrations made up of EVs [11, 12]. It is feasible to sync EV charging times with high

Fig. 9.2 Solar energy-assisted EV charging

renewable energy capacity and apply dynamic pricing or incentives for EV users to charge with renewable energy sources.

This study investigates how prepared grids are for the EV transition through benchmarking, particularly the necessity for technical updates, investments, and reforms to follow when considering an S-curve model for the adoption of EVs. Countries that are close to completing the full EV transition are highlighted as models based on their environmental policies, energy resources, grid management practices, and additional factors that may influence EV adoption. This manuscript presents as well the challenge when integrating renewable energy in charging solutions, particularly for off-grid systems that could potentially replace gas stations. Grids need to be better prepared for local challenges when integrating fast public charging stations. The USA is presented as a case study with the installed power and energy consumption, power and energy demand from the EV fleet, and the forecast of electricity usage in the transportation sector.

9.2 Increased Electricity Demand and Strain on Grids

Charging Infrastructure Planning
How much charging power should be installed to support each new EV added to the fleet?

This is commonly expressed in kilowatts (kW) of charger capacity per EV.

Rule of Thumb

- Residential/light-use scenario:

  ~1 to 2 kW/EV
- Public and commercial infrastructure:

  ~2 to 7 kW/EV
- Fast-charging focused (urban, highway):

  ~10 to 15+ kW/EV

Example (From Global Averages)

- U.S. target (DOE, NEVI plans):

  ~0.3 to 0.5 kW of public charger capacity per EV
- South Korea (2023):

  ~1.2 public chargers per EV, averaging ~7 kW/charger, so about 8.4 kW/EV (but overbuilt compared to many).

Energy Use Estimate per EV
How much energy must the grid support per EV per year (for charging)? It is a different measure:

- Average EV consumption: ~15–20 kWh/100 km
- Annual driving (global average): ~12,000–15,000 km/year
- → 1800–3000 kWh/year per EV
  So grid planning must support that much energy (not power) per EV.

For grid or station designers planning EV infrastructure, a basic planning guide might look like Table 9.1.

9.2.1 Grid Stress and Electric Vehicle Transition

Switching to non-emitting transportation could lead to higher energy demand. The increase in EVs is not necessarily an issue tied to energy. The growing number of EVs will not likely pose an energy issue but instead a capacity challenge in

Table 9.1 Common charging conditions for EVs

Use case	Power to install per EV	Notes
Home charging	2–3 kW	Overnight slow charge (Level 1/2)
Workplace/public	5–7 kW	Level 2 chargers
Fast-charging sites	10–50+ kW	DCFC or Tesla Superchargers
National Grid Avg.	~1.5–3 kW/EV	Depends on usage intensity

distribution networks, as in an extreme situation where every household charges simultaneously alongside their typical electricity usage. Simultaneous EV charging, for a given grid, could lead to peak load management challenges, particularly during high-demand periods. This kind of circumstance can intensify surges in demand. This problem will require upgrading transformers, substations, and distribution systems to manage increased loads. The implementation of sophisticated grid management systems, like real-time surveillance and predictive analysis, should also be considered.

At present, home charging is certainly the favored choice for EV users [13, 14] when it is possible. Charging an EV at home or at public stations impacts power demand and, consequently, creates strain on the grid. The charging infrastructure required to power the forthcoming EV fleet is expected to lead to significant spikes in electric load growth, and most existing grids are not equipped to meet this energy demand. To fully support the EV revolution and decrease emissions from this sector, faster and smarter grid planning is necessary.

Considering an EV's typical energy usage of 0.20 kWh for every kilometer or 0.32 kWh for each mile [15, 16], the average energy needed for the daily travel of individuals can be examined.

The U.S. Department of Transportation states that the average American drives 13,476 miles annually, which is about 36.92 miles each day. By utilizing the typical energy usage of an EV, a residential EV charger would consume approximately 11.81 kWh daily to restore its driving range. This equates to approximately 353.3 kWh each month and 4310.65 kWh annually.

Driving behaviors differ greatly among European nations, but the EU average is approximately 2814.20 miles annually, which translates to 7.7 miles daily. In this scenario, a residential EV charger would consume approximately 2.48 kWh daily, which totals 74.40 kWh monthly and 905.20 kWh annually.

Table 9.2 above compares the current installed capacity of the 20 countries with the highest EV penetration rates and the power required if all the fleet were made of EVs and would be recharged simultaneously with an average power of 7 kW/vehicle. Obviously most home chargers are between 2 and 3 kW but public level 2 chargers are up to 7 kW and some EVs, as Teslas, are charged with power over 7 kW. Only three countries, those with the highest EV penetration rate in the list, have an installed capacity more greater than that potential demand: Norway, Iceland, and Sweden. These three countries have a grid with the potential to simultaneously power the entire fleet if it were to be made exclusively of EVs. All the other countries in the list present a deficit. Norway, Iceland, and Sweden also present a very high ratio of renewable energy percentage in their total installed capacity. The USA and China present the highest power deficit.

Countries with the lowest ratio of EV charging demand vs installed power may need a strategy in case of a surge. Even though grids are being upgraded, the main problem to solve is to guarantee that the progress in the grid development is not too far behind the progress of the EV fleet and there is a real management strategy for the offer of fast public charging stations.

Table 9.2 Power demand of EVs fleets in countries with highest EV penetration [17–19]

	EV penetration rate	Installed capacity (MW)	Generation (GWh)	Vehicles (fleet)	EVs fleet	All EV (MW)	Max Evs[a] possible	Power[a] Mismatch (MW)
Norway	29%	40,540	143,383	3,469,692	900,000	24,288	5,791,429	+16,252
Iceland	18%	3006	19,865	306,563	19,215	2146	429,429	+860
Sweden	11%	52,706	173,754	5,682,251	560,000	39,776	7,529,429	+12,930
Denmark	11%	19,503	33,727	3,222,575	300,000	22,558	2,786,143	−3055
Finland	8.1%	24,784	70,162	4,221,852	83,800	29,553	3,540,571	−4769
Netherlands	8.3%	57,194	120,840	10,062,194	700,000	70,435	8,170,571	−13,241
Germany	5.4%	261,086	560,759	52,714,433	2500,000	369,001	37,298,000	−107,915
Switzerland	5.8%	24,014	58,726	5,301,789	172,000	37,113	3,430,571	−13,099
Portugal	1.87	23,316	45,543	7,198,871	111,000	50,392	3,330,857	−27,076
United Kingdom	5%	111,020	318,595	41,700,000	1,580,000	291,900	15,860,000	−180,880
France	4.1%	148,914	446,291	44,444,965	1,570,000	311,115	21,273,429	−162,201
Belgium	8.2%	26,310	90,893	6,977,664	254,240	48,844	3,758,571	−22,534
China	7.6%	2,593,589	8,881,870	440,000,000	21,800,000	3,080,000	370,512,714	−486,411
Italy	3.8%	123,327	274,164	44,888,074	220,000	314,217	17,618,143	−190,890
Spain	1.2	130,908	265,312	30,724,166	466,178	215,069	18,701,143	−84,161
Greece	2.3	22,336	49,179	5,777,241	134,484	40,441	3,190,857	−18,105
Canada	2.3%	158,973	637,996	26,302,526	550,000	184,118	22,710,429	−25,145
South Korea	2.4%	146,539	606,760	25,949,201	553,000	181,644	20,934,143	−35,105
United States	2.1%	1,201,260	4,291,954	283,400,986	4,800,000	1,983,807	171,608,571	−782,547
New Zealand	2.4%	10,412	44,178	4,794,156	102,384?	33,559	1,487,429	−23,147

[a] Remark: Max EVs possible is the maximum number of EVs that can be charged simultaneously at 7 kW under the current installed capacity. Power Mismatch is the difference between the installed capacity and the required capacity to simultaneously charge the entire fleet at 7 kW if it was made exclusively of EVs

Norway and Iceland, occupying the top two positions in the rate of EVs in new registrations, are significantly ahead of other nations in the group of countries with the highest proportions of EV uptake, as shown in Table 9.1 [11, 12]. It is probable that these two nations will achieve 100% of new EV registrations within the next few years [11, 12]. Norway and Iceland showcase characteristics that align with the EV transition. They hold the first and second positions in GDP per capita within the provided countries, second and first in lowest density of population density, and second and first in kWh per capita; they are average regarding the number of vehicles per 1000 inhabitants but are highly urbanized with extremely low population density (Table 9.1). These elements indicate that both Norway and Iceland possess the financial capacity to purchase EVs and demonstrate resilience in their energy sectors, aligning with grid preparedness or suitability for the transition to EVs (Table 9.1). Based on Table 9.1, Norway, Iceland, and Sweden are the only three countries among those listed to have an installed capacity outpacing the demand that would come from the entire fleet if it were completely electrified presently.

Conversely, countries with a dense population, low kWh/per capita production, a high vehicle count per capita, low GDP per capita, and a high kWh to liter of fuel ratio [11] will likely encounter significant difficulties in ensuring a considerable share of EVs in their fleet. These nations will require significant structural changes, especially to align the power sector with the EV transition.

Public charging stations may be a necessity to solve the problem of range anxiety but also solve the issue of grid strain, as they allow control of the maximum number of EVs that can be charged simultaneously. This is particularly the case in densely populated urban areas where home charging is not an available alternative. This can be well illustrated by the example of South Korea. As of 2024, the number of registered cars in Seoul, the capital city of South Korea, was 3,191,162. It means that Seoul citizens owned one car per 2.94 people. This is about one person different from 1.98 to 1 nationwide [20]. South Korea is gradually transitioning to electric transportation. Considering worries regarding the impending climate crisis, South Korean car manufacturers and battery companies are collaborating with the government to hasten the uptake of EVs. As a key player in the worldwide automobile industry, South Korea seeks to boost the domestic market share of EVs and deliver 4.5 million units by 2030, prioritizing eco-friendly vehicles. In 2023, South Korea saw the registration of 543.9 thousand EVs, marking a significant rise since 2013. Along with hybrid and hydrogen vehicles, the percentage of eco-friendly vehicles among all registered vehicles in South Korea was approximately 8.2 percent in 2023 [21].

In response to the problem of grid strain and the increasing demand, public charging stations seem to be an adequate solution to guarantee a successful EV transition. The development of charging infrastructure is essential for enabling the worldwide transition to electric transportation. Public charging stations are essential to the EV ecosystem, alleviating range anxiety and providing easy access to charging for EV owners who lack a charging station at home or are frequently traveling. As per the IEA, by the end of 2023, there were 2.7 million public charging stations globally, with over 900,000 established in 2022. Although home charging primarily satisfies most charging requirements, public charging infrastructure remains essential for promoting EV adoption, particularly in densely populated urban regions where home charging is significantly less common.

Faster, more powerful EV charging stations amplify the positive and negative impacts that these stations can have on the grid. One 350 kW charger impacts the grid similarly to 50 smaller 7 kW chargers operating concurrently. However, the 350 kW charger will not be allocated among different feeders or substations as 50 smaller 7 kW chargers would be. Another difference is that the individual 350 kW charger ought to be easier to manage because there is only one control point. However, the strongest EV charging stations tend to be the least flexible, making them less compatible with charging management systems. This happens as the most powerful EV charging stations, level 3, are used in situations where time is restricted. Consequently, customers using fast EV charging stations have not set aside time for a slower charge or any wait times required for grid requirements. Additionally, EV charging stations do not always operate at their full power capacity, and the charging rates for EVs vary when power sharing is enabled among nearby chargers. These are elements concerning EV charging stations that may be influenced as charging prices continue to increase.

Since 2019, China has established over a million new public charging stations, showing an unparalleled commitment to EV infrastructure compared to any other nation in recent years. South Korea is at the forefront regarding public charging stations for every 1000 EVs. As per IEA data, at the close of 2022, South Korea boasted 201,000 public chargers for its 357,000 plug-in electric passenger vehicles, which translates to 563 charging stations for every electric car, the highest density among the largest EV markets globally.

China boasts 1.76 million charging stations, making up nearly two-thirds of the global public charging network; however, because of its large EV population, the nation ranks third in chargers available per EV on the road. In that regard, the Netherlands stands in second place, boasting 235 public chargers for every 1000 EVs.

Seoul has one of the highest urban densities of population in the world (Table 9.2). The example of South Korea, with the highest number of charging stations per 1000 EVs (Fig. 9.3), is main proof that these charging infrastructures are one of the best responses to the grid stress due to the EV fleet.

Countries with the lowest density of population show a lower number of charging stations per 1000 EVs; it is the case of Norway that presents the highest EV penetration rate and one of the lowest densities of population. Home charging is the main option in Norway.

Based on Table 9.3 above and Fig. 9.3, countries with higher densities of population, particularly in urban areas, have opted for public charging stations to guarantee the EV transition. The availability and adequate distribution of public charging stations allow good control of the grid management as well and avoid any risk of power surge. Nevertheless, the challenge is to have the development of charging stations match the penetration rate of EVs, as most nations are still in the early stage of EV adoption.

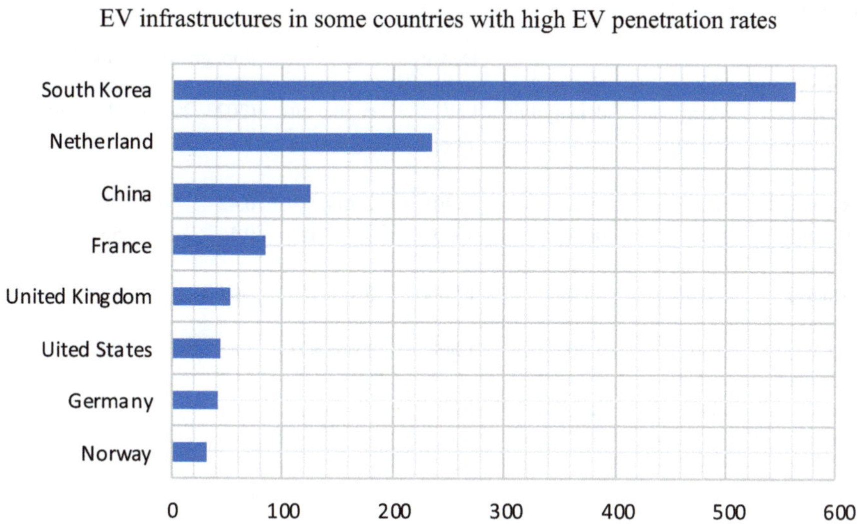

Fig. 9.3 Number of charging stations per 1000 EVs in 2022 [22]

Table 9.3 Population characteristics of capital cities and countries with highest EV penetration [23]

Capital city/ country	Size capital (km^2)	National density/km^2	Density (capital)/ km^2	Population capital city	Pop. growth (%) (2023–2024)	Number of cars per 1000 (National)
Oslo/Norway	426	18.24	1684	717,710	0.86	550
Reykjavik/ Iceland	73.1	3.733	3281	239,733	1.88	779
Stockholm/ Sweden	483	25.91	3748	1,809,412	1.12	473
Copenhagen/ Denmark	292	138.8	4717	1,378,649	0.93	472
Helsinki/ Finland	214	18.44	3148	674,500	0.88	660
Amsterdam/ Netherlands	168	533.3	6782	1,136,210	1	501
Berlin/ Germany	891	231.3	4036	3,596,999	0.81	578
Bern/ Switzerland	51.6	217.1	2655	136,988	0.54	546
Lisbon/ Portugal	84.9	112.2	6427	545,796	−0.13	549
London/ United Kingdom	1172	276	7490	8,776,535	0.74	480
Paris/France	106	121.1	20,016	2,113,705	−0.61	570
Brussels/ Belgium	33.1	383.3	5948	196,828	1.23	507

Table 9.3 (continued)

Capital city/ country	Size capital (km^2)	National density/km^2	Density (capital)/ km^2	Population capital city	Pop. growth (%) (2023–2024)	Number of cars per 1000 (National)
Beijing/ China	16,410	150	1332	18,960,744	1.27	194
Roma/Italy	1287	195.3	2140	2,754,719	0.42	682
Madrid/ Spain	606	96.32	5650	3,422,416	0.56	553
Athena/ Greece	38.96	79.40	17,000	3,146,164	−0.07	550
Ottawa/ Canada	549	4.209	1945	1,068,821	1.45	707
Seoul/ Republic of Korea	605	484.7	15,839	9,586,195	0.21	500
Washington/ United States	158	37.18	4289	678,972	0.92	850
Auckland/ New Zealand	607	19.29	2372	1,440,300	0.34	889

There is an impact of public EV charging stations on grid strain. Public charging stations, especially fast chargers, can significantly increase electricity demand, particularly during peak hours. In dense urban areas like Seoul, in South Korea, simultaneous charging by multiple EVs can strain local distribution networks, leading to voltage drops, transformer overloads, and increased grid instability. This is especially critical if the infrastructure was not originally designed to handle high, sudden loads. The challenge intensifies with high penetration of EVs and clustering of charging stations in specific zones (malls, apartment complexes). Mitigating these effects requires smart grid technologies, time-of-use pricing, demand-side management, and integration of local renewable energy and energy storage systems to reduce reliance on the main grid during high-demand periods.

As an illustration, let us consider an urban district and an evening peak load scenario. In a busy urban neighborhood, a new EV charging hub with 10 DC fast chargers (each rated at 100 kW) is installed near a shopping mall. During the evening rush hour (5–8 PM) when residents return home and electricity demand is already high, 8 out of 10 chargers are occupied simultaneously. This adds a sudden load of 800 kW to the local grid.

The area's existing transformer was originally designed for a maximum local load of 1 MW and is already operating near 90% capacity during peak hours. The added EV charging demand causes the transformer to exceed its limit, resulting in:

- Voltage fluctuations
- Overheating of grid components
- Risk of localized blackouts or the need for load shedding
- Reduced lifespan of equipment due to overuse

Cities can manage this challenge to prevent such grid strain. Utilities might implement:

- Smart charging systems that schedule EV charging during off-peak hours
- Battery energy storage at the station to buffer demand
- Time-of-use pricing to discourage peak-hour charging

Growing electricity demand and pressure on grids create challenges for the real-world application of charging solutions. There is a possibility of infrastructure being overwhelmed. Local distribution networks and transformers were not created to handle peak loads from EVs. In regions with concentrated EV usage, there is a danger of regular outages or infrastructure breakdowns. Improvements are necessary, but they come at a cost. Increasing grid capacity (such as substations, cables, and transformers) requires substantial capital investment. Utilities must seek funding or incentives to actively improve the infrastructure. In comparing charging time to grid peaks, EVs are frequently charged during evening peak times when home demand is already elevated. In the absence of load-shifting incentives, this deteriorates peak demand curves. Due to constraints in urban areas, in densely populated cities like Seoul, there is a lack of physical space available for the installation of high-capacity chargers. Renovating buildings or sites presents intricate permitting and construction difficulties. Coordination gaps and inadequate collaboration among utilities, city planners, and private charging providers hinder the rollout. Insufficient uniformity in charging infrastructure hinders grid integration. Regulatory obstacles exist; extended approval processes for grid enhancements or charger licenses hinder progress. Obsolete regulations might overlook the requirements of EV infrastructure. The use of renewable energy is still behind; EVs are best suited for the integration of renewable energy, yet numerous grids remain reliant on fossil fuels. Without pairing solar/storage, EVs can elevate carbon intensity during peak periods.

9.2.2 Renewable Energy and Off-Grid Charging Solutions

A stationary off-grid EV charging station generally employs renewable energy sources like photovoltaic solar or wind to generate electricity and store it in batteries. This configuration is perfect to reduce stress on the grid or for areas with restricted or inconsistent grid availability. This approach is eco-friendly and may be economical over time or drive subsidies if necessary. The worldwide off-grid EV charging market is projected to hit $1.2 billion by 2028 [24], fueled by increasing EV demand, investments in renewable energy, and the development of battery technology. EVs can also facilitate higher integration of renewable energy sources by using their batteries to store excess generation. This concept of Vehicle-to-Grid (V2G) [7, 8] is gaining popularity in different parts of the world. Advantages of off-grid EV charging stations comprise energy autonomy, reassurance, scalability, eco-friendliness, enhanced accessibility, readiness for emergencies, and obviously less stress on grids. Off-grid EV charging stations provide advantages that render them an appealing option for sustainable and dependable EV charging. Off-grid EV charging solutions decrease reliance on the utility grid, lessening the effects of grid-associated strain. Using renewable energy sources such as solar power for off-grid charging aids in lowering carbon emissions and conserving natural resources. Utilizing clean energy, off-grid charging stations promote a more sustainable future and support international environmental objectives.

Factors to consider encompass greater upfront investment expenses, reliance on weather, battery storage capabilities, and maintenance needs. In fact, off-grid EV charging stations provide numerous advantages, yet they also entail certain considerations and limitations. For photovoltaic energy, this may pose a restriction for regions with insufficient sunlight or during times of prolonged cloudiness. The capacity of battery storage in off-grid EV charging stations can influence both the charging speed and, obviously, the number of EVs that can be charged. The space area needed for solar panels may be a major challenge in urban areas. Despite these obstacles, off-grid EV charging stations [25] provide a dependable and sustainable option for charging electric EVs and solve the problem of grids' lack of readiness.

Standalone solar charging systems, as a developing business, are gaining popularity for single households. They remain practical to charge one vehicle per household without much challenge.

The integration of renewable energy presents challenges to powering EV charging stations. The standalone mobile EV rapid charging stations come with integrated batteries, making them ideal for charging EVs anytime and anywhere, as it is presently for gas stations and ICE vehicles. The advanced, possibly mobile, EV chargers offer unparalleled flexibility and effectiveness, allowing for a seamless, practical charging experience. The state-of-the-art charging station combines innovative DC fast charging technology with the most secure lithium-ion (Li-ion) battery chemistry, ensuring ample energy to recharge electric vehicles. These battery packs can be charged primarily using renewable energy sources, but also, if needed, the electrical grid or a DC fast charger, guaranteeing that EV owners are always ready to power up.

These stations quickly deliver fast charging to all electric vehicle models that are DC-compatible. This improves the availability of charging for various electric vehicles. Customize your charging experience by selecting various power outputs and battery sizes to meet individual requirements.

Figure 9.4 represents types of off-grid EV charging stations; with the grid as a backup, this may represent the future of electric transportation, as they allow fast charging. They alleviate grid pressure and enable opportunities for renewable energy advancements and energy storage options for an integrated decarbonization of road transport. Consideration should be given to regulation and cost. The updated subsidy guidelines may necessitate that investors work closely with decision makers to guarantee access to EV charging infrastructure and reduce range anxiety for electric vehicle owners.

In a successful transition to EVs, off-grid solar charging stations will take the place of existing gas stations due to their effective integration, particularly in densely populated areas. Nonetheless, significant challenges exist, such as cost, battery lifespan, and primarily the current energy density of Li-ion batteries, which will

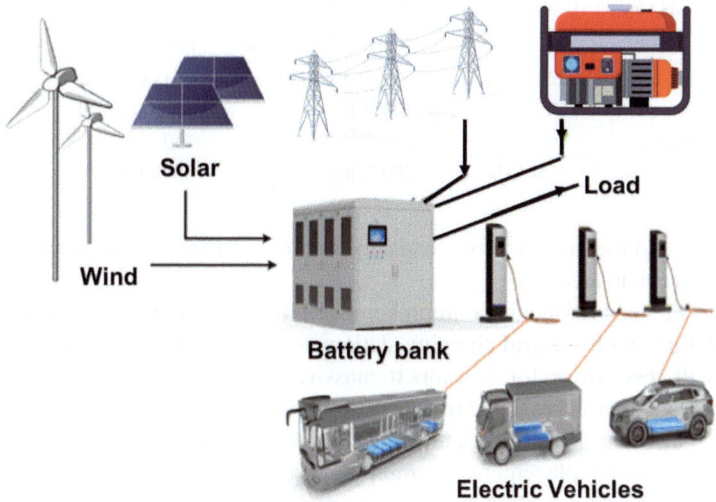

Fig. 9.4 Off-grid EV charging station with integrated battery: the developing business

require considerable space for energy storage to match the capacity of existing gas stations. The current Li-ion battery technology is projected to provide approximately 10 years of functionality before requiring replacement [26].

To achieve practical integration of battery off-grid charging stations, battery technology must improve its energy density significantly to align with the widespread adoption of EVs. A typical gasoline-powered vehicle comes with a fuel tank of 13 to 16 gallons (50 to 60 L), while an electric vehicle generally has a battery capacity of around 40 kWh on average. Gas tanks at service stations typically hold between 12,000 and 24,000 gallons, allowing them to refuel around 750 to 1500 vehicle tanks. To achieve the same capacity as an off-grid Li-ion battery station, a range of 60,000 to 120,000 kWh will be required. To achieve a density of 500 Wh/L, the total volume required will range from 120,000 to 240,000 L, or 120 to 240 m^3, which is roughly equivalent to 2–4 40 ft containers.

A more cohesive implementation of off-grid solar-powered EV charging stations will necessitate advancements in battery technology. For example, to store 45.42 m^3 (12,000 gallons) of battery energy to charge 750 electric vehicles, the energy battery density required would need to be at least around 120,000 kWh/50 m^3, which is 2400 Wh/L compared to the current density of under 1000 Wh/L. Indeed, Li-ion batteries usually possess energy densities that range from 150 to 250 Wh/kg or 300 to 700 Wh/L. The figure of 2400 Wh/L equates to almost 1000 Wh/kg.

Moreover, to power an off-grid 60,000 kWh battery system with solar photovoltaic under an average of 6 sun-hours per day using 20% efficient solar panels, it will require 10 MW of solar panels, which equals roughly 50×10^3 m^2 (=5 ha). Clearly, such projects would be quite impractical, particularly in urban areas.

Table 9.4 compares the characteristics of a gas station vs an off-grid EV charging station. The density of energy in gasoline allows for much smaller-sized stations, while an equivalent off-grid EV charging station would need much more space for

Table 9.4 Gas station vs. off-grid EV charging station

	Gas station	Off-grid EV station	Future off-grid EV station
Capacity	12,000 gal	60,000 kWh/120 m^3	60,000 kWh/45.42 m^3
Tank	60 L	40 kWh	40 kWh
Number vehicles	750	750	750
Energy density	32.4 MJ/L	500 Wh/L	2400 Wh/L

the battery, and the installed solar capacity would cover surface areas not generally available in urban areas.

There is a need for technical improvement to achieve a competitive replacement of gas stations with off-grid charging stations for EVs. The cost of Li-ion batteries as well will need a drop for investors to massively invest in the upcoming business.

There is an alternative to the replacement of gas stations. Fast-charging hubs can support long-term scalability and are ideal for highway corridors or EV-dense zones. Gas station replacements are more accessible in cities but may struggle with future demand without infrastructure upgrades. Here is a comparison between EV fast-charging hubs and traditional gas station replacements:

- Fast-Charging Hubs

  Purpose: Built from the ground up for high-throughput EV charging.
  Location: Often located at highway exits, malls, or dedicated charging zones.
  Capacity: 10–50+ chargers; supports simultaneous fast charging (50–350 kW).
  Design: Optimized for EV traffic flow, short waiting times, and grid infrastructure.
  Grid Impact: High peak demand; often equipped with energy storage or smart-grid tech to buffer strain.
  Vision: Serves as a new model for electric mobility infrastructure.

- Gas Station Replacements

  Purpose: Retrofit of existing gas stations to add EV chargers.
  Location: Urban and suburban fueling stations.
  Capacity: Typically, 2–4 chargers due to space and grid constraints.
  Design: Limited redesign; may cause bottlenecks if EV volume increases.
  Grid Impact: Lower per-site demand, but less optimized for large-scale charging.
  Vision: A transitional model to integrate EVs into existing fuel networks.

9.3 Power and Energy Demand from EV Fleet: A Case Study of the USA

EVs represent the future of transportation, provided that sufficient energy can be produced to fuel the entire new fleet. The existing charging practices for EVs will not be sustainable for the grid unless adjustments are implemented prior to achieving net-zero targets for EVs. Significant funding is required for a successful transition to EVs. The grid will undergo considerable stress without substantial investments and modifications in charging behaviors. Policy makers should urge consideration of utilities that promote daytime charging and motivate investment in charging infrastructure to transition drivers from home to their workplaces for charging. Management solutions for charging and grid enhancements to support fast EV charging stations can control the electric load that EVs contribute to the grid both in time and space. Controlled charging offers significant benefits to the utility by preventing expensive upgrades to the grid. Shifting EV load from peak periods to off-peak times will reduce the total expenses of EV-related grid enhancements. This is usually accomplished by transmitting a price signal to EV owners, which can be executed via conventional managed charging methods, unidirectional from the grid to the vehicle, or bidirectional charging approaches, occurring both from the grid to the vehicle and in the opposite direction. Managed charging strategies represent one of the easiest methods to divert EV charging from peak grid loads by making charging away from home relatively cheaper. By avoiding expensive grid upgrades, utilities can transfer those savings to EV owners, encouraging them to engage in managed charging. These incentives consist of time-of-use pricing, demand response rates, and real-time pricing, among others. The savings for EV owners could lower the overall cost of ownership, which is likely to boost EV sales.

Approximately 80% of EV charging in the U.S. takes place at home, while the remaining 20% happens at public charging stations [27]. Nevertheless, the quantity of EV charging stations has increased more than twofold since 2020. In December 2020, the Department of Energy announced that there were almost 29,000 public charging stations across the country [28]. As of February 2024, that figure had risen to over 61,000 stations [28]. More than 95% of Americans now reside in a county with at least one public EV charging station [29].

EV charging stations are easiest to reach for people living in cities: 60% of urban dwellers reside within a mile of the closest public EV charger, in contrast to 41% of suburban residents and merely 17% of rural individuals.

Nighttime home charging loses the potential of solar energy as presented in Fig. 9.5 and puts more stress on the grid, and there are limitations for an acceptable integration of wind energy in urban areas.

The grid carries significant consequences. Two significant implications stand out: firstly, the price signals do not correspond with what is optimal for the grid and for ratepayers. The second point is the need to think about investments in charging facilities for workplaces.

Daytime charging might decrease electricity demand in comparison to home charging that is mostly done at night. This requires enabling drivers to charge their EVs at their convenience outside their residence, making use of work or public charging stations, and restricting home charging. Along with minimizing grid effects, charging during the day can utilize surplus electricity generated from solar photovoltaics at peak production times. Nonetheless, this solution could render EV charging less convenient and might not be appreciated by EV owners.

A frequently employed tactic to shift the EV peak demand away from the grid's net peak is for utilities to communicate price signals to EV owners via time-of-use

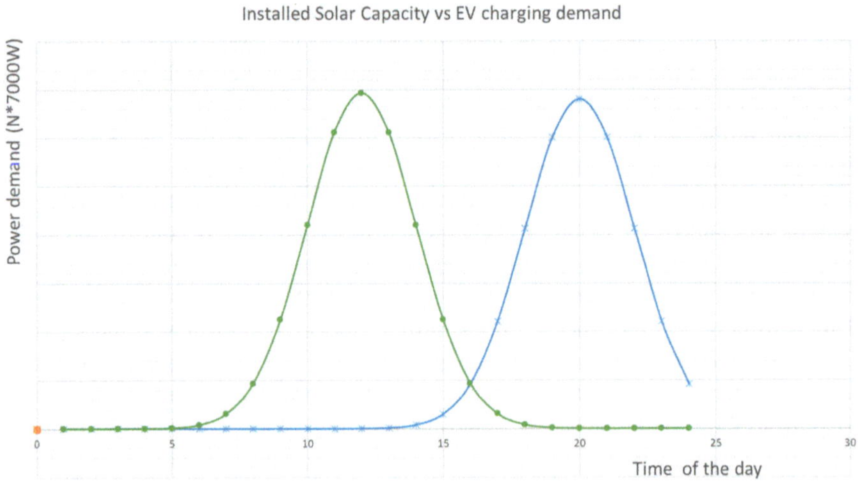

Fig. 9.5 Model solar irradiation period vs EV charging period (N is the number of EVs charged at 7000 W)

pricing or demand response rates. Real-time pricing is another possibility but has typically only been used in pilot projects so far. Such pricing strategies can motivate EV owners to recharge their vehicle during less busy peak hours.

Intelligent managed charging approaches and charging rate indicators typically depend on the EV owner's knowledge of fluctuating electricity costs. Nonetheless, there are intelligent charging methods where a controller begins charging at predetermined times based on utility pricing or grid conditions as variables.

An ideal scenario would correspond to Fig. 9.6 above; with a given proportion, the peak charging time for EVs corresponds to the peak generation of photovoltaic solar energy and gives full sense to V2G.

According to the World Economic Forum's 2021 [30] report 'Getting to Net Zero: Increasing Clean Electrification by Empowering Demand,' smart technology that enables consumers will be essential for delivering flexible, resilient, and sustainable energy. The future role of V2G, which enables the bidirectional energy flow between EVs and the grid, and vice versa. V2G will enhance the power grid's capacity during times of peak demand, alleviate local congestion, and act as a safeguard against power intermittency. It will additionally assist in decreasing the overall expense.

Readying the grid to support load growth and charging infrastructure may be difficult. The grid designed to handle load increases poses new challenges for utilities and grid operators. Currently, utilities develop infrastructure based on service requests, with some integrating broader system growth strategies that utilize conservative load growth forecasts. Nonetheless, the electric demand from EVs is anticipated to hit approximately 1500 TWh per year by 2050, about the same as the electricity usage of the whole US residential sector in 2018. Most utility load predictions and infrastructure strategies do not presently consider this increase in demand.

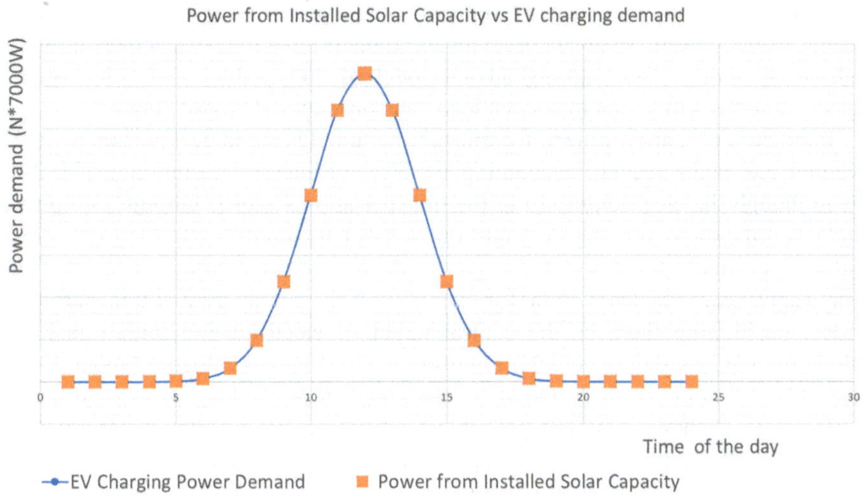

Fig. 9.6 Model of ideal EV charging habits to integrate solar photovoltaic energy (N is the number of EVs charged at 7000 W)

Moreover, the distinct traits of EV charging, particularly the substantial and immediate power requirements for fast chargers, set apart the load increase linked to transportation electrification from other types of load growth that utilities are familiar with managing. Utilities are utilized for constructing reactively instead of proactively. For instance, when servicing a newly constructed apartment complex, utilities have sufficient time to prepare during the construction phase, making it straightforward to forecast and control the upcoming demand. However, a swiftly expanding EV market generates a completely new demand to provide substantial amounts of power where it has not been traditionally required and with significantly shorter lead times.

To keep up with the rapid increase in EV adoption, utilities must act swiftly to ready the grid for this new kind of demand. However, conventional utility business frameworks and regulations, which effectively cater to reactive strategies for additional demands like new residential complexes, presently do not permit the forward-thinking grid investments required by EVs. The uncertainty about the exact charging requirements of EVs, such as when, where, and how much, stands in stark contrast to regulatory demands for the prudence of investments aimed at minimizing costs for ratepayers.

9.3.1 Installed Capacity and Energy Consumption in the United States

The United States is the second biggest consumer of energy overall and specifically of oil and refined products [31]. Nonetheless, the existing and projected capacity and consumption are enhancing towards a situation of reduced imports and more effective utilization of all energy resources. The large new reserves of oil and natural gas from domestic shale are dramatically transforming the perspective on future self-sustainability. The ongoing advancement of "renewable" and "alternative" energy sources will lessen the dependence on conventional "fossil" fuels.

Petroleum primarily utilized for transportation holds the largest portion at 36%. Natural gas ranks second, accounting for 32% of energy consumption [31–34]. Even though the electricity ratio in the transportation sector is currently minimal, and it was the same from the early introduction of EVs (under 0.2%) [35], EVs present an opportunity for this sector to greatly decrease petroleum use via electrification. Light-duty vehicles account for over 60% of the energy used in transportation today, positioning them as the central target of electrification initiatives [36]. Medium- and heavy-duty vehicles represent the next largest category (about 20%), and although some of these might be electrified, it is challenging to rival diesel powertrains in these uses. The probable load profile for medium- to heavy-duty vehicles is also quite uncertain, complicating the creation and integration of their corresponding load profiles into the models.

Table 9.5 above shows the very small proportion of electric energy used in the sector of transportation. Nevertheless, the most determinant factor is rather the power; a surge may occur if all EVs are simultaneously connected to the grid, as presented in Table 9.1.

The portion of electric energy used in the transportation sector is still limited, but it is likely to sharply grow in the next years and for 20 consecutive years, as shown by the S-curve in Fig. 9.7 below.

9.3.2 *Energy Demand from EV Fleet*

In 2022 transportation still accounts for less than 0.2% of the total electric energy consumed in the USA, Table 9.4. The inclusion of EVs in the fleet and the goal to substitute all ICE vehicles with EVs pose significant difficulties for grids. The challenge is more significant for power distribution compared to energy use.

The introduction of EVs as an emerging technology in a specific market can be effectively illustrated by the S-curve [11]. This model is also suitable for forecasting the shifts in overall power and energy needs resulting from an EV fleet. The model illustrates the duration required to enhance the power grid to meet the impending demand.

Table 9.5 Electricity consumption in the United States from 2016 to 2022, by sector (in terawatt-hours) [37]

	2016	2017	2018	2019	2020	2021	2022
Residential	1411.06	1378.65	1469.1	1440.3	1464.6	1470.5	1509.2
Commercial	1367.19	1352.89	1381.76	1360.88	1287.4	1328.4	1390.9
Industrial	976.72	984.3	1001.6	1002.35	959.1	1000.6	1020.5
Transportation	7.5	7.52	7.67	7.63	6.5	6.3	6.6

The rate of adoption for EVs as an emerging technology is illustrated by the S-curve as shown in the following equation:

$$R(t) = \frac{1}{\left[1+\exp(-\alpha t)\right]^{k}} \tag{9.1}$$

where α and k are derived from an empirical approach [11]. The penetration of EVs in the US market illustrated in Fig. 9.7 is defined by: $\alpha = 0.13$ and $k = 11.1$ [11].

The above Fig. 9.7 shows that it will take about 50 years in the USA for all new registrations to be made of EVs if the same trend is to continue, considering a meaningful uptake from 2013. The main problem will result from rapid growth of the EV fleet in the 20 years between the tenth year of introduction and 30 years. The first 10 years do not generate a real challenge for the grid; the progress in EV adoption is very slow, corresponding to the first phase of the S-curve. Utility companies will need strategies to promptly react in power generation capacity and management systems.

N (t) = number of vehicles
N_{ICE} (t) = ICE vehicles fleet
N_{EV} (t) = EVs fleet

$$N(t) = N_{EV}(t) + N_{ICE}(t) \tag{9.2}$$

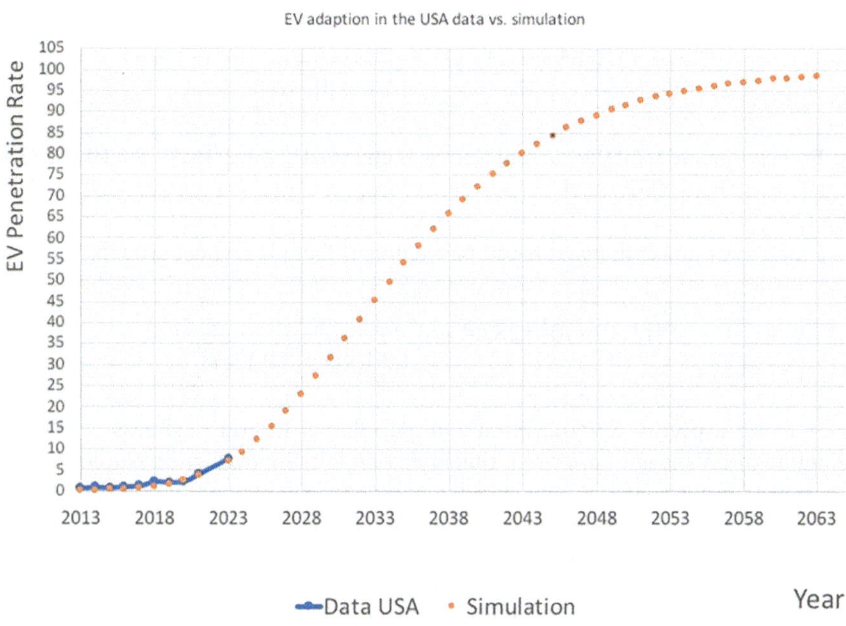

Fig. 9.7 EV penetration in the US market

$$N(t) = N_0 (1+\beta)^t \tag{9.3}$$

β is the growth rate of the total fleet:

$$N_{EV}(t) = N(t) R(t) \tag{9.4}$$

Average yearly energy demand per EV
USA: 4310.65 kWh annually
Europe: 905.20 kWh annually

Evolution in the EV related energy demand in the USA:

Regarding the US market as a reference, the US Motor Vehicle Sales Growth rate is updated monthly, available from January 1977 to December 2024, and an average growth rate of 1.8% through 2050 is considered in this manuscript.

The USA market is characterized by $\alpha = 0.13$ and $k = 11.1$ based on actual data and S-curve based simulation as presented in Fig. 9.7.

$\beta = 0.02$ is a realistic estimate.

Some 283.4 million vehicles were registered in the United States in 2022. The figures include passenger cars, motorcycles, trucks, buses, and other vehicles. The number of light trucks sold in the U.S. stood at 10.9 million units in 2022.

It can be considered that currently: $N_0 = 283.4 * 10^6$.

$$E(t) = 4310.65 * N_{EV}(t) * R(t) \quad \text{in kWh} \tag{9.5}$$

$$E(t) = 4310.65 * N_0 (1+\beta)^t * \frac{1}{\left[1+\exp(-\alpha t)\right]^k} \quad \text{in kWh} \tag{9.6}$$

Power demand if all the current fleet was changed at once: $P(t) = 7 * 10^3 * 283.4 * 10^6 = 1983.8$ GW, such a power exceeds the present power installed in the USA.

Progress of power demand based on S-curve if there is no fleet growth:

$$P(t) = 7 * 10^3 * 283.4 * 10^6 * \frac{1}{\left[1+\exp(-0.13t)\right]^{11.1}} \tag{9.7}$$

A major aspect of EV transition and the relation to grid readiness is that the adoption of EVs is not uniform in time. After an initial introduction period there is a period of fast growth before a slowdown as described by the S-curve [11]. Based on Fig. 9.8, the main challenge of EV adoption in the USA will be between 10 and 30 years after the introduction of EVs.

If the fleet remains constant, so will the power demand related to mass adoption of EVs once saturation is reached. If the fleet growth is included with a realistic rate $\beta = 0.02$ for the USA, for example, the energy and power demand from the EV fleet

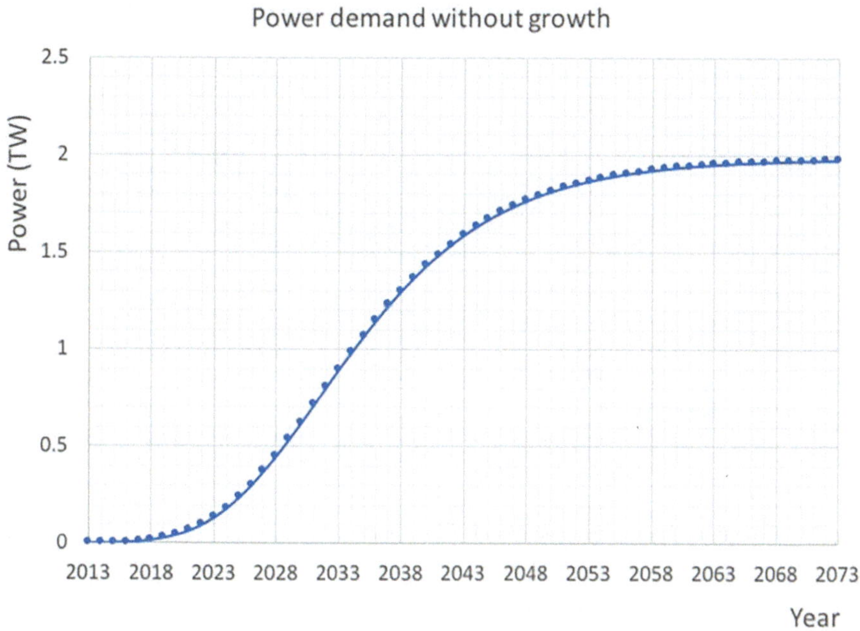

Fig. 9.8 Power demand from EV fleet from a constant fleet

will inflate as illustrated in Eqs. (9.8) and (9.9) and Fig. 9.8 with the extreme hypothesis that all EVs are charged simultaneously.

$$E(t) = 4310.65 * 283.4 * 10^6 * (1+0.02)^t * \frac{1}{\left[1+\exp(-0.13t)\right]^{11.1}} \tag{9.8}$$

$$P(t) = 7 * 10^3 * 283.4 * 10^6 * (1+0.02)^t * \frac{1}{\left[1+\exp(-0.13t)\right]^{11.1}} \tag{9.9}$$

Evolution in the EV related energy demand in the Europe:

$$E(t) = 905.20 * N_{EV}(t) * R(t) \quad \text{in kWh} \tag{9.10}$$

$$E(t) = 905.20 * N_0 (1+\beta)^t * \frac{1}{\left[1+\exp(-\alpha t)\right]^k} \quad \text{in kWh} \tag{9.11}$$

The installed capacity in the USA stands at 1201.260 GW [38]. According to Fig. 9.8 above, in approximately 50 years, the demand for power to charge EVs will equal the current capacity installed in the USA. Based on Fig. 9.9, the demand from the EV fleet may reach over 6 TW 60 years after the initial adoption of EVs.

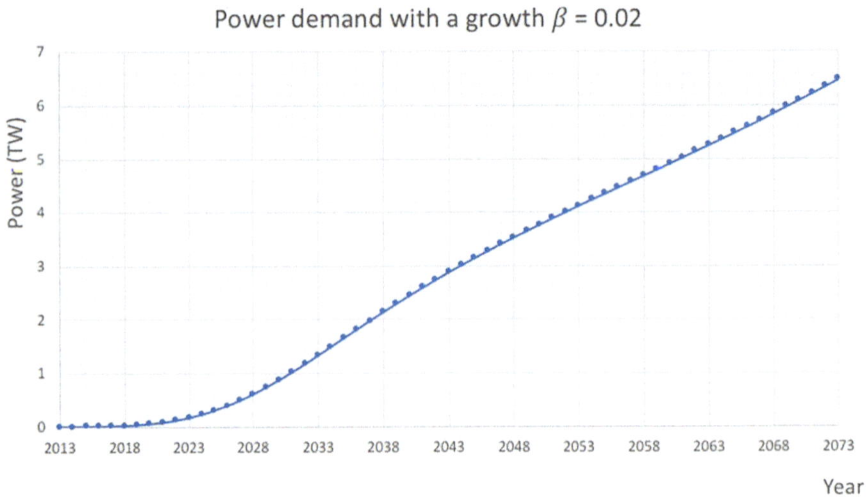

Fig. 9.9 Evolution of power demand from EV fleet in the US

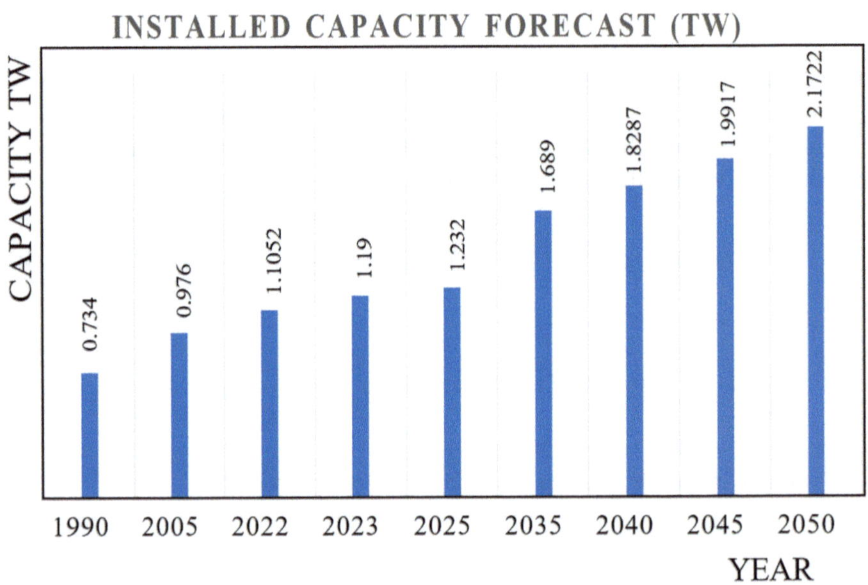

Fig. 9.10 Electric power sector capacity in the United States, with a forecast to 2050 [38]

In 2023, the electric power sector in the United States had an estimated net capacity of approximately 1.19 TW. This number is anticipated to rise by over 97 percent in the next 30 years, hitting around 2.17 TW by 2050 (Fig. 9.10).

This upgraded grid in 30 years will still be under stress if the entire fleet is replaced by EVs, as shown in Figs. 9.8 and 9.9, which estimate the power needed for a full EV fleet.

There is no doubt about the need to upgrade the power system and the whole distribution system. It is necessary to have a better management system and emphasis on level 2 or level 3 public charging stations.

9.3.3 *Forecast of Electricity Usage in the Transportation Sector*

The proportion of electricity in global primary energy usage is expected to rise from 26% in 2022 to 37% by 2050 [39]. The primary factor behind this growth will be the increase in electric transportation. EVs are progressively boosting worldwide electricity demand. EVs now account for 20% of new car sales worldwide [40]. Even though electricity continues to be a small energy source for global transportation, predictions suggest that electricity demand in transportation could rise by 10% annually up to 2050 [41]. The rising demand for electricity from EVs generates a new dynamic in energy usage and underscores the need for forward-thinking energy strategies that integrate transportation and electricity planning.

According to Figs. 9.8 and 9.9, the primary challenge for the USA during the next two decades will stem from the fact that this timeframe aligns with the phase of rapid EV adoption as presented in the S-curve, Fig. 9.7. The fast growth period is between the tenth and the 30th year of adoption of the EV in the case of the USA, as shown in Fig. 9.7. The transition to electric power for vehicles will inherently create a new need for electricity generation and energy management. Understanding load profiles will be essential. Load shapes can be described as the hourly distribution of loads based on a specific set of variables, including equipment, operational traits, and other influences like weather and demographics. EV load profiles are crucial for utility distribution and system planning and are becoming essential as more EVs are charged on the electric grid. If the effects of EV charging are not included in utility planning, the industry faces the danger of inadequate capacity in the distribution system and supply as an increasing number of customers embrace EVs. It is crucial to grasp the overall background information on EVs, clarify the significance of EV load shapes, outline the factors and choices that need to be addressed for a study to generate EV load shapes, and offer instances of essential research questions that ought to be contemplated in any EV load investigation.

For sound proactive measures, regional hourly demand trends for EVs are needed for the years through 2050 to support the analysis of a situation with considerable renewable energy output. To estimate the electrical energy needs for transportation, some input variables should be considered, such as the regional population forecast data [42], historical vehicle per capita statistics [43], and growth functions for market penetration to estimate the number of EVs in each analysis region. Upon the examination of a limiting case from reference [44], a unique market saturation situation where 50% of sales are EVs consuming an average of about 6 kWh per day indicates the necessary adjustment. It is essential to comprehend load profiles for the transportation sector. Hourly load profiles for EVs have been recorded in numerous sources [45–47]. Regarding the integration of renewable energy, three scenarios can be used [48, 49]. In the initial situation, the consumer can connect and charge right after the vehicle finishes its last trip of the day. This scenario is known as "no utility control" since the vehicle load depends on consumer demand and charges until the battery reaches full capacity or the consumer starts a new trip. The second scenario is called "opportunity," which assumes that charging infrastructure is extensive, and the consumer chooses to plug in anytime the vehicle is parked, no matter how long it is stationary. This scenario leads to significantly higher fuel savings, but it also increases daytime EV demands, total energy needs, and potential battery deterioration. The availability of a renewable energy power source is not a requirement to charge the vehicle. Even if it may have the advantage of a good match with the potential of solar photovoltaic energy on the condition that the infrastructure is available. A third scenario is called a "valley filling" approach. The "valley-filling" strategy encourages EV charging during low-demand periods, maximizing the utilization of surplus grid capacity and enhancing stability. At first, this scenario was structured to allow the utility to completely oversee the delivery of daily energy needs in accordance with the evening decline of a typical daily utility load curve. However, in the context of a high renewable energy ratio, it may be more beneficial for a utility to charge vehicles at different times throughout the day. As a result, the "valley filling" load curve is used to define an energy demand that must be met during the day, although there are no restrictions on the timing of when this needs to happen. Both the "no utility control" and "opportunity" scenarios assumed 110-V, 1.4-kW charging rates, whereas the "valley filling" curve allowed for 3-kW charging to best match the energy needs with the utility valley profile.

For an adequate integration of renewable energy, managed vehicle charging can be advantageous for utilities and system operations, and as a result, drivers would be encouraged to engage in a regulated charging program. Users would charge at home without managing utility resources, but as public infrastructure improves and consumers learn to optimize their vehicle technology investments, charging opportunities will expand. The combination of hourly load profile results over a given period can lead to the generation of a cumulative load profile for each vehicle.

The entire load pattern exclusively demonstrates the static load profile. The dynamic segment overseen by the utility, "valley filling," grows from 0% of the demand in 2010 to roughly 45% of the total demand by 2050 [50].

Figure 9.11 shows a model of load profile for the entire EV fleet presently in the USA, the current 3.3 million EVs, the potential double 6.6 million EVs, and the triple 9.9 million EVs, with an average charging power of 7 kW, 80% of vehicles charging at home with a peak around 9 PM, and 20% charging at work with a peak at around 11 AM. The power demands, as shown in Fig. 9.11, remain in the potential of the grid in general.

Charging at home during the night is the most effective method for reducing peak demands in the early evening; however, a preference for charging at work better aligns EV charging requirements with the availability of solar energy.

If the whole fleet consists of EVs, 283.4 million EVs, Fig. 9.12 indicates utilizing 20% daytime charging with photovoltaic solar energy and 80% overnight home

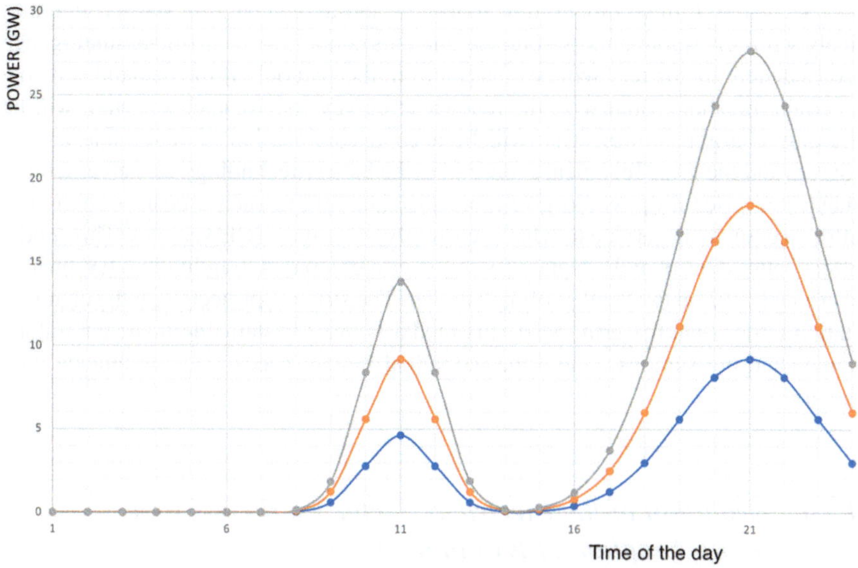

Fig. 9.11 Model shape and transition of load profile for 3.3 million EVs, the double, and the triple

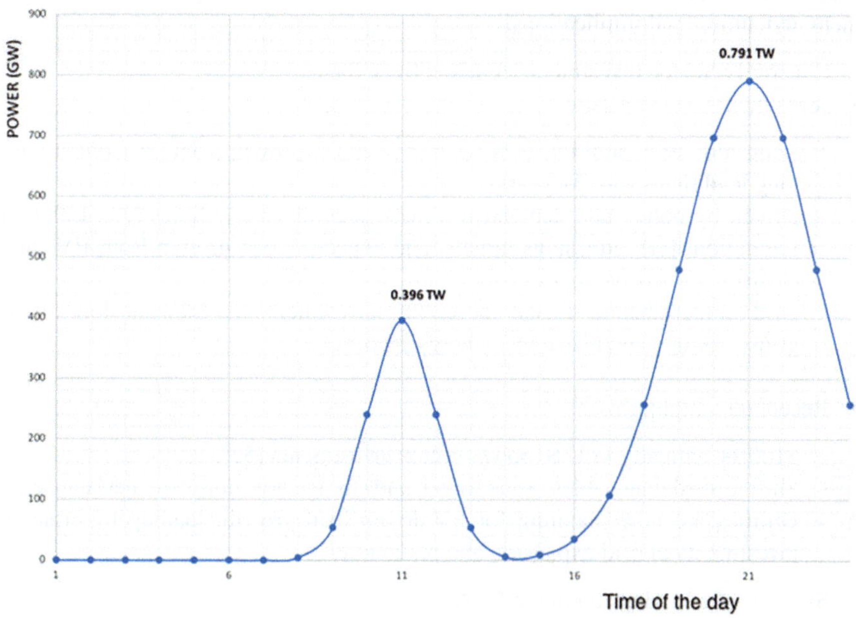

Fig. 9.12 Model shape and transition of load profile for the entire fleet made of EVs

charging. Night charging leads to a peak of 0.791 TW at 9 PM, with the charging distributed over nearly 10 h. 0.791 TW stays within the grid's capacity but could pose a significant challenge for it.

The "valley filling" model may apply to stand-alone charging stations cumulatively with renewable energy sources. Such a system may be artificial intelligence (AI) assisted; the battery will be charged from the grid only if the demand is below a certain threshold; otherwise the renewable energy source, solar or wind, will be the only charging power. This will be a good combination of renewable energy sources and the grid. Technical evolution in the battery technology as well as a more substantial price drop would give more sustainability to such a business model to offer an equivalent service to what is currently offered for ICE vehicles. This model offers better control and adaptability of the grid compared to individual EV owners charging based on the "valley filling" model. The stress is projected on the charging station rather than the grid. Also, this model is more sustainable than stand-alone home charging stations, where the individual system needs to communicate with the grid.

9.3.4 Evolution of the Electricity Sector Vs. the S-Curve in the Adoption of EVs in the USA

A comparison of the evolution in the electricity sector and power plants vs. the S-curve in the adoption of EVs in the USA is necessary, focusing on timeline alignment, dynamics, and implications.

- Electricity Sector Investment Trends in the USA
- Investment Characteristics:
  - Long-term & capital intensive: power plants require upfront investment with long lead times (5–15+ years).
  - Driven by policy and demand forecasts: federal/state policies (e.g. Inflation Reduction Act), emissions targets, and expected load growth from EVs and electrification drive investment.
  - Grid modernization focus: increasing investments in transmission, smart grids, storage, and distributed energy resources.

- Recent data highlights:
  - $100B+ annually in U.S. power sector investments [52].
  - Shift toward renewables: Solar, wind, and storage now dominate new capacity.
  - Utilities are now planning for EV-driven load growth, leading to strategic investments in grid expansion and resiliency.

- S-Curve of EV adoption in the USA
- S-Curve Basics:
  - Slow Start → Rapid Growth → Saturation.
  - EV adoption driven by price drops (esp. batteries), policy incentives, infrastructure, and consumer behavior.

- Current Phase (2025):
  - Inflection point/early majority (Fig. 9.7):

    EVs are ~10–12% of new vehicle sales (2024 data).
    Used EV market growing.
    Infrastructure (chargers) lagging behind but accelerating.

- Some of the key drivers:
  - Federal policies: IRA (Inflation Reduction Act) subsidies, tax credits for EVs & chargers.
  - State mandates: e.g. California's ZEV (Zero Emission Vehicle) mandates.
  - Private investment in charging networks (Tesla, Electrify America, etc.).

Table 9.6 summarizes the comparison between the evolution of the EV fleet and the development of the power sector in the USA.

- Interconnection & implications
  - EVs increase electricity demand → Need for new generation capacity and grid upgrades.
  - EV adoption accelerates grid modernization: more flexible, digital, and distributed systems.
  - The mismatch in timelines: EV growth is fast after the inflection point in the curve (Fig. 9.7), but utility investment cycles are slower. There is a risk of grid unpreparedness without coordination.
  - Opportunity: EVs can be grid assets (V2G, smart charging), but it requires integrated planning.

Table 9.6 Comparison between the power sector and the evolution in the EV fleet

Aspect	Electricity sector & power plants	EV adoption (S-curve)
Time Horizon	Long (10–40 years lifespan for assets)	Medium (10–15 years to full market maturity)
Investment dynamics	Policy & forecast-driven, steady capital deployment	Consumer behavior-driven, nonlinear growth
Growth curve	Linear or incremental (with policy shocks)	Exponential (S-curve)
Tech disruption	Gradual transition (e.g. coal → renewables)	Rapid with tipping points (battery breakthroughs, mandates)
Policy role	Crucial (e.g. clean energy standards, tax credits)	Crucial (e.g. EV tax credits, bans on ICE sales)
Infrastructure bottleneck	Grid interconnection delays, transmission siting issues	Charging infrastructure, battery supply chains
Market actors	Regulated utilities, IPPs, federal/state governments	Automakers, consumers, startups, federal/state regulators

- Strategic insight
- To ensure synchronized energy transition, the U.S. must:
  - Align EV adoption trajectories with utility infrastructure planning.
  - Accelerate permitting and interconnection for power projects.
  - Scale charging infrastructure with high geographic equity.
  - Enable data-sharing between utilities and automakers.

U.S. EV fleet growth versus installed electric power capacity between 2006 and 2024, along with projections through 2050.

Here is a comprehensive comparison of U.S. EV fleet growth versus installed electric power capacity between 2006 and 2024, along with projections through 2050.

EV Fleet Evolution: 2006–2024 and Future Outlook

- In 2006, EVs were virtually negligible in the U.S. fleet—fewer than ~10,000 plug-in vehicles. Low-speed EVs numbered around 60–76,000, but mass-market highway-capable EVs were nearly absent [53].
- By 2020, cumulative plug-in electric vehicles sold since 2010 reached ~1.8 million, with roughly 2 million EVs on the road (about 1% of ~290 million total vehicles) [54].
- 2023 saw cumulative plug-in sales reach ~4.68 million units. The EV market share of new car sales rose to 9.1%, and total EVs represented ~1.4% of all vehicles (~4 million EVs) [55, 56].
- In 2024, new EV registrations were around 1.1 million (~9.2% of new registrations), aligning with Experian's estimate of ~4 million EVs on the road out of 292 million (~1.4%) [56].

Projections to 2050

- EY (pre-pandemic) projected EVs would rise from ~2 million in 2020 (~2%) to ~88 million (approx. 65% share) by 2050 in U.S. fleet stock [57].
- IEA/EIA scenarios suggest 11–26% of light-duty vehicle stock in 2050 will be plug-in vehicles depending on assumptions (~up to 65% under aggressive adoption) [58].
- Based on slower turnover, even if all new sales became electric, it would likely take until the 2040s or 2050s before EVs constitute a majority of the total fleet (~60–70%) [59].

Installed Electric Generating Capacity: 2006–2024 and Trajectory

- As of 2024, U.S. total installed generating capacity stood at approximately 1250 GW (~1.25 TW) [60].
- Significant capacity additions are occurring, with about 35 GW added in 2024 [61].

Projections to 2050

- According to EIA's Annual Energy Outlook Reference Case, total capacity doubles (2.1–2.2 TW) by 2050, roughly a 2× from the current ~1.1 TW in 2022 [62].

Table 9.7 summarizes a quantitative comparison between the evolution of the EV fleet and the development of the power sector in the USA between 2006 and 2024 and projections to 2050.

Interlinked Trends & Implications

- The electrification of transportation will significantly increase electricity demand—some forecasts project demand growth of 25% by 2030 and ~78% by 2050 [63], driven by EVs, AI, electrification of heating, and industrialization [64, 65].
- Replacing fossil fuels and supporting EV load means capacity must expand not only to meet demand but also to replace retiring coal and gas plants, and firm capacity must ensure reliability.
- Battery storage growth (~doubling in 2024 alone) is vital to support intermittent solar and wind power [66].

Key Takeaways

- The EV fleet grew from virtually zero in 2006 to ~4 million by 2024 (~1–1.4% of total vehicles).
- If adoption accelerates, EVs could make up 25–65% of U.S. vehicles by 2050, depending on turnover and policy.
- Installed generating capacity has grown to ~1.25 TW by 2024 and is projected to double or even triple by 2050, with solar and wind dominating the expansion.
- The shift is not just growth; it's a transformation of the energy system, coupling clean generation growth with massive new electricity demand from electric transport.

This dual transformation reflects a system-wide shift: transportation electrification and clean power expansion are advancing in tandem. But the scale-up must

Table 9.7 Summary of the evolution of the EV fleet and installed capacity

Metric	~2006	~2020	2023/24	2050 projection
EV fleet (plug-in stock)	~10k	~2 M (1%)	~4 M (~1.4%)	~65 M–90 M (~25–65%) depending on scenario
Total vehicle stock	~250 M	~290 M	~292 M	~350–400 M
Installed capacity (GW)	~800–900 GW (est)	~1150 GW	~1250 GW	~2100–2500 GW (2 × –3 × current)

accelerate; EV adoption needs sustained policy support and turnover of the existing fleet, while electricity capacity growth must align with clean energy goals and grid reliability.

9.4 Conclusion

The S-curve is a common model used to describe the adoption of new technologies, including EVs. It shows how adoption typically starts slowly, accelerates rapidly once a tipping point is reached, and then levels off as saturation approaches. Understanding this curve is essential when considering how the electrical grid must evolve to support mass EV adoption.
Key points on EV transition and grid readiness include:

1. S-Curve of EV Adoption
   - Early stage corresponding to slow growth: High costs, limited models, and range anxiety dominate.
   - Inflection point corresponding to the beginning of rapid growth: Cost parity with internal combustion engine vehicles, improved infrastructure, and policy support fuel acceleration.
   - Mature stage, corresponding to plateau: market saturation leads to slower growth.
2. Implications for the grid involve:
   - Load growth: As EV adoption increases, electricity demand rises significantly, particularly in residential areas.
   - Peak demand stress: Unmanaged EV charging, especially in the evenings, could exacerbate peak load issues and would require a behavior shift.
   - Infrastructure needs: grid modernization, more transformers, distribution upgrades, and smart meters will be necessary.
   - Smart charging: time-of-use pricing, V2G, and managed charging can mitigate stress and improve efficiency.
   - Renewable integration: as EVs can act as distributed storage, they can help stabilize the grid when paired with intermittent renewables like solar and wind.
3. Timing is critical.

In fact, the steep part of the S-curve can sneak up on regulators and utilities. If grid planning and upgrades lag behind EV growth, it could lead to localized blackouts or grid reliability issues.

The rapid growth of the EV fleet may pose serious challenges to grids in all countries with a high ratio of electrified fleets. Norway, Iceland, and Sweden are the countries with the highest EV penetration rate worldwide, with the potential for Norway and Iceland to have all new registrations made of EVs in the next coming years. These three countries are the only countries within those with a high EV penetration rate that have an installed electric capacity greater than the potential demand from the entire fleet if it were made of EVs that would be charged simultaneously. The biggest threat does not come from the energy demand but most likely from the installed power capacity and distribution system. The other major challenge is likely to come from the phase of rapid growth of the EV fleet; this growth, as described by the S-curve, is likely to outpace the necessary upgrade of grids. New investments are needed to get grids ready for a full EV transition. Reforms are needed also as for incentives regarding tariff rearrangements so that charging habits align with the potential of renewable energy, particularly photovoltaic solar energy. Public charging stations look to be an adequate solution to solve the problem of grid anxiety and control the rhythm of charging habits, particularly in densely populated regions where home charging is not an available option, due mostly to types of housing. The integration of renewable energy in off-grid public charging structures may be a solution despite challenges, but to be a credible alternative to current gas stations, it will be necessary to progress in battery technology with energy densities that may need to double compared to what is available presently. To overcome the problem of the rapid growth of the EV fleet, microgrids may be an appropriate solution. Microgrids present a viable option for boosting resilience to prevent major power failures from overload-related disasters. The concept of a microgrid presented is a technique to consistently incorporate distributed energy resources, like energy storage systems and controlled loads. A microgrid can connect with and disconnect from the larger utility grid, enabling it to operate in either grid-connected or standalone mode. In contrast to conventional centralized grids, microgrids are primarily structured to handle and integrate distributed energy resources, allowing them to satisfy local demand without needing costly transmission infrastructures. On the other hand, and as an alternative, a smart grid [51] describes a type of electrical system that employs two-way flows of power and information, ensures secure communication technology, and provides computational intelligence from generation to consumption to create an enhanced energy network, as it can be needed from the EV fleet demand.

References

1. Abo-Khalil AG, Abdelkareem MA, Sayed ET, Maghrabie HM, Radwan A, Rezk H, Olabi AG (2022) Electric vehicle impact on energy industry, policy, technical barriers, and power systems. Int J Thermofluids 13:100134
2. Zhu Y, Choma EF, Wang K, Wang H (2023) Electric vehicle adoption delivers public health and environmental benefits. Eco Environ Health 2:193–194
3. Tan VKM, Yong JY, Ramachandaramurthy VK, Mansor M, Teh J, Guerrero JM (2023) Factors influencing global transportation electrification: comparative analysis of electric and internal combustion engine vehicles. Renew Sust Energ Rev 184:113582
4. Liu H, Wang DZW (2017) Locating multiple types of charging facilities for battery electric vehicles. Transp Res B Methodol 103:30–55
5. Yoldaş Y, Önen A, Muyeen SM, Vasilakos AV, Alan İ (2017) Enhancing smart grid with microgrids: challenges and opportunities. Renew Sustain Energy Rev 72(C):205–214

6. Khan AA, Khadijatul Kubra M, Rafiqul Islam M, Rahman A, Hassan M (2024) Microgrid-based operational framework for grid resiliency enhancement: A case study at KUET campus in Bangladesh. Energy Rep 11:1752–1765
7. Kumar P, Channi HK, Kumar R, Rajiv A, Kumari B, Singh G, Singh S, Dyab IF, Lozanović J (2025) A comprehensive review of vehicle-to-grid integration in electric vehicles: powering the future. Energy Convers Manage X 25:100864
8. Owens J, Miller I, Gençer E (2022) Can vehicle-to-grid facilitate the transition to low carbon energy systems? Energy Adv 1:943–1072
9. Mathisen S, Zeyringer M, Haaskjold K, Löffler K, Mata É, Sandvall A, Andersen KS, Vågerö O, Wolfgang O (2024) The REPowerEU policy's impact on the Nordic power system. Energ Strat Rev 54:101454
10. Vezzoni R (2023) Green growth for whom, how and why? The REPowerEU Plan and the inconsistencies of European Union energy policy. Energy Res Soc Sci 101:103134
11. Diouf B (2024) The electric vehicle transition. Environ Sci Adv 3:332
12. Diouf B (2025) Modeling the used vehicle market share in the electric vehicle transition. World Electr Veh J 16:29
13. Visaria AA, Jensen AF, Thorhauge M, Mabit SE (2022) User preferences for EV charging, pricing schemes, and charging infrastructure. Transp Res A Policy Pract 165:120–143
14. Dini P, Saponara S, Colicelli A (2023) Overview on battery charging systems for electric vehicles. Electronics 12:4295
15. Weiss M, Winbush T, Newman A, Helmers E (2024) Energy consumption of electric vehicles in Europe. Sustainability 16:7529
16. https://evbox.com/en/ev-home-charger-electricity-usage
17. https://roadgenius.com/cars/ev/statistics/sales-by-country/
18. https://en.wikipedia.org/wiki/Electric_car_use_by_country#/media/File:PEVs_in_use_Top_countries_&_regional_markets_2020.png
19. https://en.wikipedia.org/wiki/Electric_car_use_by_country
20. https://www.mk.co.kr/en/society/10921207
21. https://www.statista.com/statistics/1097976/south-korea-total-registration-number-of-electric-vehicles/
22. https://www.statista.com/chart/30897/countries-with-the-best-ev-charging-infrastructure/
23. https://www.citypopulation.de/en/norway/cities/
24. Ahmad F, Iqbal A, Ashraf I, Marzband M, Khan I (2022) Optimal location of electric vehicle charging station and its impact on distribution network: a review. Energy Rep 8:2314–2333
25. https://yocharge.com/faq/what-is-off-grid-ev-charging-station/
26. Yang M, Sun X, Liu R, Wang L, Zhao F, Mei X (2024) Predict the lifetime of lithium-ion batteries using early cycles: a review. Appl Energy 376:124171
27. Carlton GJ, Sultana S (2022) Electric vehicle charging station accessibility and land use clustering: a case study of the Chicago region. J Urban Mobility 2:100019
28. https://www.pewresearch.org/data-labs/2024/05/23/electric-vehicle-charging-infrastructure-in-the-u-s/
29. https://inframanage.com/proximity-to-charging-station-correlates-with-ev-switch-us-survey/
30. https://www.weforum.org/stories/2023/01/electric-cars-charging-habits/
31. Ahmad T, Zhang D (2020) A critical review of comparative global historical energy consumption and future demand: the story told so far. Energy Rep 6:1973–1991
32. U.S. Energy Information Administration (2022) Monthly Energy Review, April
33. https://www.eia.gov/energyexplained/us-energy-facts/
34. https://www.e-education.psu.edu/ebf301/node/457
35. Rigogiannis N, Bogatsis I, Pechlivanis C, Kyritsis A, Papanikolaou N (2023) Moving towards greener road transportation: a review. Clean Technol 5:766–790
36. Barman P, Dutta L, Bordoloi S, Kalita A, Buragohain P, Bharali S, Azzopardi B (2023) Renewable energy integration with electric vehicle technology: a review of the existing smart charging approaches. Renew Sust Energ Rev 183:113518
37. https://www.statista.com/statistics/560927/us-retail-electricity-consumption-by-sector/

38. https://www.statista.com/statistics/504369/electric-power-sector-capacity-in-the-us-forecast/
39. Gielen D, Boshell F, Saygin D, Morgan, Bazilian D, Wagner N, Gorini R (2019) The role of renewable energy in the global energy transformation. Energ Strat Rev 24:38–50
40. Nogueira T, Sousa E, Alves GR (2022) Electric vehicles growth until 2030: impact on the distribution network power. Energy Rep 8(Supplement 3):145–152
41. McKinsey & Company (2024) Global energy perspective 2024. McKinsey Global Energy and Materials Practice. https://www.mckinsey.com/industries/energy-and-materials/our-insights/global-energy-perspective
42. Hamed MM, Ali H, Abdelal Q (2022) Forecasting annual electric power consumption using a random parameters model with heterogeneity in means and variances. Energy 255:124510
43. Paul J, Blumenberg E (2023) Vehicle ownership rates: the role of lifecycle, period, and cohort effects. Transport Res Interdiscip Perspect 21:100892
44. Buhmann KM, Criado JR (2023) Consumers' preferences for electric vehicles: the role of status and reputation. Transp Res Part D: Transp Environ 114:103530
45. Wang H, Guo F, Sivakumar A (2025) Analysing the impact of electric vehicle charging on households: an interrelated load profile generation approach. Energ Buildings 335:115558
46. Andersen FM, Jacobsen HK, Gunkel PA (2021) Hourly charging profiles for electric vehicles and their effect on the aggregated consumption profile in Denmark. Int J Electr Power Energy Syst 130:106900
47. Baek K, Lee E, Kim J (2024) A dataset for multi-faceted analysis of electric vehicle charging transactions. Sci Data 11:262
48. Manousakis NM, Kanellos FD (2023) Integration of renewable energy and electric vehicles in power systems: a review. PRO 11:1544
49. Amiruddin A, Dargaville R, Liebman A, Gawler R (2024) Integration of electric vehicles and renewable energy in Indonesia's electrical grid. Energies 17:2037
50. https://www.pnnl.gov/main/publications/external/technical_reports/PNNL-23491.pdf
51. Fang X, Misra S, Xue G, Yang D (2012) Smart grid—the new and improved power grid: a survey. IEEE Commun Surv Tutorials 14(4):944–980
52. https://www.ft.com/content/051f2361-de09-4744-814f-0de4b9b6c0d0
53. https://about.bnef.com/insights/clean-transport/electric-vehicle-fleet-set-to-hit-100-million-by-2026-but-stronger-push-needed-to-stay-on-track-for-net-zero/
54. https://www.reddit.com/r/OptimistsUnite/comments/1h06svr/california_hits_264_ev_market_share_in_q3_2024/
55. https://www.spglobal.com/mobility/en/research-analysis/electric-vehicle-adoption-trends-its-not-all-grim.html
56. https://www.experian.com/blogs/ask-experian/how-many-evs-are-in-us/
57. https://www.ey.com/en_us/insights/automotive/five-hypotheses-shaping-the-electric-vehicle-market
58. https://evtcp.org/country/united-states-of-america/
59. https://time.com/6280316/ev-impact-aging-gasoline-cars/
60. https://natlawreview.com/article/evolving-mix-energy-sources-uss-electric-generating-capacity
61. https://www.eia.gov/energyexplained/electricity/electricity-in-the-us-generation-capacity-and-sales.php
62. https://www.foley.com/insights/publications/2024/09/the-evolving-mix-of-the-energy-sources-for-the-u-s-s-electric-generating-capacity/
63. https://www.utilitydive.com/news/icf-sees-25-load-growth-by-2030-up-to-40-price-increase/748711/
64. https://en.wikipedia.org/wiki/Plug-in_electric_vehicle
65. https://www.investors.com/news/oil-prices-trump-drill-baby-drill-evs-oil-stocks-demand
66. https://www.reddit.com/r/EnergyAndPower/comments/14atgki/solar_wind_storage_account_for_82_of_new_us/

Chapter 10
A Way Forward: A Full Transition to Electric Vehicles

Contents

10.1 Introduction

Human society, the environment, and future generations are interconnected. Maintaining harmony between humans and nature is of prime importance [1]. Human needs should align with sustainable development tenets, which focus on balancing environmental, social, and economic aspects. Roads have played a vital role for freight movement, local and rural connectivity and will remain important in the future. Transitioning to electric vehicles (EVs) is pivotal to mitigating greenhouse gas (GHG) emissions and averting climate crises [2].

Transportation has always been a major driver behind human advancement. From the invention of the wheel to steam power, from aviation to space exploration,

R. Pode, B. Diouf, *Decarbonization of Road Transportation*, Green Energy and Technology, https://doi.org/10.1007/978-3-032-11388-7_10

transportation technologies have shaped societies, boosted economies, and connected different cultures. These innovations, fuelled by curiosity and desire to explore new horizons, have not only revolutionized how we travel but also redefined how we live, trade, connect, and think.

The Otto combustion engine was developed in 1876, and early oil discoveries in the U.S., along with developments in refining, provided a plentiful supply of affordable fuel [3]. These technologies made a new mode of transportation possible, with the first gasoline-powered cars appearing in Germany in 1887, followed by those in the U.S. in 1893. Initially, gasoline cars competed with electric and steam-powered alternatives, and it took about a decade for them to become cost-effective and capable enough to be truly disruptive. Cars also offered advantages in speed, cargo capacity, and daily travel distance.

In 1900, however, people doubted automobiles, viewing them as expensive, unreliable, and dangerous [4]. Horses and carriages, in contrast, were familiar, trusted, and even cherished. Most could not imagine a world without them. However, as cars became more visible and reliable, skepticism shifted to curiosity and eventually desire. Meanwhile, the trusted horse began to be viewed as outdated, dirty, and obsolete. This change in public perception acted as a powerful accelerator of change, even in the face of opposition from established incumbents.

In the current scenario, EVs, autonomous systems, smart infrastructure, and digital logistics are no longer just concepts; they are now part of our everyday life. These developments aim to make transportation smarter, more environmentally friendly, and more inclusive while tackling challenges such as urban congestion, carbon emissions, and equitable access to transportation.

This chapter discusses the unexpected consequences of 100% EV surface transportation, including disruption to the oil sector, disruption in the high end of the gasoline vehicle markets, changes in the car, bus, delivery van, and truck markets, effects on employment, AVs, EV-based Transportation-as-a-Service (TaaS), and future emerging technologies in road transport. By 2050, road transport is believed to be influenced by automation, artificial intelligence (AI), electrification, and the adoption of more sustainable and efficient systems.

10.2 Importance of Gasoline Fuel Energy

Fossil fuel use remains crucial for the world's future [5]. Currently, the five key sectors of the global economy, such as information, energy, transport, food, and materials, are undergoing disruption at an unprecedented speed and scale. These changes have significant consequences for the broader economy, societies, and our civilization as a whole. Alex Epstein et al. noted that today fossil fuels continue to be the dominant source of energy around the world, providing 80% of the world's energy, and their use is still increasing [6]. In fact, fossil fuels have helped to reduce global poverty to an all-time low. Over the past 170 years, fossil fuels have contributed to a 1-degree rise in global temperature. Furthermore, climate-related deaths are at all-time lows due to development fueled by fossil fuels [6]. The authors also noted that some scientific findings could only be accepted by certain journals if they produced the "right" results. Some organizations were pressured to sell their oil stocks. Opposing oil became a moral mission, a virtual dogma, for anti-oil groups. Further, they argued that scientific data had been often misrepresented. Certain individuals and organizations maintain that fossil fuels do not contribute to climate change. Some groups with specific interests are producing and disseminating this information at different levels. Epstein disputes all the criticisms and presented a set of guidelines for evaluating climate-related claims [6].

Alex Epstein et al. argue that current policies targeting a rapid phase-out of fossil fuels could lead to catastrophic consequences, potentially creating a world that is poorer, more dangerous, and a more miserable place for people [6]. Even partial restrictions on fossil fuel use, without full elimination, might negatively impact the quality of life and exacerbate suffering for a sizable global population, especially in the most vulnerable regions. A complete removal of fossil fuels from the energy portfolio appears to be unrealistic, due to the stated plans of China, Russia, and India to expand their fossil fuel consumption [7]. Displacing fossil fuels completely with new energy sources could be disruptive and challenging [8].

10.2.1 Fossil Fuel in Road Transportations

The profile of vehicles on the road includes passenger cars, buses, trucks, and motorcycles. The road transportation sector has traditionally depended heavily on gasoline and diesel, and that remains largely true, as the system was designed around petroleum-derived fuels [9]. In 1990, motor gasoline was the most dominant fuel in Europe, making up 54.7% of road transport fuel, followed by gas/diesel oil at 43.8% [10]. Over time, diesel gained more prominence after 1998, though motor gasoline has recently regained some share. In many regions, gasoline and diesel oil account for over 90% of energy consumption in road transport [11]. According to KPMG, oil products supply nearly 91% of the transport sector's final energy [12]. However, it is gradually shifting towards alternative energy sources. Recent years have seen some tangible efforts to diversify fuel options and reduce reliance on fossil fuels, driven by environmental concerns and policy initiatives [13].

The transportation industry is experiencing a gradual shift towards environmentally friendly alternatives, including electric vehicles and hybrid technologies. Still, gasoline and diesel will remain important for transportation for the foreseeable future. The existing infrastructure, reliable supply chains, and practicality of these fuels ensure their continued use in many indispensable sectors [14]. As a result, *gasoline and diesel fuels, at present, remain the primary energy sources for powering vehicles and transporting goods around the world.*

10.3 A Full Transition to New Energy Vehicles

The transition to zero-energy transportation is certainly a great step in reducing emissions from road transport. To reach net zero by 2050, the world needs to adopt zero-emission transportation as rapidly as possible. Businesses, commercial vehicles, passenger vehicles, and public transportation all play a significant role in lowering global transportation emissions. Multiple studies indicate that increasing adoption of EVs is the key strategy to reduce GHG emissions from transportation [15]. Moving toward trains, trams, cars, buses, and bikes powered by renewable electricity is necessary in achieving total decarbonization of the sector [16]. A complete transition to NEVs, such as EVs, is intended to replace conventional internal combustion engine (ICE) vehicles with alternatives that use cleaner fuels [17].

A full shift (100%) to new energy vehicles (NEVs) is an ambitious yet increasingly feasible objective. This involves: (i) the complete phasing out of ICE vehicles; (ii) ensuring that all new vehicle sales are NEVs; (iii) retiring existing ICE fleets; and (iv) extensive upgrades to charging and fueling infrastructure. The transition to 100% NEVs is a complex process with substantial environmental, economic, and social consequences. While challenges persist, the momentum towards cleaner and more sustainable transportation is evident. Realizing a 100% transition to NEVs requires several critical strategies and considerations. Key aspects of this transition are outlined in Table 10.1 [18–24]. Collectively, these approaches can support a sustainable and environmentally friendly transportation system.

In 2024, China's heavy vehicle sector demonstrated significant advancements in electrification, showing its continued leadership in the global shift to zero-emission commercial transportation [25]. Data from the China Association of Automobile

Table 10.1 Major aspects of the transition to 100% NEVs [19–24]

Aspects	Programs to achieve goals	References
Targets and timeline	Many countries and companies are advocating for the gradual phasing out of the sale of ICE vehicles, with a goal of achieving 100% EV sales by a specific timeframe	[19, 20]
Infrastructure & development	Expanding the charging infrastructure is a key focus and a significant challenge	[21, 22]
Policy support	Various stakeholders are enacting policies, including both monetary and non-monetary subsidies as well as regulations, to promote EV adoption and facilitate the transition	[23, 24]
Technological advancements	Continuous advancements in battery technology, charging speeds, and vehicle design are important for the successful transition	[21]
Global collaboration	Programs such as EV100, a global initiative led by the Climate Group, unite businesses committed to transitioning their fleets to EVs	[21]
Global progress	China: Over 30% of new car sales are NEVs, with a target of 100% by 2035	
	Norway: The goal is to end ICE vehicle sales by 2025, as electric cars already represent over 80% of new car sales	

Manufacturers (CAAM) and industry sources such as China Trucks indicate a notable increase in the adoption of NEVs in heavy vehicle segments, particularly battery-electric vehicles emerging as the leading technology.

10.3.1 Challenges to 100% NEVs Transition

Challenges to a 100% NEV transition include limitations in battery technology, inadequate charging infrastructure, high upfront costs, potential strain on the electricity grid, and concerns about raw material availability and environmental impact [26, 27]. A comprehensive overview of these challenges is described in Table 10.2 [28–32].

In addition to these significant challenges, numerous other factors, including *technical*, *logistical*, and *social obstacles*, are very vital to address in order to achieve a complete transition to NEVs without adversely affecting any sector of society.

Table 10.2 Main challenges to fully transition to NEVs

Challenges			
Battery technology, production, and environment	Charging infrastructure	Cost	Grid capacity and energy sources, and Other challenges
• The anxiety regarding battery power depletion before reaching a charging station poses a major psychological obstacle for prospective EV buyers • The manufacturing of EV batteries necessitates critical raw materials such as lithium, cobalt, and nickel, which may result in considerable environmental impacts related to their mining and processing • Therefore, it is vital to ensure sustainable sourcing of these materials and to promote the development of more environmentally sustainable battery technologies	• The limited availability and accessibility of public charging stations discourage potential long-distance travelers • The existing infrastructure is insufficient to support the demands of a fully electric fleet • Concerns exist regarding charging speed. EVs require significantly more time to charge compared to refuelling conventional gasoline vehicles • The expansion and improvement of charging infrastructure, particularly by incorporating fast-charging options, are essential	• EVs are often more expensive compared to their gasoline counterparts • While operating costs, such as electricity and maintenance, tend to be lower, the upfront cost can discourage potential buyers • The price disparity is noticeable in the heavy-duty segment, where electric trucks cost more than their diesel counterparts	• As the EV adoption rate increases, it is necessary to upgrade the existing electricity grid to accommodate the growing demand • To maximize the environmental advantages of EVs, the additional demand of electricity should be met through renewable energy sources

Transitioning to a 100% NEV fleet presents multiple technical challenges that must be resolved to ensure feasibility and sustainability [30]. These are primarily categorized as limitations in battery technology [28], insufficient charging infrastructure [28], and grid integration complexities [33, 34], as elucidated in Table 10.3.

Certainly, these technical challenges are not merely obstacles; they represent the framework that must be addressed to make complete NEV adoption a reality. They are systemic issues that directly influence the pace and scope of our progress. Overcoming them is essential for paving the way to a cleaner, smarter, and more resilient transportation future.

The logistical challenges associated with a full transition to an EV fleet extend beyond manufacturing and sales. These challenges include fleet electrification and operational management, the establishment of charging infrastructure, the necessity for global coordination and policy support across different countries and cities, as well as workforce training and specialized skill development, among other factors. Table 10.4 provides a comprehensive illustration of the intricate *logistical hurdles* involved. Although technical innovation often gets the spotlight, logistics serves as an essential backbone, underscoring its importance in the transition process. These logistical hurdles are the unseen mechanisms that can either facilitate or impede the complete shift to EVs. Essentially, these hurdles form the structural framework behind the scenes. Without resolving them, even the most advanced EV tech can come to a standstill. Tackling these challenges means building a resilient, adaptable, and future-oriented ecosystem for NEVs.

Table 10.3 A detailed technical challenges to complete transition to NEV fleet

Battery technology limitations	Charging infrastructure	Grid integration
* **Energy density**: Present-day lithium-ion batteries fall short of gasoline in terms of energy density, limiting the driving range of EVs * **Charging time**: While fast-charging technology is improving, it still doesn't offer the same level of convenience as refuelling a gasoline car * **Degradation**: Batteries gradually lose their capacity over time, adversely impacting vehicle performance and resale value. External factors like temperature, usage pattern, and charging practices can also influence battery longevity and overall performance * **Raw materials**: Scarcity of lithium, cobalt, and nickel poses potential risks to the supply chain	* **Insufficient coverage and inadequate charging stations**: Many regions, especially rural ones, lack sufficient public charging stations. The existing charging infrastructure is insufficient to support a large volume of EVs * **Grid capacity**: Widespread adoption of EVs could strain local and national electricity grids * **Standardization**: The absence of universal charging standards creates challenges for interoperability between different brands and regions	* **Increased load**: A large-scale transition to EVs will substantially raise the demand on the electrical grid * **Grid stability and management**: Integrating a large number of EVs can affect grid stability and necessitate the use of advanced grid management technologies to maintain a reliable power supply

Table 10.4 Logistical challenges to fully transitioning to new energy vehicles

Logistical challenges	Details
Fleet electrification and operational management	• **High costs of vehicles**: Electric trucks and vans may incur costs up to three times more than their diesel counterparts
	• **Range constraints**: Managing delivery routes with limited EV range necessitates careful planning and could lead to diminished operational flexibility
	• **Driver training**: EVs require unique driving practices to optimize battery performance and range, which adds complexity to fleet operations
Charging infrastructure deployment	• **Installation expenses**: Establishing charging stations, especially for large fleets, requires substantial financial investment and infrastructure works
	• **Charging logistics**: Managing charging schedules for multiple vehicles while maintaining smooth operations presents a considerable logistical challenge
	• **Power demand**: High-capacity charging stations need robust grid connections and effective energy management systems
Global coordination and policy support	• **Fragmented regulations**: Diverse EV policies across different countries and cities present obstacles for international logistics operations
	• **Incentive gaps**: Lack of reliable financial incentives and inadequate infrastructure support hinder EV adoption rates in emerging markets
Workforce training and expertise	• **Specialized maintenance and repair**: EVs need specialized maintenance and repairs, which require technicians and mechanics to undergo specific training
	• **Route planning and operations**: Fleet managers and drivers need to be trained on the specific operational needs of EVs, which encompass route planning, charging management, and range optimization

Finally, shifting to a complete EV fleet goes beyond a technological or economic challenge; it is inherently a social issue as well. This transition affects communities, livelihoods, and cultural practices. The primary social and societal challenges are outlined in Table 10.5. Resolving these social challenges is crucial for achieving a successful and fair transition to NEVs. A comprehensive strategy involving government, industry, and communities is necessary to address barriers and make sure that the benefits of NEVs are accessible to everyone.

Indeed, the social and societal challenges are central to the EV transition, holding comparable significance to technology and logistics considerations. Without community support, cultural synergy, and equitable access, even the most cutting-edge NEVs risk being ineffective. Essentially, the shift to EVs revolves around humans, not just about machines. Addressing these societal issues requires fostering trust, inclusivity, and a sense of shared purpose. This is the pathway for the transition from innovation to meaningful transformation.

Table 10.5 Social and society challenges to 100% transition to new energy vehicles

Social and society challenges	Details
Equity and access	• **Affordability gap**: NEVs are still more expensive than internal combustion vehicles, limiting their accessibility to low-income households
	• **Charging inequality**: Urban areas get more charging stations, while rural and underserved communities lag behind
	• **Energy justice**: Communities historically reliant on fossil fuel jobs may perceive exclusion from the green transition
Community disruption	• **Loss of identity**: Regions built around coal, oil, or gas industries encounter cultural and economic disruption as these sectors decline
	• **Social fragmentation**: Changes in employment and migration patterns may weaken local social networks and institutions, such as schools, churches, and religious organizations
Workforce transition	• **Job displacement**: Workers in fossil fuel industries may struggle to find comparable employment in the NEV or renewable energy sectors
	• **Skills mismatch**: Retraining programs are often underfunded or misaligned with actual job market demands
	• **Union concerns**: Labor organizations worry about losing bargaining power in emerging, decentralized industries
Global inequality and perception	• **North-South divide**: Developing nations argue that wealthy countries are pushing green tech agendas without acknowledging historical emissions or developmental needs
	• **Cultural resistance**: In certain regions, NEVs are seen as symbols of elite or foreign influence, fostering social discord
Behavioral and cultural shifts	• **Consumer habits**: People are used to fast refueling, long driving ranges, and low maintenance costs. EVs require a mindset shift
	• **Trust and misinformation**: Doubts regarding battery safety, range, and environmental impact can slow adoption
	• **Lifestyle changes**: Owning an EV may require new routines, like planning charging stops or installing home chargers

In summary, tackling these challenges necessitates a multifaceted strategy that includes technological innovations, infrastructure development, supportive policies, industry collaboration, and active involvement from community stakeholders.

10.4 Implications of Transitioning to New Energy Vehicles

Earlier, McKinsey analysis in 2022 showed that the shift could create opportunities for companies across the automotive value chain around the world, which include manufacturing EV batteries and fuel cells and producing the materials needed to make these essential components; building, making, and operating infrastructure for charging BEVs and refueling FCEVs; and creating digital solutions to integrate the new vehicle energy infrastructure with the power grid [35]. Potential benefits of a complete transition to NEVs are displayed in Fig. 10.1. Switching to electric vehicles can significantly reduce GHG emissions from road transport, which currently contributes to more than 20% of global CO_2 emissions [36]. Lower tailpipe emissions lead to better air quality in urban areas and overall improvements in public health. The widespread adoption of electric cars will lead to increased electricity consumption, necessitating upgrades to the grid and the implementation of smart charging solutions. This demand will drive the development of a new type of grid capable of managing the increased load [37]. Integrating renewable energy sources is essential for creating a cleaner grid. EVs only reduce emissions when they are powered by low-carbon energy sources [38]. Additionally, the transition to electric cars will open up new avenues in battery tech, charging infrastructure, and employment in the green energy sector.

However, such a complete transition is not without risk, despite its environmental benefits. Lower-income households would naturally be more affected by the relatively high cost of acquisition and would recoup it only over the life of the vehicle. Highly promoted renewable energy sources have contributed to steeply rising electricity prices and more frequent power outages. Furthermore, there is a risk of market monopolies arising as a few dominant companies control software, repair

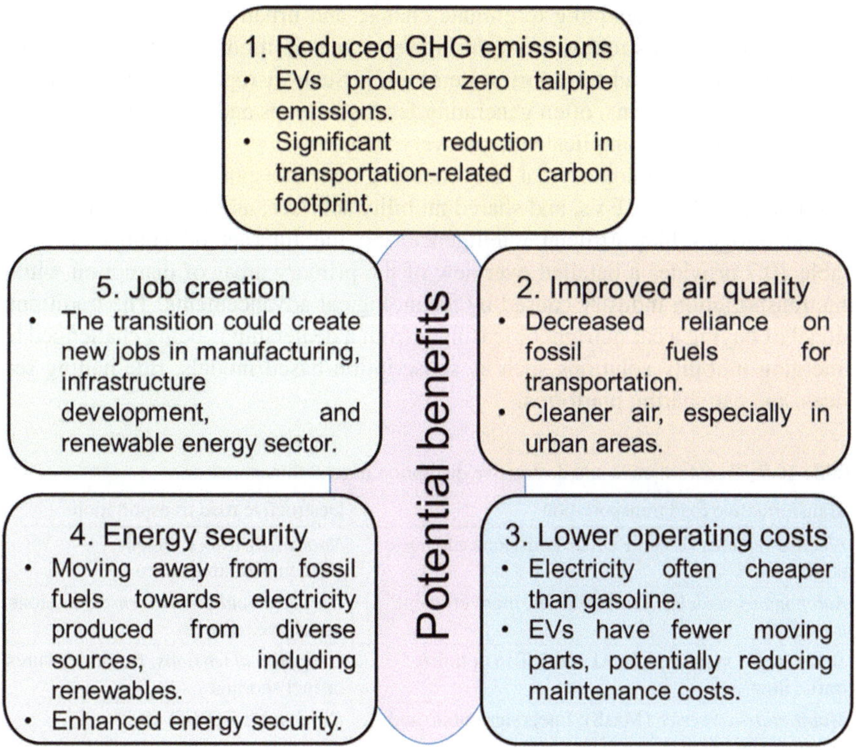

Fig. 10.1 Potential benefits of a 100% transition to new energy vehicles

ecosystems, and EV charging networks. The large-scale production of NEVs could exert pressure on natural resources and may inadvertently shift pollution rather than mitigating it. Additionally, jobs in traditional auto sectors, particularly in internal combustion engine manufacturing, might decrease, leading to workforce displacement. Nevertheless, these risks do not diminish the benefits of the transition; they simply require careful planning, innovation, and global collaboration.

10.5 Disruptions in Road Transportation Due to New Energy Vehicles

Disruptive road transportation refers to substantial alterations or interruptions in conventional road transport systems, resulting from technological innovations, environmental influences, or socio-political developments. Such disruptions can produce both beneficial (positive or transformative) and adverse (negative or destructive) outcomes [39]. Table 10.6 illustrates a breakdown of both aspects. Positive disruptions lead to innovations and drive systemic change that improves and enhances the existing infrastructure, fostering transport systems' resilience and supporting cities in adapting to climate change and urban expansion [40, 41]. In contrast, negative disruptions arise from events or conditions that impede the normal operations of road transport systems [42]. Such disruption poses challenges within complex systems, often generating feedback loops and unanticipated consequences, which can complicate the recovery efforts.

Disruptive technologies that are changing the transportation sector include autonomous vehicles, EVs, and shared mobility services, as well as the integration of technologies like artificial intelligence and the Internet of Things [43, 44]. Table 10.7 provides a detailed overview of the primary areas of disruption within the transportation industry caused by technological advancements. The traditional model of buying a car outright or leasing it from a dealership is being challenged by emerging mobility solutions such as subscription-based models, ride-hailing services, and car-sharing platforms.

Table 10.6 Transformative and destructive disruption in road transportation

Transformative road transportation	Destructive road transportation
Electric Vehicles (EVs): Reduce emissions and noise pollution	*Natural Disasters*: Floods, earthquakes, and storms
Autonomous vehicles: Promise safer, more efficient transport	*Accidents and breakdowns*: Collisions or vehicle failures
Smart traffic systems: Use AI and IoT to optimize traffic flow	*Geopolitical tensions*: Border closures or fuel shortages
Mobility-as-a-Service (MaaS): Integrates public and private transport options	*Pandemics*: COVID-19
Green infrastructure: Investment in bike lanes, pedestrian zones, and EV-friendly roads	*Infrastructure failures*: Aging bridges, roads, or tunnels

Table 10.7 Major areas of disruption in transportation due to technologies

Autonomous vehicles (AVs)	Electric vehicles (EVs)	Shared mobility service	Intelligent transportation systems (ITS)
• Autonomous vehicles, including cars, trucks, and drones, are poised to revolutionize the future of transportation • Offer advantages such as increased safety, reduced traffic congestion, and greater efficiency • The widespread adoption of AVs could present challenges concerning infrastructure, cybersecurity, and potential job loss within the transport industry • AVs are classified according to their level of automation, ranging from level 0 (no automation) to level 5 (full automation)	• EVs are gaining popularity and offering a sustainable option to traditional ICE vehicles • The rise of EVs will affect transportation infrastructure. Some existing facilities, like gas stations, could become outdated • EVs are set to transform the automotive industry by potentially creating new markets for charging infrastructure and battery technology	• Ride-sharing, bike-sharing, and other forms of shared mobility services are becoming popular, providing convenient and affordable transportation solutions • These services are made possible through technologies such as online booking platforms, mobile apps, and payment systems • Utilizing shared services can lead to more efficient and sustainable transportation systems, especially in urban environments	• AI, big data, and the Internet of Things are revolutionizing the management and optimization of transportation networks • ITS technologies facilitate real-time traffic flow monitoring, enhance traffic management, and optimize route planning • These systems can enhance resource distribution, alleviate congestion, and improve overall transportation efficiency

Certainly, the rise of electric and autonomous vehicles, the expansion of transport-as-a-service models, and the integration of smart infrastructure are driving the progressive advancements in the transportation sector [45]. According to Sarah Schmidt of Market Research, the industry is anticipated to undergo a dramatic transformation due to revolutionary technologies such as self-driving robo-taxis, autonomous trucks, delivery robots, and EVs, including cars and buses [46]. Ride-hailing services have seen substantial growth in adoption since Uber's introduction in 2009 [47]. The advent of shared mobility services, including Uber, Lyft, and Zipcar, is transforming conventional transportation business models.

A report from the World Intellectual Property Organization (2025) indicates that advancements in solid-state battery (SSB) technologies and platooning communication systems may have a considerable influence on industries, societies, and economies in the near future [48]. SSBs reported a notable breakthrough in battery technology, offering several advantages over conventional batteries. Similarly, platooning also demonstrates significant improvement in road transportation by allowing multiple vehicles to travel in close proximity at high speeds, facilitated by automated driving systems and vehicle-to-vehicle (V2V) communication. At present, China and Japan are at the forefront of global innovation in road transportation technologies, while Germany and Sweden demonstrate high levels of specialization in this field. Additionally, China and India are experiencing rapid growth within this sector.

Disruptive technologies have significantly transformed the transportation industry, which is closely connected to our everyday lives. These technology innovations are revolutionizing the way we commute, travel, and transport goods. With advancements like self-driving vehicles, drones, electric cars, and advanced AI, the transportation landscape is undergoing a revolution. These technologies are not only enhancing efficiency and sustainability but also making strides in improving safety and reducing costs.

10.5.1 Disruption in the Automotive Industry

The automotive industry, long dominated by legacy ICE car manufacturers, is experiencing a seismic shift driven by the rise of EVs. EV technologies are significantly disrupting the automotive industry by changing manufacturing processes, sales models, and the overall competitive landscape [49]. The shift from fossil fuel-based combustion engines to electric technologies is impacting the transitional automotive supply chains [50]. New types of companies, equipped with EV technologies, are challenging the status quo, disrupting traditional markets, and accelerating the transition to sustainable mobility. They are transforming the design of cars, manufacturing, the sales business model, and the integration of automobiles into a future characterized by connected, electric, and autonomous mobility. The rise of EV technologies has substantially disrupted fossil fuel-based automotive industries in various regions around the globe. The most impacted areas are categorized into four distinct groups, as illustrated in Table 10.8 [51].

The rise of EV technologies has dramatically disrupted China's fossil fuel-based automotive industry, transforming both domestic and international markets [52, 53]. In July 2025, new energy vehicles (NEVs) reached a 51.1% retail penetration rate, officially surpassing traditional fuel-powered cars as the mainstream choice. This is

Table 10.8 Disruption in fossil fuel-based automotive sectors due to the rise of EV technologies

China	European Union	USA	South East Asia
• Accounting for 60% of worldwide EV sales • Chinese EVs now represent over 20% of the EU's EV sales, up from just 3% in 2020	• EV sales still rose by 17% between 2022 and 2023 • Facing competitive pressure from Chinese imports	• Experiencing rapid growth in EV adoption • Traditional automakers like GM and Ford are heavily investing in EV production • The situation may change in 2025 due to changed EV policies	• Undergoing a major shift, with EVs expected to make up 38% of the market by 2030 • Thailand, Indonesia, and Vietnam are investing in local EV manufacturing and battery supply chains

impacting oil demand and potentially altering global market dynamics [54]. The surge in EV adoption is putting pressure on the traditional automakers who rely on fossil fuel vehicles. A key factor driving this disruption is the ability of Chinese companies to bypass conventional stages of automotive development and rapidly adopt advanced technologies [55].

The fossil fuel-based automotive sector in the European Union (EU) is undergoing a major disruption due to the rapid rise of electric vehicle (EV) technologies [56]. This transition is driven by environmental objectives, regulatory mandates, and competitive pressures from both internal and external sources. Legacy automakers are retooling factories for EV production. The transition to zero-emission mobility is a critical opportunity for the EU to achieve its climate goals and maintain its industrial leadership.

The rise of EV technology is significantly disrupting the US fossil fuel automobile industry [57]. The transition toward EVs is expected to alter the transportation sector, with potential implications for employment, infrastructure, and global oil demand [58]. In the USA, the car industry played a more direct role in the growth of the middle class by raising incomes. The disruption of transportation could drive the resale value of ICE cars, trucks, buses, and vans down to zero or even negative territory. A reduction of one percentage point in resale value may result in financial losses amounting to hundreds of millions of dollars for automobile manufacturers. A significant decline in resale value could cause liquidity problems, which again would have implications for jobs and credit markets. EVs are increasingly software-driven, enabling automakers to generate new revenue streams through services like in-car subscriptions and data analytics [59]. Additionally, technologies such as generative AI are being used to optimize design, manufacturing, and customer interaction processes [60]. Alongside, U.S. automakers are facing increasing competition from Chinese and European EV manufacturers.

The fossil fuel-based automotive sector in Southeast Asia is being rapidly disrupted by the rise of EV technologies. While the region has long been dominated by ICE vehicles, especially from Japanese manufacturers, EVs are now reshaping the landscape due to policy shifts, foreign investment, and changing consumer preferences. Thailand aims for EVs to make up 50% of new car sales by 2030, offering subsidies, tax breaks, and import duty exemptions to attract EV manufacturers [61]. Indonesia is positioning itself as a regional hub for EV and battery production, leveraging its vast nickel reserves. In this region, Japanese ICE brands face competition from Chinese EV makers [62]. By 2030, EVs (including hybrids and plug-ins) are expected to account for 38% of new vehicle sales in Southeast Asia [61].

Traditional automakers must pivot to NEV production, impacting supply chains and labor markets. Reduced demand for gasoline and diesel could disrupt oil refining and petrochemical industries [38].

The transition to EV technologies affects the automotive sector in three ways: (i) it increases competition among assembly plants at the intra-group level for the awarding of new models; (ii) it continues to reconfigure supply chains, threatening the viability of certain traditional providers of components and giving rise to new players, battery manufacturers, that may challenge the chain hierarchy; and (iii) the transition to electric vehicles has led to a reduction in employment volume in the chain as a whole [63].

In summary, as the EV transition is an ongoing process, it will continue to pose significant challenges for multiple actors involved in the automotive industry.

10.5.2 Disruption in the Oil Industry

For decades, oil and gas served as the backbone of global transportation, fueling the majority of vehicles and driving oil demand continuously more and more. As a result, extensive infrastructure was developed to support extraction, refining, and distribution processes. However, growing concerns regarding emissions and climate change have accelerated the pursuit of cleaner and sustainable alternatives. EVs have emerged as a pivotal factor in this transition. The rising prevalence of EVs on roads has had a considerable impact on the oil industry. For better understanding of the impact of EVs on the oil industry, a comprehensive analysis is presented in Table 10.9.

IEA says that electric vehicles slashed oil demand by over 1.3 million barrels per day (mb/d) in 2024 [64]. The IEA estimates that by 2030, EVs could displace over five million barrels of oil per day globally (equivalent to a nearly 5% reduction in total demand), with China responsible for half that amount [65]. One EV can displace roughly 15 barrels of oil annually [66]. According to BloombergNEF's Economic Transition Scenario, EVs could displace over 20 million barrels per day by 2040 [67]. Demand for road fuels is projected to peak at 49 million barrels per day in 2027. Subsequently, a structural decline is anticipated, with demand decreasing to 35 million barrels per day by 2040. These forecasts indicate that the oil industry must adapt to an impending structural decline in demand.

Despite EVs accounting for about 90% of new car sales in Norway, fuel demand has declined by only 10% between 2017 and 2023, showing relative stability [68].

Table 10.9 Impact of transition to EV technologies on the oil industry

Reduced oil demand	Shift in transportation energy demand	Adaption and diversification	Potential for further disruption
• Directly displace gasoline and diesel consumption • Result in decline in demand of these oil products • Projected significant reduction in oil demand by 2040 • Likely to lead to lower global oil prices	• Shifts towards EVs represent a fundamental change in the energy landscape of the transportation sector	• The oil and gas industry is facing the challenge of adapting to this changing landscape • Some refiners and producers are diversifying portfolios to include hydrogen, biofuels, and carbon capture technologies	• The speed of EV adoption and the resulting impact on oil demand will likely vary across different regions

Additionally, Norway's road transportation oil demand decreased by 12% from 2021 to 2024. As EVs replace ICE vehicles, a reduction in domestic oil consumption for transportation is expected in China. The adoption of EVs is also accelerating the integration of renewable energy into China's grid power, thereby reducing reliance on coal and oil. Furthermore, total oil demand may be nearing its peak. Notably, China's crude oil imports declined by nearly 2% in 2024, representing the first annual drop in two decades [69].

In the US, the transition to EV is expected to impact the oil and gas industry, as well as the broader automotive supply chain, potentially leading to job displacement and new economic opportunities [57]. Indeed, the rise of EVs is disrupting the traditional oil and gas sector [70, 71].

10.5.3 Disruption in Labour Market

The adoption of EV technologies is significantly disrupting the global labor market, creating both job losses in traditional automotive manufacturing and new opportunities in related sectors [72, 73]. The skills needed for EV manufacturing differ from those for ICE vehicles, requiring workers to adapt to new technologies like battery production, electric motor assembly, and software integration [74, 75].

Furthermore, this transition is expected to reduce maintenance costs by up to 40% compared to conventional vehicles due to the simpler design and fewer moving parts in EVs. The rise of EVs will necessitate new areas of expertise, such as battery diagnostics, charging infrastructure installation, and software updates, creating new job opportunities for trained technicians [76]. Additionally, augmented focus will be directed toward software and electronics due to their bigger role in car functionality. This transition underscores the necessity of enhancing the skills of existing technicians in these domains to leverage emerging opportunities within the automotive industry, in accordance with the changing dynamics of electric mobility.

An EV-only approach could lead to half a million job losses in the EU, particularly in the ICE powertrain component production sector if technology is phased out by 2035, according to CLEPA (the European Association of Automotive Suppliers) [77]. The transition necessitates reskilling and upskilling of the workforce to meet the demands of the EV industry and other related sectors (software development and charging infrastructure management, etc.), avoiding potential chaos in the society [78].

Governments and industry are investing in training and reskilling programs to equip workers with the new skills needed for the EV sector [79]. Policies are needed to support workers displaced from traditional automotive jobs, including financial assistance, retraining opportunities, and job placement services, and also to support the economy of the region [80].

In summary, workers in traditional automotive and oil-related sectors are being retrained or displaced as EV production ramps up. New jobs could emerge in areas like battery manufacturing, EV charging infrastructure, and software development.

10.6 Emerging Technologies in the Automotive Sector

Emerging technologies refer to new and innovative technological developments in their initial phases of creation or implementation, frequently representing progress with the potential to substantially influence industries, societies, or economies. Advancements in automotive technology address common industry challenges related to sustainability, safety, and connectivity [81]. AI, robotics, big data, IoT, and more are the emerging technologies transforming the automotive sector [82]. The EVs' advanced mobility reduces reliance on fossil fuels, while hydrogen fuel cell technology offers an alternative for zero-emission transportation [83]. Advanced driver-assistance systems (ADAS) integrate AI-powered sensors to enhance vehicle safety and performance by enhancing real-time environmental awareness [84, 85]. Vehicle-to-everything (V2X) communication connects cars to pedestrians, other vehicles, traffic lights, and even city infrastructure like road signs and traffic management systems [86, 87]. This technology improves traffic management and reduces accidents. Autonomous driving technology, incorporating machine learning and real-time data processing, improves the development of fully self-driving cars, shaping the future of transportation [88, 89]. As the automotive industry progresses toward a technologically advanced and sustainable future, stakeholders must adopt these innovations to maintain competitiveness.

The future of automotive technology is characterized by electrification, autonomy, connectivity, and cutting-edge software integration. Current technological trends show that EVs and hydrogen-powered transportation will dominate the road transportation sector in the foreseeable future [90, 91]. Whereas, the epoch of autonomous vehicle adoption is still some time away [92].

10.6.1 Autonomous Electric Vehicles

The combination of two groundbreaking technologies, electric vehicles and autonomous driving technology, promises to redefine the future of transportation. Autonomous electric vehicles (AEVs) are powered by electric motors and are capable of navigating and operating with little to no human intervention [93]. The integration of AI in the automotive industry has undergone dramatic evolution, progressing from basic automation to fully intelligent mobility systems. Table 10.10 presents a chronological overview of the transformative impact of AI on the automobile sector [94, 95]. AI systems are enhancing vehicle safety by equipping cars with superior intelligence, faster reaction, and increased awareness of their environments beyond the capabilities of human drivers. These innovations are greatly contributing to the future of mobility. In 2025, AEVs have gained substantial traction with advancements from companies like Tesla, BYD, and other Chinese automakers [96]. Tesla's full self-driving (FSD) system, Alphabet's subsidiary Waymo robotaxi service, GM's Cruise unit, Navya's autonomous electric shuttles, and Ford's

Table 10.10 Evolution of AI integration in automobile over the time

Early 2000s (Foundations of Automation)	2010–2015 (Rise of Smart Features)	2016–2020 (Autonomous Aspirations)	2021–2025 (Full Ecosystem Intelligence)	Future outlook (2025–2030 & Beyond)
* *Basic Driver Assistance*: Cruise control, anti-lock braking systems (ABS), and early forms of lane-keeping assistance established the foundation for intelligent driving * *Navigation Systems*: GPS-based routing emerged, offering real-time traffic updates and route optimization	* *Advanced Driver Assistance Systems* (ADAS): Technologies like lane-keeping assist, adaptive cruise control, and automatic emergency braking were developed * *Voice Recognition*: Early infotainment systems started using basic voice command functionalities	* *Self-Driving Prototypes*: Companies like Tesla, Waymo, and Uber initiated the testing of autonomous vehicles using the application of deep learning techniques and sensor fusion methodologies * *Predictive Maintenance*: AI algorithms started analyzing vehicle data to forecast potential component failures prior to their occurrence * *Personalization*: Cars have started learning the process of acquiring knowledge of driver preferences, including seat positions, climate settings, and music choices	* *Connected Vehicles*: AI facilitates vehicle-to-everything (V2X) communication, enabling vehicles to interact with infrastructure, other vehicles, and pedestrians * *Emotion-Responsive Systems*: AI is now capable of detecting driver fatigue, stress, or distraction through the use of facial recognition technology and biometric sensors * *Digital Twins*: Virtual models of vehicles are used for real-time diagnostics and optimization of design * *Sustainability*: AI helps to the reduction of fuel consumption, the optimization of logistics, and the minimization of material waste	* *Multimodal AI*: Integration of voice, gesture, and gaze control mechanisms for seamless interaction * *AI-Powered EVs*: Electric vehicles will use AI to optimize battery usage and route planning, thereby improving energy efficiency * *Autonomous Fleets*: The extensive deployment of self-driving taxis and delivery vehicles is anticipated * *Insurance & Risk Assessment*: Real-time driving behavior analysis will revolutionize insurance models

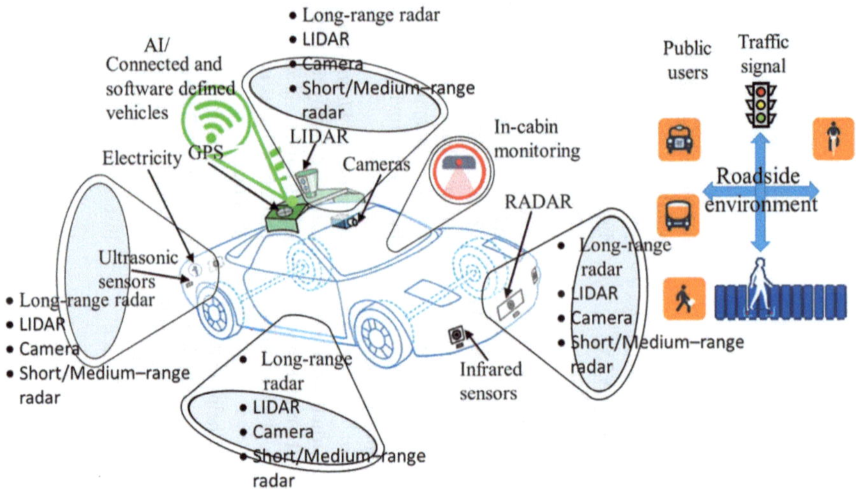

Fig. 10.2 Future automotive technology—Autonomous electric vehicles

BlueCruise technology are at the forefront of AEVs. They offer a wide range of benefits across safety, sustainability, and efficiency. Figure 10.2 shows the integration of various technologies with the electric vehicles. Earlier, the findings of Ercan et al. indicated that the integration of autonomous electric vehicles could potentially reduce GHG emissions by as much as 34% of the total transportation-related emissions by 2050 [97, 98].

In 2023, researchers have shown that developing a control structure model of the existing road transport system serves as a foundational step in identifying potential gaps in control mechanisms and adaptive feedback functions necessary for sustaining safety over time [99]. Unaddressed gaps could lead to the materialization of risks linked to the integration of future technologies, including advanced automated vehicles. Tesla's data indicate their EVs with Autopilot engaged are now between six and nine times safer than the average human-driven car in America.

The emergence of connected, autonomous, and electric vehicles is expected to accelerate the growth of the automotive sector. With the advent of revolutionary new technologies, the transportation industry is anticipated to transform significantly in the near future. As a result, the transportation industry is encountering notable changes, with various technological developments occurring in the short and long term. Companies that can adjust to these trends and anticipate future shifts will have a better chance to succeed in this dynamic and quickly evolving market. However, there are some challenges and considerations to this AEV technology, which are displayed in Table 10.11 [100–105].

Autonomous vehicles are becoming a reality in personal transportation. Equipped with advanced sensors, complex algorithms, and high-performance processing capabilities, these vehicles can operate without human intervention. They promise the potential to radically reduce traffic congestion, emissions, and traffic accidents caused by human error. Additionally, autonomous driving technologies and vehicle

Table 10.11 Challenges in adoption of AEV technologies

Challenges to AEV adoption		References
Public trust	Fostering public trust in the safety and reliability of autonomous vehicles is essential for their widespread adoption	[100]
Regulations and laws	It is important to establish clear regulatory frameworks to govern the operation of AEVs and address liability concerns	[101]
Infrastructure development	Investments in charging infrastructure and communication networks are necessary to support the growth of AEVs	[102, 103]
Ethical considerations	The development and deployment of AEVs pose ethical questions regarding decision-making in complex driving situations	[104, 105]

electrification are significant trends influencing the automotive industry. As a result, AEVs are rapidly evolving, with notable progress in both technology and legislation.

10.6.2 Electric Vehicle-Based Transportation-as-a-Service

Transportation as a Service (TaaS) represents an evolving model in which vehicle ownership is not required for consumers. Rather, individuals utilize electric cars operated by a fleet management entity. TaaS will create an entirely new business model and is likely to disrupt and transform our transportation industries.

The TasS transportation model is illustrated in Fig. 10.3. Instead of owning a car, individuals and businesses can access transportation on demand through fleets of electric vehicles. EV TaaS is a model where EVs are offered as a service; users pay for access rather than ownership. Transportation modes include ride-sharing, car-sharing, subscription-based services, autonomous EV fleets, and two-wheeler sharing systems, such as bicycles and motorcycles. The advantages of the EV-based TaaS model are shown in Fig. 10.4. TaaS is expected to offer cheaper transportation per mile than buying a new car.

With the implementation of TaaS, 95% of passenger miles traveled in the U.S. will be served by on-demand, autonomous EVs operated by fleets by 2030. The impacts of TaaS include an 80% reduction in transportation energy demand, a 90% reduction in tailpipe emissions, about $1 trillion in household savings, and more than 200 million cars taken off American roads.

The economic consequences of this model of transportation are far-reaching. In the first phase of disruption, manufacturing of vehicles with ICE will collapse because of the decline in cost of EVs. In the second phase of disruption, autonomous EVs will support TaaS (transport mode as a service). This emerging business model has the potential to change the transportation landscape, ending individual vehicle ownership and reshaping the world's energy economy.

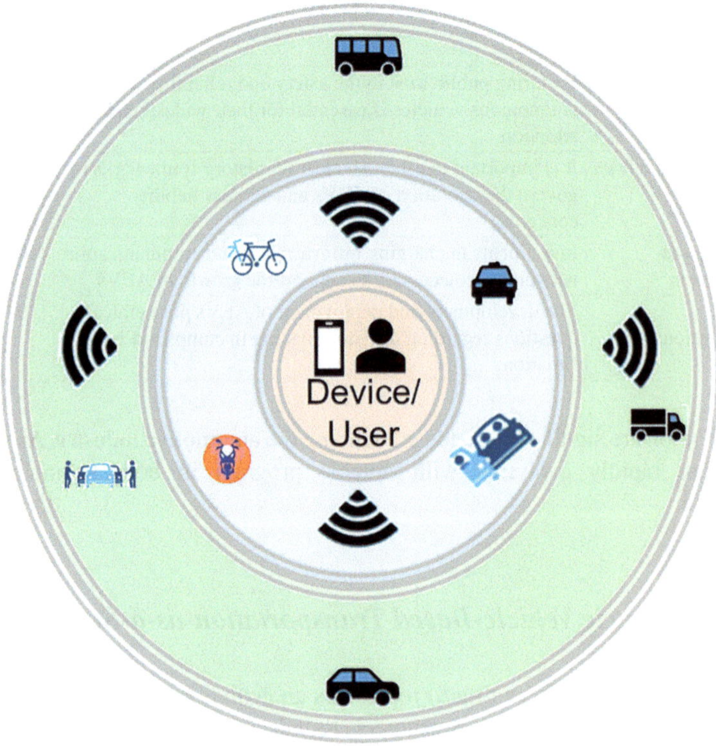

Fig. 10.3 Transportation as a service model

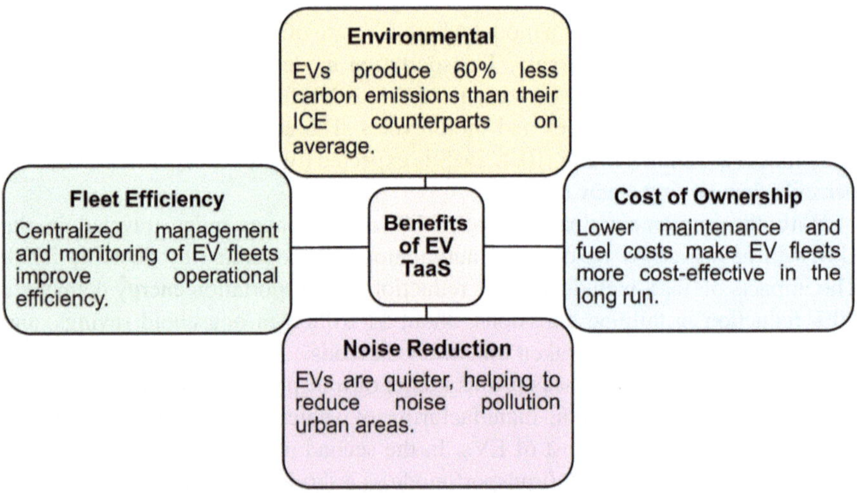

Fig. 10.4 Benefits of EVs for Transportation-as-a-Service (TaaS)

10.7 Next Level of Disruptive Technologies in Road Transportation

The transportation sector is currently undergoing the most significant transformation in more than a century. This shift is primarily driven by the convergence of artificial intelligence (AI), EVs, and sustainability initiatives. The future of transportation is characterized by innovation, sustainability, and integrated multimodal transportation systems with seamless connections between modes of travel and connectivity. Disruptive technologies such as EVs, *autonomous vehicles with an ability to interact with drivers using natural language*, *AI-driven dialogue systems* and *adaptive interfaces*, urban air mobility, and hyperloop systems are reshaping how people and goods move around the world. Several automotive companies (Volkswagen Group, Zoox (Amazon), Pony.ai, Aptiv (via Motional), Ford (with Argo AI), May Mobility, Tesla, etc.) are integrating AI, machine learning, and natural language processing to develop vehicles that are not only self-driving but also capable of self-explanation, self-adjustment, and even self-regulation. Real-time traffic management will emerge as a promising technology. Connected vehicles, sensors, and smart navigation systems will enable efficient traffic management, routing both freight and passenger flows in real-time.

As we embrace these transformative trends, it's essential to prioritize safety, sustainability, and equity to ensure that the future of transportation benefits all stakeholders. By leveraging technology, fostering collaboration, and embracing a forward-thinking mindset, we can create a transportation system that is safer, more efficient, and more inclusive for generations to come.

The future of transportation is expected to increasingly incorporate autonomous technologies, offering improvements in efficiency, safety, and accessibility. Although challenges persist, the potential benefits of autonomous transportation systems are vast and potentially transformative.

10.7.1 Road Transportation in 2050

In 2050, road transportation is expected to be significantly transformed by automation, electrification, and a shift towards more sustainable and efficient systems [106, 107]. It will be equipped with AI-powered personalization and comfort features. Materials will be sustainable, and focus will be on the safety and seamless integration into daily life. An augmented reality (AR) display projected onto the windshield will provide drivers with navigation, hazard alerts, and other information in an intuitive way. Ownership models may shift towards shared mobility systems. EVs are expected to become predominant in both public transportation and the private sector as well. Vehicle fleets, in the form of car-sharing services, will be the norm. Cars, trucks, and delivery vehicles will be autonomous electric vehicles that communicate with one another to optimize traffic flow and ensure safety [48].

Mobility as a Service (MaaS) will gain more popularity because it offers a seamless, flexible, and user-centric approach to transportation [108]. By integrating various modes of transit, such as buses, trains, ride-hailing, bike-sharing, and even car rentals, into a unified digital platform, MaaS enables people to plan, book, and pay for journeys through a single smartphone application.

By 2050, automobiles are expected to become intelligent, interconnected mobile ecosystems. Autonomous driving will be the norm, with vehicles featuring sensors, LIDAR for creating 3D maps, and AI systems. EVs will utilize sustainable energy sources, eliminating range anxiety. Cars are anticipated to communicate with each other and with infrastructure to optimize traffic flow and enhance safety. Advanced AI systems are expected to predict traffic light changes and facilitate communication with other vehicles. Autonomous semi-trucks, vans, electric robotaxis, electric powertrains, and MaaS will enable the transportation of people around streets and in underground tunnels. These changes may result in a decrease of private vehicle ownership, a shift toward shared mobility options, and a more seamless, personalized transportation experience [109].

10.8 Conclusion

The global road transportation sector has long depended mainly on gasoline and diesel fuels and continues to do so, as it is primarily built around petroleum-derived fuels. In many regions, gasoline and diesel oil account for more than 90% of energy used in road transport. Transitioning to zero-energy transportation is an important step toward reducing carbon emissions in this sector. To be net-zero emissions by 2050, a worldwide shift to zero-emission vehicles must happen as fast as possible. A complete transition to NEVs is an ambitious but increasingly achievable target. Concurrently, the growing adoption of EVs is changing the traditional oil and gas industries. Corrective measures should be taken to assist the affected sectors and communities.

Advances in EVs, autonomous systems, smart infrastructure, and digital logistics could lead to smarter, greener, and more inclusive transportation. These developments may help to address challenges such as urban congestion, carbon emissions, and fair access to transport in the future.

The future of transportation is undoubtedly moving towards more autonomous vehicles, potentially improving efficiency, safety, and accessibility. Although challenges exist, autonomous transportation systems offer significant benefits. New technologies and innovative methods are changing how people and goods are transported, with an emphasis on sustainability.

In summary, the future of global road transportation will be significantly different from today. Technological advancements and innovation will continue to influence both demand for and supply of transportation options and services. Vehicles are expected to go electric and fully autonomous, operating without human intervention. Autonomous electric vehicles (AEVs), the rise of a new form of transport known as TaaS, and emerging technologies are expected to dominate in the automotive sector. These technological innovations may kill the traditional gasoline vehicle market, individual car ownership, and even the oil industry that supports them.

References

1. Horton P, Horton BP (2019) Re-defining sustainability: living in harmony with life on Earth. One Earth 1(1):86–94. https://doi.org/10.1016/j.oneear.2019.08.019
2. Kumar R, Kanwal A, Asim M, Pervez M, Mujtaba MA, Fouad Y, Kalam MA (2024) Transforming the transportation sector: mitigating greenhouse gas emissions through electric vehicles (EVs) and exploring sustainable pathways. AIP Adv 14(3):035320. https://doi.org/10.1063/5.0193506
3. Benson AK (2023) Otto's practical internal combustion engine. https://www.ebsco.com/research-starters/history/ottos-practical-internal-combustion-engine#summary-of-event
4. Bladh M (2019) Origin of car enthusiasm and alternative paths in history. Environ Innov Soc Trans 32:153–168. https://doi.org/10.1016/j.eist.2018.09.003
5. Ritchie H, Rosado P (2017) Fossil fuels. OurWorldinData.org. https://ourworldindata.org/fossil-fuels
6. Epstein A (2022) Fossil future: why global human flourishing requires more oil, coal, and natural gas—not less. Portfolio, New York
7. Stanway D (2023) China climate envoy says phasing out fossil fuels 'unrealistic'. Reuters. https://www.reuters.com/sustainability/climate-energy/china-climate-envoy-says-phasing-out-fossil-fuels-unrealistic-2023-09-22/
8. Soeder DJ (2025) Replacing fossil fuels. In: Energy futures. Springer, Cham, pp 143–183. https://doi.org/10.1007/978-3-031-83603-9_6
9. IEA (2023) Road transport. IEA, Paris. https://www.iea.org/reports/road-transport
10. Final energy consumption in transport—detailed statics, Statistics explained. Eurostat (Sept 2024). https://ec.europa.eu/eurostat/statistics-explained/SEPDF/cache/131944.pdf
11. IEA (2023) Energy consumption in transport by fuel in the net zero scenario, 1975–2030. IEA, Paris. https://www.iea.org/data-and-statistics/charts/energy-consumption-in-transport-by-fuel-in-the-net-zero-scenario-1975-2030
12. Evolution of transport fuels (2024) KPMG International. https://assets.kpmg.com/content/dam/kpmg/xx/pdf/2024/06/evolution-of-transport-fuels-report.pdf#:~:text=The%20global%20transport%20sector%20is%20at%20a%20criticalneed%20for%20a%20transformative%20shift%20to%20alternative%20fuels
13. IEA (2023) Net zero roadmap: a global pathway to keep the 1.5 °C goal in reach. IEA, Paris. https://www.iea.org/reports/net-zero-roadmap-a-global-pathway-to-keep-the-15-0c-goal-in-reach
14. Wei X (2023) The transportation industry: powerhouses in the modern world with a gasoline and diesel fuels. J Pet Environ Biotechnol 14:523
15. Evro S, Oni BA, Tomomewo OS (2024) Global strategies for a low-carbon future: lessons from the US, China, and EU's pursuit of carbon neutrality. J Cleaner Prod 461:142635. https://doi.org/10.1016/j.jclepro.2024.142635
16. Scott S, Burkart K (n.d.) The global transition to 100% renewable transportation. One Earth. https://www.oneearth.org/renewable-transport/
17. Making electric transport the new normal by 2030. Climate Group Inc. https://www.theclimategroup.org/OLDev100
18. IEA (2025) Global EV outlook 2025. IEA, Paris. https://www.iea.org/reports/global-ev-outlook-2025
19. Udendhran R, Roosefert Mohan T, Babu R, Uthra RA, Anupama CG, Selvakumarasamy S, Dinesh G, Mukhopadhyay M, Saraswat V, Chakraborty P (2025) Transitioning to sustainable

E-vehicle systems—global perspectives on the challenges, policies, and opportunities. J Hazardous Mater Adv 17:100619. https://doi.org/10.1016/j.hazadv.2025.100619
20. Sustainable transport and accelerating the transition to electric vehicles. AstraZeneca (2025). https://www.astrazeneca.com/media-centre/articles/accelerating-transition-electric-vehicles.html#:~:text=As%20part%20of%20our%20Ambition%20Zero%20Carbon%20programme%2Cincluding%20in%20the%20EU%2C%20Japan%20and%20the%20US
21. Sankhwar P (2024) Evaluation of transition to 100% electric vehicles (EVs) by 2052 in the United States. Sustain Energy Res 11:35. https://doi.org/10.1186/s40807-024-00128-w
22. SME Communications (2024) Emerging trends in electric vehicle technology and infrastructure. SME. https://www.sme.org/sme-blog/posts/emerging-trends-in-electric-vehicle-technology-and-infrastructure/
23. Nyaz N (2025) Fleet electrification incentives and subsidies around the world. Fleetyr, Brisbane. https://fleetyr.com/resources/blog/fleet-electrification-incentives-and-subsidies/
24. Government subsidies for EV fleet transitions. FreightAmigo, Hong Kong. https://www.freightamigo.com/blog/government-subsidies-for-ev-fleet-transitions/
25. Barnard M (2025) China's Electrification of Trucks, Buses, & Construction Vehicles Accelerates Rapidly. Clean Technica. https://cleantechnica.com/2025/04/22/chinas-electrification-of-trucks-buses-construction-vehicles-accelerates-rapidly/
26. Qadir SA, Ahmad F, Mohsin A, Al-Wahedi AB, Iqbal A, Ali A (2024) Navigating the complex realities of electric vehicle adoption: a comprehensive study of government strategies, policies, and incentives. Energ Strat Rev 53:101379. https://doi.org/10.1016/j.esr.2024.101379
27. Kumar A (2025) 10 biggest EV challenges, industry research 2025 in go-to-market challenges in the EV and auto industry. Growth Jockey, Gurugram. Updated on 25 Feb 2025. https://www.growthjockey.com/blogs/challenges-in-ev-industry
28. Alanazi F (2023) Electric vehicles: benefits, challenges, and potential solutions for widespread adaptation. Appl Sci 13(10):6016. https://doi.org/10.3390/app13106016
29. Zhao P, Zhang S, Santi P, Cui D, Wang F, Liu P, Zhang Z, Liu J, Wang Z, Ratti C, Wu Y (2024) Challenges and opportunities in truck electrification revealed by big operational data. Nat Energy 9:1427–1437. https://doi.org/10.1038/s41560-024-01602-x
30. Eyes B (2024) Technical challenges of the battery electric vehicle transition: emissions, energy, and policy implications, 1st edn. CRC Press. https://doi.org/10.1201/9781003502746
31. Fulton L, Schäfer AW, Sperling D (2024) Transitions to deeply decarbonized transportation and energy systems around the world: challenges and solutions. Transp Res Part D Transp Environ 134:104219. https://doi.org/10.1016/j.trd.2024.104219
32. Lee H, Clark A (2018) Charging the future: challenges and opportunities for electric vehicle adoption. Environment and Natural Resources Program, Belfer Center for Science and International Affairs Harvard Kennedy School, Cambridge, MA. https://projects.iq.harvard.edu/files/energyconsortium/files/rwp18-026_lee_1.pdf
33. Global electric vehicle charging station market size and forecast 2025–2033. Renub Research. https://www.renub.com/electric-vehicle-charging-station-market-p.php
34. Santana-Méndez I, García-Afonso Ó, González-Díaz B (2024) Full road transport sector transition towards 100% autonomous renewable energy supply in isolated systems: Tenerife Island test case. Appl Sci 14(21):9734. https://doi.org/10.3390/app14219734
35. Hannon E, Krishnan M, Patel J, Sahdev S (2022) Mobility's net-zero transition: a look at opportunities and risks. McKinsey & Company, Automotive & Assembly. https://www.mckinsey.com/industries/automotive-and-assembly/our-insights/mobilitys-net-zero-transition-a-look-at-opportunities-and-risks#/
36. Sen A, Teter J, Miller J (2025) Vision 2050 update on the global zero-emission vehicle transition in 2024, ICCT report. International Council on Clean Transportation. https://theicct.org/wp-content/uploads/2025/01/ID-260-%E2%80%93-Vision-2050-update_report_final.pdf
37. Impact of Electric Vehicles on the Grid, Report to Congress June 2024. Office of Electricity, U.S. Department of Energy. https://www.energy.gov/sites/default/files/2024-10/Congressional%20Report%20EV%20Grid%20Impacts.pdf

38. Donahue L, Scott V (2024) Electric vehicles are key to the energy transition—but the switch must be sustainable. Here's why World Economic Forum. https://www.weforum.org/stories/2024/04/electric-vehicle-evs-energy-transition-demand/
39. Seizing the moment of opportunity: supercharging the new energy era of renewables, efficiency, and electrification. New York (2025). https://www.un.org/sites/un2.un.org/files/un-energy-transition-report_2025.pdf
40. Disruptive technologies in transportation. International Finance Corporation, World Bank Group. https://www.ifc.org/en/what-we-do/sector-expertise/infrastructure/transportation/disruptive-technologies-in-transportation
41. Transport System Resilience (2024) Roundtable report, international transport forum. OECD Publishing, Paris. https://www.itf-oecd.org/sites/default/files/docs/transport-system-resilience.pdf
42. Rodrigue J-P (2024) Transport planning and policy. In: The geography of transport systems, 6th edn. Routledge, London, p 402. https://doi.org/10.4324/9781003343196
43. Dia H, Bagloee S, Ghaderi H (2021) Technology-led disruptions and innovations: the trends transforming urban mobility. In: Augusto JC (ed) Handbook of smart cities. Springer, Cham, pp 1163–1198. https://doi.org/10.1007/978-3-030-69698-6_51
44. Disruptive technologies and the changing nature of work in the transportation sector (Report). SOM Steering Committee on Economic and Technical Cooperation (SCE), Transportation Working Group (TPTWG), Asia-Pacific Economic Cooperation (2021). https://www.apec.org/docs/default-source/publications/2021/3/disruptive-technologies-and-the-changing-nature-of-work-in-the-transportation-sector/221_tpt_disruptive-technologies-and-the-changing-nature-of-work-in-the-transportation-sector.pdf?sfvrsn=
45. Disruptive technology in the transport sector. Public-Private Partnership, World Bank Group. https://ppp.worldbank.org/disruptive-technology-transport-sector
46. Schmidt S (n.d.) 5 disruptive transportation industry trends to watch in 2024. MarketResearch.com. https://blog.marketresearch.com/5-disruptive-transportation-industry-trends-to-watch
47. Clewlow RR, Mishra GS (2017) Disruptive transportation: the adoption, utilization, and impacts of ride-hailing in the United States. Institute of Transportation Studies, UC Davis. https://escholarship.org/uc/item/82w2z91j
48. WIPO (2025) The future of transportation. WIPO technology trends report 2025. World Intellectual Property Organization (WIPO). https://www.wipo.int/web-publications/wipo-technology-trends-future-of-transportation/en/5-emerging-technologies-in-transportation.html
49. How electric vehicles are disrupting the automotive Industry. Disrupt Magazine (2025). https://disruptmagazine.com/how-electric-vehicles-are-disrupting-the-automotive-industry/
50. Industry report: the EV transition—challenges and opportunities for companies, SMEs, and the workforce. DTO Consulting GmbH, Dusseldorf (2025). https://dto-research.com/images/downloads/automotive/industry_report-EV-transition.pdf
51. Hilfinger SH, Parins RL, Strom JH, Johnson KA (2024) EV and EV-infrastructure deployment encounters regulatory, political, and market challenges in 2024: hitting speed bumps or shifting into reverse? National Law Review's (NLR's), Highland Park, IL. https://natlawreview.com/article/ev-and-ev-infrastructure-deployment-encounters-regulatory-political-and-market
52. Ou S, Hsieh I-YL, He X, Lin Z, Yu R, Zhou Y, Bouchard J (2021) China's vehicle electrification impacts on sales, fuel use, and battery material demand through 2050: optimizing consumer and industry decisions. iScience 24(11):103375. https://doi.org/10.1016/j.isci.2021.103375
53. Competition-fueled Chinese EV boom to benefit world market, speed up decarbonization. Xinhua, China Daily (26 Apr 2024). https://www.chinadaily.com.cn/a/202404/26/WS662b7a03a31082fc043c446c.html
54. Quinn R (2025) Electric trucks and the future of Chinese oil demand. Rhodium Group. https://rhg.com/research/electric-trucks-and-the-future-of-chinese-oil-demand/

55. Wang H, Kimble C (2011) Leapfrogging to electric vehicles: patterns and scenarios for China's automobile industry. Int J Automot Technol Manage 11(4):312–325. https://doi.org/10.1504/IJATM.2011.043164
56. Prado VD, Fabry E, Laya AG, Köhler-Suzuki N, Lamy P, Praetorius S (2025) The road to a new European automotive strategy: trade and industrial policy options. Rapport n°129. Notre Europe, Jacques Delors Institute, Paris School of International Affairs, Paris. https://institut-delors.eu/content/uploads/2025/04/REUOBQT.pdf
57. Colato J, Ice L (2023) Charging into the future: the transition to electric vehicles. In: Beyond the numbers: employment & unemployment, 12(4). U.S. Bureau of Labor Statistics. https://www.bls.gov/opub/btn/volume-12/charging-into-the-future-the-transition-to-electric-vehicles.htm
58. Kitti T (2025) EVs on the rise: the disruption of the auto industry. Electrik Living, Wyoming. https://electrikliving.com/how-electric-vehicles-are-disrupting-the-auto-industry/
59. Galer S (2023) How electric vehicles are disrupting automobile supply chains. Forbes Media. https://www.forbes.com/sites/sap/2023/04/04/how-electric-vehicles-are-disrupting-automobile-supply-chains/
60. Nolan S (2024) Disruption as opportunity: transforming the automotive Indus. EV Magazine. https://evmagazine.com/charging-and-infrastructure/disruption-as-opportunity-transforming-the-automotive-indus
61. Yuan W, Wong T (2024) Navigating electric vehicle disruption in Southeast Asia. Roland Berger, GmbH, Munich. https://www.rolandberger.com/en/Insights/Publications/Navigating-electric-vehicle-disruption-in-Southeast-Asia.html
62. Nishino K (2024) Global strategic studies institute monthly report, Mitsui & Co. Japan. Outlook on automotive electrification in Southeast Asia: the need to prepare for escalating EV supply. https://www.mitsui.com/mgssi/en/report/detail/__icsFiles/afieldfile/2024/05/21/2404i_nishino_e.pdf
63. Ramos MR, Ruiz-Gálvez ME (2024) The transformation of the automotive industry toward electrification and its impact on global value chains: Inter-plant competition, employment, and supply chains. Eur Res Manage Bus Econ 30(1):100242. https://doi.org/10.1016/j.iedeen.2024.100242
64. Saptakee S (2025) How EV adoption is reshaping global oil demand: IEA's 2025 outlook and 2030 forecast. Carbon Credits. https://carboncredits.com/how-ev-adoption-is-reshaping-global-oil-demand-ieas-2025-outlook-and-2030-forecast/
65. IEA (2025) Slowly backtracking on peak oil demand by 2030 narrative. Organization of the Petroleum Exporting Countries, Vienna. https://www.opec.org/pn-detail/1414266-16-may-2025.html
66. Dogar A (2025) The global shift to EVs: what it means for oil companies and their future. Dosolar.io. https://dosolar.io/the-global-shift-to-evs-what-it-means-for-oil-companies-and-their-future/
67. Electric cars have dented fuel demand. By 2040, They'll Slash It. BloombergNEF (2023). https://about.bnef.com/insights/clean-transport/electric-cars-have-dented-fuel-demand-by-2040-theyll-slash-it/#:~:text=Oil%20consumption%20displaced%20by%20EVs%20rises%20to%20overmore%20oil%20than%20the%20US%20consumed%20last%20year
68. Rystad Energy (2023) Is Norway's love for EVs enough to put a dent in fuel demand? OilPrice.com. https://oilprice.com/Energy/Gas-Prices/Is-Norways-Love-For-EVs-Enough-To-Put-A-Dent-In-Fuel-Demand.html
69. Paddison L, Nilsen E (2025) China made a bet decade ago because it couldn't compete with the US on cars. That bet is paying off big. CNN. https://edition.cnn.com/2025/01/23/climate/china-evs-growth-oil-market/index.html
70. How the EV boom is disrupting the oil & gas market. GEP (25 Aug 2022). https://www.gep.com/blog/mind/how-the-ev-boom-is-disrupting-oil-and-gas
71. The impact of electric vehicles on oil & gas demand. Share Vault (5 Aug 2025). https://sharevault.com/blog/thought-leadership/the-impact-of-electric-vehicles-on-oil-gas-demand

72. Charette RN (2023) How EVs are reshaping labor markets. IEEE Spectrum. https://spectrum.ieee.org/the-ev-transition-explained-2658797703
73. Saha D, Shrestha R, Hunt N, Kim E (2024) Navigating the EV transition: 4 emerging impacts on auto manufacturing jobs. World Resources Institute, Washington, DC. https://www.wri.org/insights/ev-transition-auto-manufacturing-jobs
74. Oya C, Sher G, Topalova P, Zhou J (2023) Cars and the green transition: challenges and opportunities for European workers. IMF Working Paper 23/116. International Monetary Fund, Washington, DC. 2 Jun. https://www.imf.org/en/Publications/WP/Issues/2023/06/02/Cars-and-the-Green-Transition-Challenges-and-Opportunities-for-European-Workers-534091
75. Cotterman T, Fuchs ERH, Whitefoot KS, Combemale C (2024) The transition to electrified vehicles: evaluating the labor demand of manufacturing conventional versus battery electric vehicle powertrains. Energy Policy 188:114064. https://doi.org/10.1016/j.enpol.2024.114064
76. Albatayneh A (2024) The electric cars era transforming the car repairs and services landscape. Adv Mech Eng 16(7):1–12. https://doi.org/10.1177/16878132241266536
77. An electric vehicle-only approach would lead to the loss of half a million jobs in the EU, study finds. CLEPA Communications, the European Association of Automotive Suppliers, Brussels (6 Dec 2021). https://www.clepa.eu/insights-updates/press-releases/an-electric-vehicle-only-approach-would-lead-to-the-loss-of-half-a-million-jobs-in-the-eu-study-finds/
78. Szabó J, Newell P (2024) Driving towards a just transition? The case of the European car industry. Energy Res Social Sci 115:103649. https://doi.org/10.1016/j.erss.2024.103649
79. OECD Education and Skills Today (2024) Why governments need to do more to support the transition to electric vehicles. https://oecdedutoday.com/why-governments-need-to-do-more-to-support-the-transition-to-electric-vehicles/
80. The future of work in the automotive industry, Technical meeting on the future of work in the automotive industry (Geneva, 15–19 February 2021). Note on the proceedings. International Labour Office, Sectoral Policies Department, ILO, Geneva (2021). https://www.ilo.org/sites/default/files/wcmsp5/groups/public/@ed_dialogue/@sector/documents/meetingdocument/wcms_821994.pdf
81. Neeley K (2024) Top 10 emerging technologies in the automotive industry. Auto-Journals. https://www.auto-journals.com/top-10-emerging-technologies-in-the-automotive-industry/
82. Top emerging technologies revolutionizing automotive manufacturing in 2025. Project Consultancy Services (NPCS), Delhi. https://www.npcsblog.com/top-emerging-technologies-revolutionizing-automotive-manufacturing-in-2025/
83. Lechowski G, Krzywdzinski M, Humphrey J, Pardi T (2025) Global shifts in the automotive industry. Outlining the new arrangements between technology, policy, firms, and markets. In: Krzywdzinski M, Lechowski G, Humphrey J, Pardi T (eds) Global shifts in the automotive sector, Palgrave studies of internationalization in emerging markets. Palgrave Macmillan, Cham, pp 345–365. https://doi.org/10.1007/978-3-031-80641-4_14
84. Chengula TJ, Mwakalonge J, Comert G, Sulle M, Siuhi S, Osei E (2024) Enhancing advanced driver assistance systems through explainable artificial intelligence for driver anomaly detection. Mach Learn Appl 17:100580. https://doi.org/10.1016/j.mlwa.2024.100580
85. Kathiresh S, Poonkodi M (2025) Artificial intelligence-powered advanced driver assistance systems in vehicles. In: Sivakumar PD, Ramachandran R, Pasupathi C, Balakrishnan P (eds) Computing technologies for sustainable development. IRCCTSD 2024, Communications in computer and information science, vol 2361. Springer, Cham, pp 131–146. https://doi.org/10.1007/978-3-031-82383-1_11
86. Walden D (2024) V2X communication technology: the future of connected vehicles. Bell Engineering, Mansfield, TX. 20 Sept. https://bellengineering.net/automobile-news/v2x-communication-connected-vehicles/#:~:text=With%20V2X%2C%20a%20car%20can%20communicate%20with%20othereven%20paves%20the%20way%20for%20fully%20autonomous%20vehicles

87. Zhang X, Li J, Zhou J, Zhang S, Wang JY, Yuan Y, Liu J, Li J (2025) Vehicle-to-everything communication in intelligent connected vehicles: a survey and taxonomy. Automot Innov 8:13–45. https://doi.org/10.1007/s42154-024-00310-2
88. Daniel Nareshkumar M, Reddy DN, Sowmiya A, Pandi VS, Sateesh R, Hindumathy N (2025) Autonomous vehicles: steering the future of transportation with self-driving technology and advanced AI systems. In: 2024 5th IEEE Global Conference for Advancement in Technology (GCAT). IEEE, Bangalore, pp 1–6. https://doi.org/10.1109/GCAT62922.2024.10924066
89. Bachute MR, Subhedar JM (2021) Autonomous driving architectures: insights of machine learning and deep learning algorithms. Mach Learn Appl 6:100164. https://doi.org/10.1016/j.mlwa.2021.100164
90. Hossain MS, Kumar L, El Haj AM, Alayi R (2022) Advancements and future prospects of electric vehicle technologies: a comprehensive review. Complexity (Hindawi) 2022:1–21. https://doi.org/10.1155/2022/3304796
91. Farooq A, Cora ÖN (2024) Current status and future perspectives for mobility options toward sustainable transportation with a focus on fuel cells. WIREs Energy Environ 13:e539. https://doi.org/10.1002/wene.539
92. Dimitrov D (2025) The future of transportation: AI, connectivity and safety in 2025. Forbes. https://www.forbes.com/councils/forbestechcouncil/2025/02/10/the-future-of-transportation-ai-connectivity-and-safety-in-2025/
93. Lu H, Zhu M, Yang H (2025) Human-like driving technology for autonomous electric vehicles. Nat Rev Electr Eng 2:218–219. https://doi.org/10.1038/s44287-025-00155-9
94. Shuliak M (2025) AI in automotive industry: use cases & real-world applications. Acropolium. https://acropolium.com/blog/ai-in-automotive-industry-use-cases-and-real-world-applications/
95. Wallner S (2025) AI in automotive: a strategic guide for industry leaders (2025–2030). StartUs Insights. https://www.startus-insights.com/innovators-guide/ai-in-automotive/
96. Lee W-I (2025) Driving the future of autonomous electric vehicles in Korea: strategic foresight, technology, and innovation cluster. Asia-Pac J Convergent Res Interchange 11(2):125–144
97. World Economic Forum (2025) When will autonomous vehicles and self-driving cars hit the road. https://www.weforum.org/stories/2025/05/autonomous-vehicles-technology-future/
98. Ercan T, Onat NC, Keya N, Tatari O, Eluru N, Kucukvar M (2022) Autonomous electric vehicles can reduce carbon emissions and air pollution in cities. Transp Res Part D: Transp Environ 112:103472. https://doi.org/10.1016/j.trd.2022.103472
99. Read GJM, McLean S, Thompson J, Stanton NA, Baber C, Carden T, Salmon PM (2023) Managing the risks associated with technological disruption in the road transport system: a control structure modelling approach. Ergonomics 67(4):498–514. https://doi.org/10.1080/00140139.2023.2226850
100. Pocuca R (2025) Autonomous electric vehicles in 2025: Tesla, China, and the future of self-driving EVs. Motorwatt. https://motorwatt.com/ev-news/autonomous-electric-vehicles-in-2025
101. Electric cars and autonomous vehicle regulations (2024) Cyber Switching, SanJose, CA. https://cyberswitching.com/electric-cars-and-autonomous-vehicle-regulations/
102. Todorovic M, Aldakkhelallah A, Simic M (2023) Managing transitions to autonomous and electric vehicles: scientometric and bibliometric review. World Electric Vehicle J 14(11):314. https://doi.org/10.3390/wevj14110314
103. Mateus (2025) How AI-driven autonomous EVs are revolutionizing urban transport. The Next Avenue. https://thenextavenue.com/2025/07/24/how-ai-driven-autonomous-evs-are-revolutionizing-urban-transport/
104. Dogan E, Costantini F, Boennec RL (2020) Ethical issues concerning automated vehicles and their implications for transport. In: Milakis D, Thomopoulos N, van Wee B (eds) Advances in transport policy and planning, vol 5. Academic Press, pp 215–233. https://doi.org/10.1016/bs.atpp.2020.05.003

105. Gros C, Kester L, Martens M, Werkhoven P (2025) Addressing ethical challenges in automated vehicles: bridging the gap with hybrid AI and augmented utilitarianism. AI Ethics 5:2757–2770. https://doi.org/10.1007/s43681-024-00592-6
106. Williams MS (2021) Life in 2050: a glimpse at transportation in the future. Interesting Engineering. https://interestingengineering.com/innovation/life-in-2050-a-glimpse-at-transportation-in-the-future
107. Zhou WD (2021) Vision 2050 transportation & mobility pathway: we can keep on moving, smarter. World Business Council for Sustainable Development (WBCSD), Geneva. https://www.wbcsd.org/news/vision-2050-transportation-mobility-pathway/
108. Musa S (2024) MaaS global is dead: long live MaaS. ZAG Daily. 15 Mar. https://zagdaily.com/trends/maas-global-is-dead-long-live-maas/
109. Road transport vision 2050 adopted. Plenary-24-06-2024, ERTRAC. European Union's Horizon Europe research and innovation programme under grant agreement No 101096253. https://www.ertrac.org/wp-content/uploads/2024/06/ERTRAC-Vision-2050-adopted-ERTRAC-Plenary-24-06-2024.pdf

MIX
Papier aus verantwortungsvollen Quellen
Paper from responsible sources
FSC® C105338

If you have any concerns about our products,
you can contact us on
ProductSafety@springernature.com

In case Publisher is established outside the EU,
the EU authorized representative is:
Springer Nature Customer Service Center GmbH
Europaplatz 3, 69115 Heidelberg, Germany

Printed by Libri Plureos GmbH
in Hamburg, Germany